T0310164

AN INTRODUCTION TO THERMODYNAMIC CYCLE SIMULATIONS FOR INTERNAL COMBUSTION ENGINES

AN INTRODUCTION TO THERMODYNAMIC CYCLE SIMULATIONS FOR INTERNAL COMBUSTION ENGINES

Jerald A. Caton

Department of Mechanical Engineering
Texas A&M University
College Station, TX, USA

Registered office
John Wiley & Sons Ltd, The Atrium, Southern Gate, Chichester, West Sussex, PO19 8SQ, United Kingdom

For details of our global editorial offices, for customer services and for information about how to apply for permission to reuse the copyright material in this book please see our website at www.wiley.com.

Library of Congress Cataloging-in-Publication Data
Caton, J. A. (Jerald A.)
  An introduction to thermodynamic cycle simulations for internal combustion engines / Jerald A Caton.
    pages cm
  Includes bibliographical references and index.
  ISBN 978-1-119-03756-9 (cloth)
  1. Internal combustion engines–Thermodynamics–Computer simulation.   2. Internal combustion
engines–Thermodynamics–Mathematical models.   I. Title.
  TJ756.C38 2015
  629.25001'5367–dc23

                                                                                    2015022961

A catalogue record for this book is available from the British Library.

ISBN: 9781119037569

Cover image: teekid/Getty

Set in 10/12 pt Times LT Std by Aptara Inc., New Delhi, India

Printed in Singapore by C.O.S. Printers Pte Ltd

1   2016

To my wife, Roberta,
our children, Jacob, Lewis and Kassandra, and
our grandchildren

Contents

Preface   **xiii**

1   Introduction   **1**
1.1   Reasons for Studying Engines   1
1.2   Engine Types and Operation   2
1.3   Reasons for Cycle Simulations   3
       1.3.1   Educational Value   3
       1.3.2   Guide Experimentation   3
       1.3.3   Only Technique to Study Certain Variables   4
       1.3.4   Complete Extensive Parametric Studies   4
       1.3.5   Opportunities for Optimization   4
       1.3.6   Simulations for Real-time Control   4
       1.3.7   Summary   5
1.4   Brief Comments on the History of Simulations   5
1.5   Overview of Book Content   6

2   Overview of Engines and Their Operation   **9**
2.1   Goals of Engine Designs   9
2.2   Engine Classifications by Applications   10
2.3   Engine Characteristics   11
2.4   Basic Engine Components   12
2.5   Engine Operating Cycles   12
2.6   Performance Parameters   12
       2.6.1   Work, Power, and Torque   12
       2.6.2   Mean Effective Pressure   15
       2.6.3   Thermal Efficiencies   16
       2.6.4   Specific Fuel Consumption   17
       2.6.5   Other Parameters   17
2.7   Summary   18

3   Overview of Engine Cycle Simulations   **19**
3.1   Introduction   19
3.2   Ideal (Air Standard) Cycle Analyses   19
3.3   Thermodynamic Engine Cycle Simulations   21

3.4   Quasi-dimensional Thermodynamic Engine Cycle Simulations   22
3.5   Multi-dimensional Simulations   23
3.6   Commercial Products   24
        3.6.1   Thermodynamic Simulations   24
        3.6.2   Multi-dimensional Simulations   25
3.7   Summary   26
Appendix 3.A: A Brief Summary of the Thermodynamics of the "Otto" Cycle Analysis   29

4   **Properties of the Working Fluids**   **37**
4.1   Introduction   37
4.2   Unburned Mixture Composition   37
        4.2.1   Oxygen-containing Fuels   40
        4.2.2   Oxidizers   41
        4.2.3   Fuels   41
4.3   Burned Mixture ("Frozen" Composition)   42
4.4   Equilibrium Composition   43
4.5   Determinations of the Thermodynamic Properties   46
4.6   Results for the Thermodynamic Properties   47
4.7   Summary   61

5   **Thermodynamic Formulations**   **63**
5.1   Introduction   63
5.2   Approximations and Assumptions   64
5.3   Formulations   65
        5.3.1   One-Zone Formulation   65
        5.3.2   Two-Zone Formulation   67
        5.3.3   Three-Zone Formulation   72
5.4   Comments on the Three Formulations   77
5.5   Summary   77

6   **Items and Procedures for Solutions**   **79**
6.1   Introduction   79
6.2   Items Needed to Solve the Energy Equations   79
        6.2.1   Thermodynamic Properties   79
        6.2.2   Kinematics   80
        6.2.3   Combustion Process (Mass Fraction Burned)   82
        6.2.4   Cylinder Heat Transfer   85
        6.2.5   Mass Flow Rates   86
        6.2.6   Mass Conservation   89
        6.2.7   Friction   89
        6.2.8   Pollutant Calculations   94
        6.2.9   Other Sub-models   94
6.3   Numerical Solution   94
        6.3.1   Initial and Boundary Conditions   95
        6.3.2   Internal Consistency Checks   96
6.4   Summary   96

7	**Basic Results**	**99**
7.1	Introduction	99
7.2	Engine Specifications and Operating Conditions	99
7.3	Results and Discussion	101
	7.3.1 Cylinder Volumes, Pressures, and Temperatures	102
	7.3.2 Cylinder Masses and Flow Rates	106
	7.3.3 Specific Enthalpy and Internal Energy	108
	7.3.4 Molecular Masses, Gas Constants, and Mole Fractions	110
	7.3.5 Energy Distribution and Work	114
7.4	Summary and Conclusions	116

8	**Performance Results**	**119**
8.1	Introduction	119
8.2	Engine and Operating Conditions	119
8.3	Performance Results (Part I)—Functions of Load and Speed	119
8.4	Performance Results (Part II)—Functions of Operating/Design Parameters	129
	8.4.1 Combustion Timing	129
	8.4.2 Compression Ratio	131
	8.4.3 Equivalence Ratio	133
	8.4.4 Burn Duration	135
	8.4.5 Inlet Temperature	135
	8.4.6 Residual Mass Fraction	136
	8.4.7 Exhaust Pressure	136
	8.4.8 Exhaust Gas Temperature	140
	8.4.9 Exhaust Gas Recirculation	142
	8.4.10 Pumping Work	145
8.5	Summary and Conclusions	149

9	**Second Law Results**	**153**
9.1	Introduction	153
9.2	Exergy	153
9.3	Previous Literature	154
9.4	Formulation of Second Law Analyses	154
9.5	Results from the Second Law Analyses	158
	9.5.1 Basic Results	158
	9.5.2 Parametric Results	163
	9.5.3 Auxiliary Comments	174
9.6	Summary and Conclusions	176

10	**Other Engine Combustion Processes**	**179**
10.1	Introduction	179
10.2	Diesel Engine Combustion	179
10.3	Best Features from SI and CI Engines	180
10.4	Other Combustion Processes	181
	10.4.1 Stratified Charge Combustion	181
	10.4.2 Low Temperature Combustion	181

10.5   Challenges of Alternative Combustion Processes   182
10.6   Applications of the Simulations for Other Combustion Processes   183
10.7   Summary   184

11   Case Studies: Introduction   187
11.1   Case Studies   187
11.2   Common Elements of the Case Studies   188
11.3   General Methodology of the Case Studies   189

12   Combustion: Heat Release and Phasing   191
12.1   Introduction   191
12.2   Engine and Operating Conditions   191
12.3   Part I: Heat Release Schedule   191
        12.3.1   Results for the Heat Release Rate   197
12.4   Part II: Combustion Phasing   205
        12.4.1   Results for Combustion Phasing   206
12.5   Summary and Conclusions   221

13   Cylinder Heat Transfer   225
13.1   Introduction   225
13.2   Basic Relations   226
13.3   Previous Literature   227
        13.3.1   Woschni Correlation   228
        13.3.2   Summary of Correlations   229
13.4   Results and Discussion   230
        13.4.1   Conventional Engine   230
        13.4.2   Engines Utilizing Low Heat Rejection Concepts   241
        13.4.3   Engines Utilizing Adiabatic EGR   247
13.5   Summary and Conclusions   250

14   Fuels   253
14.1   Introduction   253
14.2   Fuel Specifications   254
14.3   Engine and Operating Conditions   255
14.4   Results and Discussion   255
        14.4.1   Assumptions and Constraints   255
        14.4.2   Basic Results   255
        14.4.3   Engine Performance Results   259
        14.4.4   Second Law Results   266
14.5   Summary and Conclusions   268
Appendix 14.A: Energy and Exergy Distributions for the Eight Fuels at the Base
        Case Conditions (*bmep* = 325 kPa, 2000 rpm, ϕ = 1.0 and MBT timing)   269

15   Oxygen-Enriched Air   275
15.1   Introduction   275
15.2   Previous Literature   276

15.3   Engine and Operating Conditions   277
15.4   Results and Discussion   277
        15.4.1   Strategy for This Study   278
        15.4.2   Basic Thermodynamic Properties   278
        15.4.3   Base Engine Performance   280
        15.4.4   Parametric Engine Performance   283
        15.4.5   Nitric Oxide Emissions   289
15.5   Summary and Conclusions   291

16   Overexpanded Engine   295
16.1   Introduction   295
16.2   Engine, Constraints, and Approach   296
        16.2.1   Engine and Operating Conditions   296
        16.2.2   Constraints   296
        16.2.3   Approach   296
16.3   Results and Discussion   297
        16.3.1   Part Load   297
        16.3.2   Wide-Open Throttle   304
16.4   Summary and Conclusions   309

17   Nitric Oxide Emissions   311
17.1   Introduction   311
17.2   Nitric Oxide Kinetics   312
        17.2.1   Thermal Nitric Oxide Mechanism   312
        17.2.2   "Prompt" Nitric Oxide Mechanism   312
        17.2.3   Nitrous Oxide Route Mechanism   313
        17.2.4   Fuel Nitrogen Mechanism   313
17.3   Nitric Oxide Computations   313
        17.3.1   Kinetic Rates   315
17.4   Engine and Operating Conditions   316
17.5   Results and Discussion   317
        17.5.1   Basic Chemical Kinetic Results   317
        17.5.2   Time-Resolved Nitric Oxide Results   320
        17.5.3   Engine Nitric Oxide Results   324
17.6   Summary and Conclusions   329

18   High Efficiency Engines   333
18.1   Introduction   333
18.2   Engine and Operating Conditions   334
18.3   Results and Discussion   336
        18.3.1   Overall Assessment   336
        18.3.2   Effects of Individual Parameters   343
        18.3.3   Emissions and Exergy   347
        18.3.4   Effects of Combustion Parameters   351
18.4   Summary and Conclusions   353

19   Summary: Thermodynamics of Engines   **355**
19.1   Summaries of Chapters   355
19.2   Fundamental Thermodynamic Foundations of IC Engines   356
        Item 1: Heat Engines versus Chemical Conversion Devices   356
        Item 2: Air-Standard Cycles   357
        Item 3: Importance of Compression Ratio   357
        Item 4: Importance of the Ratio of Specific Heats   359
        Item 5: Cylinder Heat Transfer   360
        Item 6: The Potential of a Low Heat Rejection Engine   360
        Item 7: Lean Operation and the Use of EGR   361
        Item 8: Insights from the Second Law of Thermodynamics   361
        Item 9: Timing of the Combustion Process   362
        Item 10: Technical Assessments of Engine Concepts   362
19.3   Concluding Remarks   362

Index   **363**

Preface

The use of engine cycle simulations is an important aspect of engine development, and yet there is limited comprehensive documentation available on the formulations, solution procedures, and detailed results. Since beginning in the 1960s, engine cycle simulations have evolved to their current highly sophisticated status. With the concurrent development of fast and readily available computers, these simulations are used in routine engine development activities throughout the world. This book provides an introduction to basic thermodynamic engine cycle simulations and provides a substantial set of results.

This book is unique and provides a number of features not found elsewhere, including:

- comprehensive and detailed documentation of the mathematical formulations and solutions required for thermodynamic engine cycle simulations;
- complete results for instantaneous thermodynamic properties for typical engine cycles;
- self-consistent engine performance results for one engine platform;
- a thorough presentation of results based on the second law of thermodynamics;
- the use of the engine cycle simulation to explore a large number of engine design and operating parameters via parametric studies;
- results for advanced, high efficiency engines;
- descriptions of the thermodynamic features that relate to engine efficiency and performance;
- a set of case studies that illustrate the use of engine cycle simulations—these case studies consider engine performance as functions of engine operating and design parameters;
- a detailed evaluation of nitric oxide emissions as functions of engine operating parameters and design features.

Although this book focuses on the spark-ignition engine, the majority of the development and many of the results are applicable (with modest adjustments) to compression-ignition (diesel) engines. In fact, the major difference between the two engines relates to the combustion process, and these differences are mostly related to the details and not the overall process. But to be consistent, extrapolations to compression-ignition engines are largely avoided.

The examples and case studies are based on an automotive engine, but the procedures and many of the results are valid for other engine classifications. In addition, the thermodynamic simulation could be used for these other applications. Many of the results are fairly general and would be applicable to most engines. For example, results highlighting the difficulty of converting thermal energy into work (a consequence of the fundamental thermodynamics) applies to all engines.

Although the main purpose of the writing of this book was to document the development and use of thermodynamic engine cycle simulations, a secondary purpose was to stimulate the interest and excitement of using fundamental thermodynamic principles to understand a complex device. As the following pages will demonstrate, many phenomena related to engine operation and design may be understood in a more complete fashion by focusing on the fundamental thermodynamics.

The work of Professor John B. Heywood needs to be acknowledged as a major part of the foundations of the material in this book. These foundations are recognized in the book by numerous citations to the work of Professor Heywood, his colleagues, and his students.

The author has enjoyed his work on this topic and writing this book. He hopes that the reader will gain insight into engine design and operation, and be stimulated to use engine cycle simulations to answer his/her own questions. Although this presentation and these results have been examined by many reviewers, any mistakes remaining are the sole responsibility of the author. Notification of the author of these mistakes and suggestions for improvements would be greatly appreciated.

1

Introduction

The internal combustion (IC) engine is a spectacular, complex device that has been an unqualified success. The IC engine is probably best known as the power plant for vehicles, but, of course, is also successfully used in a variety of other applications. These other applications include, for example, simple garden equipment, stationary electrical power generation, locomotives, and ships. A powerful approach to aid in the design and understanding of these engines is through the use of mathematical simulations.

Engine cycle simulations have been developed and used to study a variety of features and issues relative to IC engines since the 1960s. In the beginning, engine cycle simulations were fairly elementary, and were limited by both computing capabilities and a lack of knowledge concerning key sub-models. In time, these simulations have become more complete and more useful.

Today, engine cycle simulations are sophisticated, complex computer programs that provide both global engine performance parameters as well as detailed, time-resolved information. Many of these simulations contain advanced and detailed sub-models for the fluid mechanics, heat transfer, friction, combustion, and chemical kinetics. The most advanced simulations include calculations in three dimensions. Some of these simulations are grouped in the general category of computational fluid dynamics (CFD). Some comments on the early history (pre-1990) of the development of engine simulations may be found in References 1–3.

1.1 Reasons for Studying Engines

As mentioned above, IC engines have been an unqualified success in several major economic markets. Certainly, as the propulsion unit for light duty vehicles, the IC engine has been a significant accomplishment. The number of such vehicles and their engines is estimated at one billion throughout the world, and is expected to be about two billion by 2020. For a rather complex, major device, these are exceptional numbers. Other applications of IC engines include stationary power generation, marine propulsion, small utility, off-road, and agriculture.

The reasons for the success of the IC engine have been well documented (e.g., References 2, 4, and 5). These reasons include relatively low initial cost, high power density, reasonable

An Introduction to Thermodynamic Cycle Simulations for Internal Combustion Engines, First Edition.
Jerald A. Caton © 2016 John Wiley & Sons, Ltd. Published 2016 by John Wiley & Sons, Ltd.

driving range (say, more than 200 miles for a standard fuel tank size), able to refuel on the order of minutes at many locations, robust and versatile, reasonably efficient, able to meet regulated emission limits, and well matched to available fuels. This last item is particularly important and results in some of the other favorable features.

Liquid hydrocarbon fuels (such as gasoline and diesel) possess relatively high energy densities, are relatively safe and stable, and (currently) are widely available. In addition, these fuels possess excellent characteristics for combustion processes utilized by spark-ignited and compression-ignited engines.

Current (2015) engine technology spans a wide range from fairly basic to relatively advanced. Some engines are still based on the use of carburetors, mechanical valve trains, and large displacements. More advanced engine designs include direct fuel injection, variable valve timings, turbocharging, and the capability to deactivate some cylinders for part load operation. Most spark-ignition engines are designed for operation at or near stoichiometric with compression ratios less than about 11 (to avoid spark knock).

The demise of the IC engine is often a popular topic in the lay press due to the perception that it is based on "old" technology. Despite this perception, the IC engine remains a successful device. Alternative power plants for light-duty vehicles include electric motors operated with batteries or fuel cells. Some advances have been accomplished regarding these technologies, but these alternatives are still many years away from displacing the IC engine. Especially considering the long time frame for replacement of existing vehicles in the current fleet, IC engines are expected to be the dominant power plant for many decades into the future.

1.2   Engine Types and Operation

Several versions of the IC engine exist. The two major categories are spark-ignition engines and compression-ignition (diesel) engines. The spark-ignition engine is based largely on a (nearly) homogeneous mixture of fuel and air, and on a more-or-less organized flame propagation. To satisfy this type of ignition and combustion process, the fuel must vaporize relatively easily and resist autoignition. Fuels with these characteristics include gasoline, natural gas, propane, and alcohols. The spark-ignition engine is often restricted to moderate compression ratios to avoid spark knock. Almost all spark-ignition engines for today's light-duty vehicles operate with stoichiometric mixtures and utilize three-way catalyst systems to meet emission regulations.

The compression-ignition engine, on the other hand, is based on the injection of the fuel into a cylinder with air, and on the self-ignition of the fuel due to the temperature of the compressed air. For the compression-ignition engine, combustion occurs in various locations throughout the cylinder with no organized flame propagation. To satisfy this type of ignition and combustion process, the compression ignition engine must utilize a fuel that can readily self-ignite. This fuel is typically a diesel fuel, but jet fuel and other oils can be used. The compression-ignition engine generally requires a relatively high compression ratio to generate sufficiently high temperature air for the auto-ignition process. These engines typically must operate with excess air (fuel lean) to ensure all the fuel is burned. In many applications, the compression-ignition engine uses intake air compression (turbochargers and superchargers) to increase its power density.

Another important classification of engines is the number of strokes the engine uses per power event. The four-stroke cycle engine uses four (4) strokes (two revolutions) for each cycle, while the two-stroke cycle engine uses two (2) strokes (one revolution) for each cycle. Almost all engines for light-duty vehicles use four-stroke cycle engines. Utility engines and some engines for small scooters and motorcycles use two-stroke cycle engines. This book will consider only the four-stroke cycle engines.

The IC engine is not a heat engine,[1] and is more accurately described as a chemical conversion device. This means that the "Carnot limitation" is not applicable. In fact, the IC engine may potentially approach 100% efficiency and still be consistent with the first and second laws of thermodynamics [2, 6]. This book will provide quantitative information on this feature of IC engines.

The IC engine consists of a series of processes which include induction, compression, combustion, expansion and exhaust. Each of these processes (and their sub-processes) are subject to real effects (irreversibilities and energy losses) which prohibit the engine from achieving 100% efficiency. For example, heat transfer from the cylinder gases is an energy loss that reduces the maximum possible efficiency. The implications of minimizing heat losses for higher efficiency are described in subsequent parts of this book.

Another, more severe, limitation is the irreversibility associated with the combustion process. For an adiabatic conversion of chemical energy to thermal energy, energy losses can approach zero. But the second law of thermodynamics describes that this process is highly irreversible and results in lower grade energy, and a loss of the potential to produce work. This, then, is the key thermodynamic fact that limits IC engine efficiency (and not the "Carnot limitation").

1.3 Reasons for Cycle Simulations

The focus of this book is the development and use of zero-dimensional, thermodynamic engine simulations. The overall goal of such thermodynamic engine simulations is to mathematically simulate the significant processes, and to predict engine performance and engine operation details including certain emission parameters. Benefits of developing and using these engine simulations are multiple. Examples of these benefits are described next.

1.3.1 Educational Value

The development of engine cycle simulations forces engineers to understand the fundamentals of engine operation, and to recognize the complex interaction of the various processes. The engineers obtain a much deeper appreciation for the role of thermodynamics, fluid flow, heat transfer, and combustion on engine performance and emissions.

1.3.2 Guide Experimentation

Since experimentation is a highly intensive activity involving many people, facilities, and funds, methods to minimize unnecessary experiments and to more efficiently conduct the

---

[1] Further discussion of the "heat engine" concept relative to IC engines is provided in Section 19.2.

experiments are important. Although thermodynamic engine cycle simulations may never completely eliminate experimentation, such simulations do help guide experimentation. This is accomplished in a number of ways, but the main result is that less experimentation will be needed. By calibrating an engine cycle simulation, fairly detailed and reliable results may be obtained. The use of these types of simulations and careful, efficient experimentation are a highly effective and systematic way to develop engines, and engine concepts and technology.

1.3.3 Only Technique to Study Certain Variables

Certain variables are difficult to study experimentally either because access is difficult or the very act of measurement will change the variable's value. For example, the overall spatial-average gas temperature (particularly during combustion) is difficult if not impossible to obtain. The cylinder pressure, however, may be measured and with the proper algorithm, the average gas temperature may be obtained using this pressure. Another example is the completion of second law analyses which are dependent on detail thermodynamic parameters.[2]

1.3.4 Complete Extensive Parametric Studies

To complete cost-effective, extensive parametric studies, the use of thermodynamic simulations is a proper choice. Since multiple sets of engine operating conditions are of increasing interest, experimentation could be prohibitively expensive. As an alternative, thermodynamic simulations can be used in a cost-effective manner to examine extensive sets of parameters.

1.3.5 Opportunities for Optimization

The use of the cycle simulations is ideal for determining optimal combinations of engine parameters for a given set of operating variables such as load, speed, spark timing, equivalence ratio, and so forth. Either a simple manual search or a more complex algorithm can be used to find the optimum set of parameters for a given set of constraints.

1.3.6 Simulations for Real-time Control

Versions of engine cycle simulations may be used as algorithms for engine control. An advantage of this approach is that the engine control will be able to accommodate a much wider range of variables. The control algorithm can employ a degree of intelligence in selecting the combination of operating parameters to meet specific goals. In addition, the controls and diagnostics can be integrated into a single, common system.

---

[2] Some authors have reported experimentally based second law analyses, but in these cases the authors have used their experimental data as input to analyses which contain at least some of the portions of thermodynamic cycle simulations.

1.3.7   Summary

The above are examples of the motivations which drive the interest in cycle simulations. From a practical perspective, the use of engine cycle simulations offers shorter engine development times, reduces development costs, minimizes emissions, maximizes fuel efficiency and performance, and provides a data base for current and future development efforts.

The development of engine cycle simulations is a challenging task largely because the IC engine is a complex device. The characteristics of IC engines which contribute to this complexity include turbulent, unsteady flow, non-uniform mixture composition, highly exothermic chemical reactions, two or three phase compositions, and pollutant species (typically with low relative concentrations). In addition, the important time scales have a large dynamic range of between 1 μs and 1 s, and the important length scales range roughly between 1 μm and 1 m.

1.4   Brief Comments on the History of Simulations

Chapter 3 provides a detailed description of the evolution of engine cycle simulations, but for completeness, this introductory chapter will summarize this evolution (e.g., References 1–3.) From the beginning of reciprocating engine development, engineers and scientists have attempted to model the overall engine operation to help understand and improve the technology. The earliest such models were the ideal air standard cycles which were first presented in the late 1800s. Although these were very crude, they helped the early engineers understand engine operation and actually provided some trend-wise insight. The history of engine model development shows a continual improvement of these models.

These ideal (air-standard cycle) models are based on the simple application of the first law of thermodynamics, and a set of simplifying assumptions and approximations. Improvements to the air standard cycle analysis were adopted in the 1930s. These improvements included using more realistic thermodynamic properties based on more realistic mixtures. Although this was a much more realistic model for the working gases, the predictions were improved only in a modest manner since the other assumptions and approximations were still present.

A great improvement relative to the air-standard cycle models was the development of thermodynamic engine cycle simulations. These developments have been reported since the early 1960s. These types of cycle simulations could not be developed prior to this time due to the lack of sufficient computer capability and, to a lesser degree, the lack of understanding of the important engine processes. The original simulations were based on one-zone and two-zone formulations, and then these types of simulations were extended to include three or more zones.

Beginning in the late 1970s, thermodynamic engine cycle simulations were more complete and many included predictions of emissions, particularly nitric oxides. An important improvement in these simulations was a more complete description of the combustion process using three or more zones. Several researchers have mentioned the importance of multiple zones for predicting nitric oxide emissions.

Thermodynamic engine simulations can provide important and detailed information on engine operation. Although the simplest thermodynamic simulations generally provide no detail spatial information, they do provide time (or crank-angle) varying quantities such as cylinder gas pressure and temperature, heat release, heat loss, intake and exhaust flow rates, and cylinder mass. In addition, these simulations are typically more practical than the multi-dimensional simulations for extensive parametric studies and other related investigations.

In summary, the development of engine cycle simulations may be described as continual improvements in two related aspects: (i) the detail descriptions of the thermodynamic properties, and (ii) the descriptions of the processes. These improvements continue as more understanding is obtained on the various engine processes. In addition, these simulations continue to be extended to include other features (such as second law analyses), and continue to find new applications (such as onboard engine control). Chapter 3 will expand on the above history for thermodynamic simulations, and also include some comments about multidimensional simulations.

1.5   Overview of Book Content

This book has been divided into a number of chapters to cover the formulation of engine simulations, required items and procedures for solutions, detailed and performance results, and case studies. The following are descriptions of each chapter:

Chapter 1: Introduction. This chapter is a brief summary of the role of engine simulations relative to designing and understanding IC engines.

Chapter 2: Overview of Engines and Their Operation. This chapter provides a brief overview of engine fundamentals, terminology, components, operation, and performance parameters.

Chapter 3: Overview of Engine Cycle Simulations. This chapter provides a description of the evolution of thermodynamic engine cycle simulations. The description begins with the simple ideal (air-standard) cycle analyses, and continues with the historical development of the thermodynamic simulations. Brief comments are included on quasi-dimensional thermodynamic simulations, and multi-dimensional (CFD) simulations.

Chapter 4: Properties of the Working Fluids. This chapter is a fairly comprehensive presentation on the development of the algorithms needed to determine the composition of the unburned and burned mixtures. Chemical equilibrium compositions for combustion products are described. The chapter ends with results for various properties for typical engine conditions.

Chapter 5: Thermodynamic Formulations. This chapter includes the details for developing the governing thermodynamic relations for the cylinder contents. The chapter ends with a concise summary of a set of differential equations for the cylinder pressure, zone temperatures, zone volumes and zone masses.

Chapter 6: Items and Procedures Required for Solutions. This chapter outlines all the required items to solve the governing differential equations developed in Chapter 5. These items include cylinder volumes, fuel burning rates, heat transfer, flow rates, friction, and other sub-models as needed. The chapter ends with comments on solution procedures and convergence to final answers.

Chapter 7: Basic Results. This chapter starts the second part of the book which is focused on presenting results using the engine cycle simulation. This chapter presents detailed, time resolved results for the base case condition. Included are results for pressures, temperatures, species, and residual fractions.

Chapter 8: Performance Results. This chapter provides a fairly complete presentation of overall global performance metrics as functions of engine operating and design parameters. Overall performance metrics include power, torque, thermal efficiency, and mean effective pressure. Engine operating and design parameters include inlet pressure, speed, load, combustion variables, compression ratio, equivalence ratio, EGR, and combustion timing.

Chapter 9: Second Law Results. This chapter includes the development of the aspects needed to complete second law assessments of IC engines. This includes the property exergy which allows the irreversibilities of the engine processes to be quantified. Results for these irreversibilities are presented as functions of engine operating and design parameters.

Chapter 10–18: Case Studies. These chapters consist of a number of case studies using the engine cycle simulations. These cases include studies of combustion, cylinder heat transfer, fuels, oxygen enriched reactants, over-expanded engines, nitric oxide emissions, and high efficiency engines.

Chapter 19: Summary: Thermodynamics of Engines. This final chapter highlights the thermodynamic features of IC engines. Many of these features are quantified by the cycle simulation results reported in this book.

References

1. Mattavi, J. N. and Amann, C. A. (eds.) (1980) *Combustion Modeling in Reciprocating Engines*, Plenum Press, New York.
2. Heywood, J. B. (1988). *Internal Combustion Engine Fundamentals*, McGraw-Hill Book Company, New York.
3. Ramos, J. I. (1989). *Internal Combustion Engine Modelling*, Hemisphere Press, Inc.
4. Mayersohn, N. (2013). The combustion engine refuses to die, Nautilus, issue 7, November. http://nautil.us/issue/7/waste/the-combustion-engine-refuses-to-die (accessed September 5, 2014).
5. Perry, M. J. (2013). Gas engine stands the test of time, *Detroit News*, June. http://www.aei.org/article/energy-and-the-environment/alternative-energy/gas-engine-stands-the-test-of-time/ (accessed September 5, 2014).
6. Lauck, F., Uyehara, O. A., and Myers, P. S. (1963). An engineering evaluation of energy conversion devices, Society of Automotive Engineers, *SAE Transactions*, **71**, 41–50.

2

Overview of Engines and Their Operation

Before discussing the details of engine cycle simulation development, the overall reciprocating engine design and operation will be briefly reviewed. This chapter will include descriptions of the goals of engine designs, engine classifications, engine characteristics, engine components, engine operating cycles, and engine performance parameters. The discussion of performance parameters includes descriptions of torque, power, mean effective pressure, thermal efficiencies, specific fuel consumption, and volumetric efficiency. These performance parameters may be determined on a gross indicated, net indicated or brake basis. This chapter is a brief overview of engines and engine operation—much more on these topics may be found in typical engine text books (e.g., References 1–5).

2.1   Goals of Engine Designs

Today's engines must meet several simultaneous goals to be successful. These goals are emphasized in different ways for different applications, but these goals are to some degree important in all market segments. These goals include acceptable performance (power and torque), high efficiency, low initial cost, able to meet emissions regulations, acceptable reliability, and low maintenance. In addition, in some market segments, fuel adaptability is important—that is, the engine is able to operate on two or more fuels. Obviously, realizing these goals means that the design and production of successful internal combustion engines is a difficult and expensive process.

As examples of the different ways that these goals are emphasized for different applications, the following contrasts the heavy-duty truck and the light-duty automotive markets. For the heavy-duty truck market, engine performance and efficiency often dominate the other considerations, whereas for the light-duty market, the initial cost may be of

An Introduction to Thermodynamic Cycle Simulations for Internal Combustion Engines, First Edition.
Jerald A. Caton © 2016 John Wiley & Sons, Ltd. Published 2016 by John Wiley & Sons, Ltd.

most concern. Similar differences in the other market segments could be cited as well. For all applications, the applicable emission regulations must be satisfied for the engine to be sold.

2.2   Engine Classifications by Applications

Internal combustion, reciprocating engines are prevalent in many economic sectors and dominant in some of these sectors. The following is a description of some of the major engine classifications by applications. There is considerable overlap between the various classifications, but the following descriptions provide a rough idea of the breath of internal combustion engine applications.

Small engines are best exemplified by utility and gardening applications such as trimmers, edgers, and lawn mowers. These engines are typically light and inexpensive, and most often based on one cylinder. Due to the requirements for light weight and low first cost, many of these applications use a two-stroke cycle design. (Two-stroke cycle and four-stroke cycle designs are described below.) Motorcycles and scooters are equipped with the next category of engines. These engines are also light, but may be a little more expensive, may utilize more than one cylinder, and most often will be based on a four-stroke cycle design.

The next classification is light-duty vehicles. This classification includes the typical automobile, pick-up truck, utility vehicle, and even some airplanes and helicopters. This is a large (if not the largest) market segment, and the engines range from some small two-cylinder engines to large eight-(or more) cylinder engines. For a number of reasons, recent production trends have favored the smaller engines (e.g., four-cylinder engines) for light-duty automotive applications. These engines may range from basic to full featured engines. In 2015, full featured spark-ignition (gasoline) engines included items such as direct cylinder injection, variable valve lift and timing, and sophisticated electronic control schemes. Such full featured engines are able to achieve high performance, high efficiency and low emissions.

Heavy-duty on-road and off-road vehicles would be the next classification. This classification includes a range of markets including trucks, earth moving vehicles, agricultural vehicles, and military vehicles. These applications almost exclusively utilize large, diesel engines for high efficiency, long life, and high torque. The next classification would include railroad locomotives and intermediate marine. Again, large diesel engines with a dozen or more cylinders would be used in these applications.

Stationary engines, largely for electric power generation, would be the next classification. These engines could be used for continuous power, standby power, emergency power, or temporary power. The engines may be of many types, but in the larger situations, they are almost always diesel engines. These engines typically operate at a constant speed which is selected to match the generator requirements.

The largest known reciprocating engines are for large marine applications. These engines are sometimes called cathedral engines, because the structure is more like a building than a typical engine. These diesel engines are often low speed (~100 rpm), two-stroke cycle designs, and often operate with low quality diesel fuel. They are able to achieve some of the highest efficiencies known for reciprocating engines due to their large size and low speed.

As the engines described above indicate, IC engines are designed for a wide range of applications. These examples, then, are a testament to the versatility of the IC engine.

2.3 Engine Characteristics

Engines may be described in a number of ways. The following are some major characteristics that may be used in designating engines.

1. Number and arrangement of cylinders: Most engines may be described as using an inline or a "V" configuration. Therefore, engines may be designated, for example, as I-4, V-6, V-8, V-16, ... Other arrangements include the use of horizontally opposed pistons (also known as flat or boxer engines), and various versions of rotary engine designs.
2. Working cycle: Engines may be based on a two-stroke cycle or a four-stroke cycle design. The material in this book is based solely on four-stroke cycle engines.
3. Valve/port arrangement: Valves are almost always installed in the cylinder head, but past designs have included valves mounted in the block. For two-stroke cycle engines, ports may be arranged in a number of ways with the goal to provide more complete and efficient scavenging [2].
4. Valve actuation: Valves may be actuated from below using the conventional push-rod and cam-in-block approach, or from above with an overhead cam arrangement. Valve actuation may be purely mechanical, pneumatic, hydraulic, electromagnetic, or some combination of these techniques.
5. Fuel: Obviously, engines may be designated by the fuel they used. This leads to naming conventions such as gasoline engine, diesel engine, propane engine, ...
6. Mixture presentation: The way that the fuel and air enter the engine is a major characteristic. Carburetion typically means that the fuel is admitted into the air stream and this mixture then enters the cylinder. Port injection is when the fuel is injected into or near the inlet port of each cylinder. Direct cylinder injection is the term for fuel injection into the cylinder—usually after the intake valve has closed.
7. Combustion chamber design: Typically, combustion chambers are described as open (single) or divided (multiple). Most engines utilize open chambers, but some engines may include multiple chambers. One category of engines that use the multiple chamber designs is the pre-chamber diesel engine.
8. Load control: Throttling the intake mixture (typically used by spark-ignition engines) and fuel metering (typically used by compression ignition engines).
9. Cooling: Engines are typically liquid or air-cooled. The liquid is almost always a water mixture (which includes some anti-freeze liquid components).
10. Ignition: Spark ignition (forced) or compression ignition (spontaneous). This is perhaps the most defining classification. Once the ignition classification is specified, many of the other engine characteristics are fixed. For example, for spark ignition, the engine must utilize a premixed fuel and air mixture, the fuel must vaporize fairly easily and resist knock, and the combustion process must be one of continuous flame propagation. For compression ignition, on the other hand, the fuel needs to enter the cylinder near TDC, the fuel needs to be relatively easy to autoignite, and combustion will occur at various sites with limited (or no) formal flame propagation.

As can be understood from the above, many items may be used to characterize a specific engine. More importantly, the above characterizations provide a way to understand the great variety of engine designs that are possible.

2.4 Basic Engine Components

Although the objective of this book is not to provide a comprehensive introduction to engines, it is useful for some brief comments to be included on the basic engine components and operation. Basic engine components include the engine structure which is often divided into the engine block and the cylinder head. The engine block includes the engine cylinders, pistons, camshaft (if located in the block), crankshaft and crank case. The cylinder head may include the spark plugs, overhead camshaft, valves, and (for some designs) fuel injectors. Valve systems include valve stems, seats, and return springs. Pistons include the piston, rings, and piston pin attachment. Connecting rods are the links between the piston and the crankshaft. Push rods are often part of the valve actuation system if the camshaft is located in the block. Other items include bearings, water cooling components, gaskets, oil components, and fuel pumps. Although these components are different for different versions of engines, they often share some basic similarities. More details on engine components may be found elsewhere [1–5].

2.5 Engine Operating Cycles

Two operating cycles are possible for reciprocating engines. The two-stroke cycle is a design based on accomplishing all events within one revolution (two strokes) of the crank [2]. This means that many of the events must overlap. For example, as the piston is moving away from the cylinder head during the expansion stroke, the exhaust ports must open so that exhaust can be completed before the return stroke (which will be part intake and part compression stroke). In theory, the two-stroke cycle engine should produce twice as much power as the four-stroke cycle engine, but due to practical limitations (related to the overlapping events) this is not realized. In fact, most two-stroke cycle engines struggle to have the performance of a comparable four-stroke cycle engine.

The four-stroke cycle is a design based on two revolutions (four strokes). Although not precisely true, this provides adequate time for the four main processes: intake, compression, expansion, and exhaust. All the work presented in this book is based on four-stroke cycle engines.

2.6 Performance Parameters

2.6.1 Work, Power, and Torque

The fundamental objective of an engine is to provide shaft work. The time rate of work is power—so, in practice, engines are rated by power. Torque is another related performance measure that provides yet another aspect of engine output. Another measure is mean effective pressure. All of these performance parameters and others are described in the following.

The work from an engine cycle is determined from,

$$W = \int p dV \tag{2.1}$$

For an engine cycle, this would be the area within the curve of a pressure–volume diagram. Figure 2.1 shows an example of cylinder pressure as a function of cylinder volume for

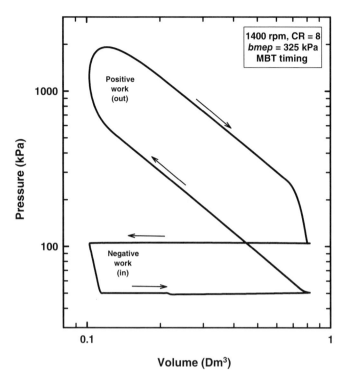

Figure 2.1   Cylinder pressure as a function of cylinder volume with the "work out" and the "work in" noted for a four-stroke cycle engine operating at part load and 1400 rpm

a four-stroke cycle engine using logarithmic scales.[1] As noted in the figure, the area of the compression-expansion (clockwise) part of the cycle is positive work out of the system, and the area of the exhaust-intake (counterclockwise) part of the cycle is negative work into the system. These two parts can be expressed as the gross indicated work and the pumping work:

$$W_{gross} = \int_{+180}^{-180} pdV \tag{2.2}$$

$$W_{pump} = \int_{+540}^{+180} pdV \tag{2.3}$$

The crank angle numbering scheme is such that "−180°aTDC" is defined to be at BDC prior to the compression stroke. This means that a crank angle position of 0.0°aTDC is TDC during the combustion phase of the process. The gross indicated work is the work associated with the compression and expansion strokes. This is called indicated work—that is, the gases are able to provide this work at the piston. If the pumping work (W_{pump}) is subtracted from the gross

---

[1] The use of log scales is particularly useful for showing the pressures during the exhaust and intake processes which are lower than the combustion pressures.

indicated work, the result is the net indicated work. Again, this is the net result of the gases acting on the piston.

$$W_{\text{net ind}} = W_{\text{gross}} - W_{\text{pump}} \qquad (2.4)$$

Finally, to get the work at the engine shaft (W_{shaft}), the friction work (W_{frict}) needs to be subtracted.

$$W_{\text{shaft}} = W_{\text{net ind}} - W_{\text{frict}} \qquad (2.5)$$

The shaft work is also known as brake or delivered work,

$$W_{\text{shaft}} = W_{\text{brake}} = W_{\text{delivered}} \qquad (2.6)$$

To get the equivalent power terms, the work values are multiplied by the engine speed. One additional consideration in doing this conversion is to note the number of revolutions per "event." Since this is a four-stroke cycle, two revolutions are needed for each event. So a conversion term is needed,[2]

$$n = 2\frac{\text{rev}}{\text{event}} \qquad (2.7)$$

Power is then found,

$$Power = \dot{W} = \int p\frac{dV}{dt} \qquad (2.8)$$

In terms of units,

$$Power = \dot{W} = \frac{W\left(\dfrac{\text{kJ}}{\text{event}}\right)N\left(\dfrac{\text{rev}}{\text{s}}\right)}{n\left(\dfrac{\text{rev}}{\text{event}}\right)} \qquad (2.9)$$

where N is the engine speed in "rev/s." This provides power with units of kJ/s or kW. As with the work terms, the power may be expressed on a "gross," "net ind," or "brake" basis.

$$Power_{\text{gross}} = \dot{W}_{\text{gross}} = \frac{W_{\text{gross}}N}{n} \qquad (2.10)$$

$$Power_{\text{net ind}} = \dot{W}_{\text{net ind}} = \frac{W_{\text{net ind}}N}{n} \qquad (2.11)$$

$$Power_{\text{brake}} = \dot{W}_{\text{brake}} = \frac{W_{\text{brake}}N}{n} \qquad (2.12)$$

---

[2] For a two-stroke cycle, the conversion term is 1 rev/event.

Torque may be determined from the power value

$$T_{\text{gross}} = \frac{Power_{\text{gross}}}{2\pi N} \tag{2.13}$$

In terms of units,

$$T_{\text{gross}} = \frac{Power_{\text{gross}}\left(\dfrac{\text{kJ}}{\text{s}}\right)}{2\pi\left(\dfrac{\text{rad}}{\text{rev}}\right)N\left(\dfrac{\text{rev}}{\text{s}}\right)} \tag{2.14}$$

This provides torque with units of "kJ" or "kN-m." Note that torque can be expressed in the three ways (as described above).

$$T_{\text{gross}} = \frac{Power_{\text{gross}}}{2\pi N} \tag{2.15}$$

$$T_{\text{net ind}} = \frac{Power_{\text{net ind}}}{2\pi N} \tag{2.16}$$

$$T_{\text{brake}} = \frac{Power_{\text{brake}}}{2\pi N} \tag{2.17}$$

2.6.2 Mean Effective Pressure

The mean effective pressure is the average cylinder pressure that provides the equivalent work of the actual cycle. Figure 2.2 shows the cylinder pressure as a function of cylinder volume for a wide open throttle condition (WOT). For this case (for one cylinder), the indicated work is 0.616 kJ/event, and the displaced volume is 0.717 dm^3. The rectangle in the figure represents the same area that applies to the compression and expansion strokes, and hence, provides the same gross indicated work value. This is accomplished with a mean pressure of 1006 kPa. Therefore, the indicated mean effective pressure is 1006 kPa. The mean effective pressure can be found from the work or power:

$$mep = \frac{W}{V_d} = \frac{\dot{W}n}{V_d N} \tag{2.18}$$

As with the other parameters, the mean effective pressure may be defined for the different cases:

gmep = gross indicated mep (2 strokes)

nmep = net indicated mep (4 strokes)

bmep = brake mep

fmep = friction mep

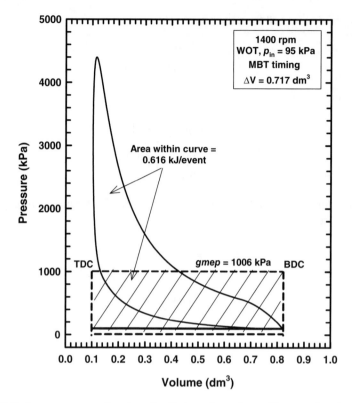

Figure 2.2   Cylinder pressure as a function of cylinder volume for a wide open throttle (WOT) condition. The gross indicated mean effective pressure (gmep) is noted

In terms of performance parameters, the mean effective pressure is a good way to compare different size engines. Small or large, the mean effective pressure is an indication of the relative power for that size engine. A high mean effective pressure indicates an engine design that is producing high output for its size.

2.6.3   Thermal Efficiencies

Thermal efficiencies are of major importance for engine designers. The thermal efficiency for engines is based on the amount of work (power) that is obtained for a certain amount of fuel energy (rate of fuel energy). The fuel energy is almost always specified by the lower heating value (LHV) of the fuel. Thermal efficiencies are given by

$$\eta_t = \frac{W}{m_{f,\text{burned}} LHV} \tag{2.19}$$

where $m_{f,\text{burned}}$ is the fuel burned. If all the fuel is not burned, then combustion efficiency is defined to reflect the amount of fuel actually burned,

$$\eta_c = \frac{m_{f,\text{burned}}}{m_{f,\text{input}}} \qquad (2.20)$$

The product of the two efficiencies is known as the fuel efficiency,

$$\eta_f = \eta_c \eta_t = \frac{W}{m_{f,\text{input}} LHV} \qquad (2.21)$$

These efficiencies, as with all the other parameters, may be expressed for the different cases (gross indicated, net indicated, and brake).

2.6.4   Specific Fuel Consumption

Another measure of the engine's efficiency is the specific fuel consumption (*sfc*) which is defined as

$$sfc = \frac{\dot{m}_{f,\text{input}}}{\dot{W}} \qquad (2.22)$$

The *sfc* is obviously related to the thermal efficiency,

$$sfc = \frac{1}{\eta_f LHV} \qquad (2.23)$$

The *sfc* may be expressed for the different cases (gross indicated, net indicated, and brake). For example, if brake specific fuel consumption (bsfc) is desired, the conversion (eq. 2.23) would use the brake fuel efficiency ($\eta_{f,brake}$).

2.6.5   Other Parameters

Another performance parameter is mechanical efficiency,

$$\eta_m = \frac{W_{\text{brake}}}{W_{\text{net ind}}} = \frac{\dot{W}_{\text{brake}}}{\dot{W}_{\text{net ind}}} = \frac{bmep}{nmep} \qquad (2.24)$$

Definitions of the mechanical efficiency vary in the literature. In some cases (e.g., [1]), the definition is based on the gross indicated work. In this case, the mechanical efficiency provides a measure of both the mechanical friction as well as the pumping work losses. Other authors (e.g., [5]) base the mechanical efficiency on the net indicated work—as defined in eq. 2.24. For this case, the mechanical efficiency provides a measure of only the mechanical friction. Values for mechanical efficiency may range between zero (idle) and about 95% (WOT).

Finally, one other performance parameter will be introduced. The volumetric efficiency is defined as the ratio of the mass of the actual air inducted and the mass of air that would occupy the displaced volume at some standard air density (ρ_{air}).

$$\eta_v = \frac{m_{air}}{\rho_{air} V_d} = \frac{\dot{m}_{air}}{\rho_{air} V_d N} \tag{2.25}$$

The volumetric efficiency is defined to provide a standard indication of how well the engine is provided air. This requires some definition for the air density (ρ_{air}) which could be defined for the ambient conditions, the inlet manifold conditions, the inlet port conditions, or some other specification. For this reason, any presentation of the volumetric efficiency should be accompanied by its definition. Typically, values of volumetric efficiency range between about 50% and over 100%. Volumetric efficiencies greater than 100% are possible due to dynamic flow charging, the definition of the air density, or other reasons.

2.7 Summary

This chapter provided descriptions of the goals of engine designs, classifications of engines, major engine characteristics, a brief list of engine components, two operating cycles, and performance parameters. The discussion of performance parameters included descriptions of torque, power, mean effective pressure, thermal efficiencies, specific fuel consumption, and volumetric efficiency. This section described that the performance parameters may be determined on a gross indicated, net indicated or brake basis. Much more detail on these topics may be found in many available engine books (e.g., References [1–5]).

References

1. Heywood, J. B. (1988). *Internal Combustion Engine Fundamentals*, McGraw-Hill Book Company, New York.
2. Heywood, J. B. and Sher, E. (1999). *The Two-Stroke Cycle Engine: Its Development, Operation, and Design*, Taylor & Francis, Philadelphia, PA.
3. Lumley, J. L. (1999). *Engines: an Introduction*, Cambridge University Press.
4. Ferguson, C. R. and Kirkpatrick, A. T. (2001). *Internal Combustion Engines: Applied Thermal Sciences*, second edition, John Wiley & Sons, Inc.
5. Pulkrabek, W. W. (2004). *Engineering Fundamentals of the Internal Combustion Engine*, second edition, Pearson Prentice Hall, New Jersey.

3

Overview of Engine Cycle Simulations

3.1 Introduction

This chapter will provide a brief overview of the various types and evolution of engine cycle simulations. Engine cycle simulations have been developed and used to study a variety of features and issues relative to internal-combustion engines for over half a century. A portion of these previous studies are mentioned in this chapter, but a comprehensive review is beyond the scope of this book. In the beginning, engine cycle simulations were fairly elementary, and were limited by both computing capabilities, and a lack of knowledge concerning key sub-processes. In time, these simulations have evolved into sophisticated tools that are used in a fairly standard fashion by all engine manufacturers. Many of the modern simulations contain advanced and detailed features for items such as fluid mechanics, combustion, chemical reactions, heat transfer, and friction.

This chapter will briefly describe the earliest attempts at engine analysis (air standard cycles), the development of thermodynamic simulations, and the related development of multi-dimensional simulations. The history of the development of engine cycle simulations is a record of continual improvement and sophistication in these models. An overview of this development history is provided next.

3.2 Ideal (Air Standard) Cycle Analyses

From the beginning of reciprocating engine development, engineers and scientists have attempted to model the overall engine operation to help understand and improve the technology. The earliest such models were based on ideal (air standard) cycles. Although these were very crude, they helped the early engineers understand engine operation and actually

An Introduction to Thermodynamic Cycle Simulations for Internal Combustion Engines, First Edition.
Jerald A. Caton © 2016 John Wiley & Sons, Ltd. Published 2016 by John Wiley & Sons, Ltd.

provided some trend-wise insight. The first of these models were developed in the late 1800s.[1] These models are based on the simple application of the first law of thermodynamics, and a set of simplifying assumptions and approximations. The most common set of assumptions and approximations are

1. the working fluid is approximated as pure air (often based on constant properties);
2. the working fluid is spatially uniform;
3. the compression and expansion processes are adiabatic and reversible;
4. the heat release due to combustion is simulated by heat addition under prescribed conditions (e.g., constant volume at TDC);
5. the exhaust blow-down process is simulated by a constant volume heat loss at BDC;
6. inlet and exhaust displacement processes are not included.[2]

Since these assumptions and approximations are based on ideal processes, the results are typically much too optimistic compared to actual engine performance. The trends of a couple of important engine characteristics, however, are predicted (this is described below).

Improvements to the air standard cycle analysis were adopted in the 1930s [2]. These improvements included using more realistic thermodynamic properties based on more realistic mixtures. For example, during the compression process, the working fluid was modeled as a mixture of fuel vapor, air, and residual gases. After combustion, the working fluid was modeled as consisting of equilibrium products for the appropriate inlet mixture. Although this was a much more realistic model for the working gases, the predictions were improved only in a modest manner since the other assumptions and approximations were still present.

This set of assumptions and approximations define the ideal (air standard) cycles. Special versions of these cycles have been given specific names:

1. "Otto" cycle—constant volume heat addition. This version is named after Nikolaus Otto (1832–1891), a German inventor who built the first practical four-stroke cycle engine in 1876. This version of the air standard cycle is often thought to best represent the spark-ignition engine.
2. "Diesel" cycle—constant pressure heat addition. This version is named after Rudolf Diesel (1858–1913). He patented (1898) an engine concept that attempted to have constant temperature combustion with the intent that the engine could approach the Carnot heat engine concept. In practice, the original versions of this engine possessed more of a constant pressure combustion event. This explains the naming of this version of the air standard cycle. Most modern diesel engines do not possess a constant pressure combustion process.
3. "Dual" cycle—combination of constant volume and constant pressure heat addition. This version of the air standard cycle might be expected to result in a better representation of the time-varying cylinder pressure for both SI and CI engines. The disadvantage is that this version requires two additional constants that must be specified.

---

[1] Direct evidence of this knowledge was not found, but indirect evidence is available (e.g., Reference 1).
[2] Some versions of the ideal cycle analyses do attempt to include some elementary models for the intake and exhaust processes [3].

4. "Atkinson" cycle—larger expansion ratio than compression ratio. This version of the air standard cycle provides an expansion stroke that is longer than the compression stroke. This cycle has the potential to supply additional work compared to a similar engine with equal compression and expansion strokes. In practice, the larger expansion ratio can reduce the effectiveness of the exhaust process. A case study on this concept using the engine cycle simulation is presented at the end of the book.

Results from these ideal cycles are widely available (e.g., Reference 3), and these results will not be repeated here. The performance parameters such as efficiency, work, torque, and power predicted by the air standard cycles are often a factor of about two too high. This defect is largely due to the assumptions of heat addition instead of actual combustion, adiabatic and reversible processes, and no gas exchange. The assumption of air as a working fluid is a more modest weakness. The deficiencies of these ideal cycles have been quantified elsewhere [4].

The simple, ideal cycle analyses do, however, provide a couple of trend-wise results that are consistent with actual practice. These are that efficiency and performance increase with (i) increasing compression ratio, and (ii) increasing ratio of specific heats. These aspects are actually quite significant, and attest to the fact that for all the simplicity, the ideal cycles do capture at least some of the important thermodynamics associated with IC engines. Appendix A is a description of the thermodynamic implications from the simple "Otto" cycle.

3.3 Thermodynamic Engine Cycle Simulations

The development and use of thermodynamic engine cycle simulations have been reported since the early 1960s. As mentioned above, these simulations were not fully possible until computer capability increased, and understanding of the various sub-processes improved. The original simulations were based on one-zone and two-zone formulations, and utilized a number of simplifying assumptions and approximations. Patterson and van Wylen [5] reported one of the first thermodynamic simulations for a spark-ignition engine in the early 1960s that included unburned and burned zones, progressive combustion, heat transfer, and flame propagation for uniform (homogeneous) charge engines. Their simulation did not, however, include flow rates, heat transfer from the unburned zone, or heat transfer during the compression process. Nevertheless, in many ways this simulation was a signal achievement, and the start of a long history of the development and use of thermodynamic engine simulations. Interestingly, this work also included one of the first presentations of the results of an exergy (availability) analysis.

McAulay et al. [6] and Krieger and Borman [7] reported in the early 1960s on the development of thermodynamic simulations for compression-ignition engines. Their simulations were fairly complete, but a major weakness was the lack of a comprehensive description of the complex diesel engine combustion process. Engine cycle simulations improved rapidly during the 1960s and early 1970s. Bracco [8] summarized some of these developments.

A study by Heywood et al. [9] in 1979 described one of the first simulations using three zones for the combustion process. For their formulation, the burned zone was divided into an adiabatic core and a boundary layer zone (in addition to the unburned zone). The introduction of the adiabatic core is especially important for the accurate prediction of nitric oxide species due to the high dependence on temperature of the chemical kinetics for the nitric oxide reactions. The use of three zones for the combustion process is employed in the current work. The details of this formulation and their results are provided in subsequent portions of this book.

Blumberg et al. [10] in 1979 published a review of these types of cycle simulations, and described the use of multiple zone simulations. In particular, they noted that the use of a boundary layer zone allows a better representation of the heat loss which is known to be confined largely to the region near the walls. James [11] in 1982 also commented on the importance of multiple zones for predicting nitric oxide emissions. In addition, Raine et al. [12] in 1995 published results from a cycle simulation using multiple zones for the burned gas. They were able to show the importance of multiple zones especially for the computation of nitric oxides. This work did not include the specific use of a boundary layer and did not include flows.

A complete citation of all investigations in this area is beyond the scope of the current work, but the above is representative of this work. A good reference for the state of the art regarding engine simulations at the end of the 1970s is a collection of papers presented at a 1978 conference [13]. Since that time, the development and use of thermodynamic engine cycle simulations has continued to make progress. The use of such simulations is now a routine aspect of engine design and development. Most major engine companies have their own "in-house" version of thermodynamic simulations. Reviews of the development of these simulations are provided by Ramos [14] and Primus [15]. A related review is supplied by Rakopolpulas and Giakoumis [16] on engine simulations for transient operation. Transient engine operation is not included in the current book. In addition, beginning in the 1980s, several companies offer commercial engine cycle simulation products. Some of these commercial simulations are described at the end of this chapter.

3.4   Quasi-dimensional Thermodynamic Engine Cycle Simulations

The basic thermodynamic engine cycle simulation has no spatial context. This aspect can be improved by developing empirical relations for the location of the flame front. By using the geometric structure of the combustion chamber and using a flame propagation velocity and geometry, the flame may be identified and assigned a spatial location. These models are often called "quasi-dimensional" since there is a sense of dimensionality, but these models do not have a strict mathematical dependence on any of the dimensions.

Some of these models use elaborate algorithms to predict the flame development and growth. These models may use laminar flame velocity data augmented with turbulence considerations. Other models may use a turbulent entrainment submodel with a laminar burning process based on a characteristic length scale. Several references are available that review the status of these models [17,18].

Often these quasi-dimensional engine cycle simulations will assume a spherical flame shape. This provides a way to "locate" the flame relative to combustion chamber features such as spark plug, piston top, or cylinder head. Many of these models need sophisticated geometric sub-routines to compute the complex intersections of the spherical flame front and the various boundaries.

In general, the quasi-dimensional engine cycle simulations have proven successful in predicting engine performance for spark-ignition engines. The level of empirical input, however, is often quite high and the simulation may be "tuned" to provide good agreement with experimental information. Although these simulations may provide some added benefits, they are also less fundamental and may mask some of the fundamental thermodynamic features of the engine. For the current purposes, the results presented in this book are limited to zero-dimensional, thermodynamic simulations.

3.5 Multi-dimensional Simulations

The development of full multi-dimensional engine simulations has a long and successful history. To a large degree, multi-dimensional simulations have been developed in parallel with the thermodynamic simulations, and serve a different purpose. The multi-dimensional simulations, in general, are based on the full set of governing partial-differential conservation equations. Often these simulations are considered a part of computational fluid dynamics (CFD) work. To complete the solution of the three-dimensional governing equations, the simulations use sub-models for items such as the turbulence, chemical reactions, heat transfer, fuel jet behavior, and boundary layer processes. The potential of these simulations is that the detailed spatial and time information is available in an exact manner from the fundamental governing equations. The obvious disadvantages are that the computations are time-consuming, and that the detail sub-models are not always known.

An important aspect of these simulations is establishing the three-dimensional computational grid. For actual combustion chambers with irregular surfaces, this can be a complex procedure. Some of the issues with developing this computational grid are that different regions of the computational space need different grid resolution. For example, the region near surfaces may need a finer grid to capture the processes in the boundary layer whereas the grid in the center of the cylinder may need less resolution.

The solution of these governing equations is difficult and requires intensive computer time. The time and length scales differ significantly so that no simple computational scheme can be employed with universal success.

Obviously, the advantages of the multi-dimensional simulations are that much of the high level physics and chemistry is inherent in the results. For example, the chemistry and turbulence can be determined throughout the combustion chamber. The local temperatures are likely to be much more realistic than the simple approach used by the thermodynamic simulations. In addition, assessments of the effects of combustion geometry changes should be much more accurate.

The following are some brief comments that compare and contrast the multi-dimensional and the thermodynamic simulations. The two categories of simulations have different goals and provide different types of outputs. These two categories of simulations are often used together in a synergistic fashion—the strengths of each helping to better understand engine operation. The obvious contribution of the multi-dimensional simulations is that the spatial and temporal information is highly detailed and complete. These details may help better understand the fuel sprays, fuel vaporization, fuel vapor and air mixing, local ignition, combustion, local heat transfer, and emissions formation. Most of these items (on a detailed spatial basis) are outside the scope of the basic thermodynamic simulations. On the other hand, determining overall thermodynamic states during the engine cycle for a variety of engine design and operating conditions is readily obtained from the thermodynamic simulations and would be impractical from the multi-dimensional simulations.

The next discussion will provide a brief review of the status of multi-dimensional simulations as of 2015. This is a highly evolving field due to advances in computing processing (faster hardware), advances of computing techniques (faster software), and advances in understanding the physical and chemical processes. Shi and Reitz [19] provide a review of these types of simulations as of 2010. They summarize the basic approach of CFD type computations, and the modeling needed for turbulence, spray and evaporation, combustion, pollutant emissions, and cylinder heat transfer. They also include a discussion of approaches for computationally efficient strategies for diesel engine combustion with detailed chemistry. Shi and Reitz end this review by commenting on the various computer codes available. These include the KIVA CFD codes which are largely open-source codes. Other engine CFD codes are commercial and include CFX, CONVERGE, FIRE, FLUENT, FORTÉ, STAR-CD, and VECTIS [19]. More details on KIVA and the commercial codes are provided at the end of this chapter.

As an example, the evolution of the KIVA codes will be described next. The initial KIVA codes were developed in the early 1980s by a team at Los Alamos National Laboratory, and were initially released to a select few industrial and academic partners. The first public release of the code occurred in 1985, and the code continues to be available to the public. Much of the success of the KIVA code has been attributed to the close collaboration with the users in industry, national laboratories and academia [20]. The KIVA codes continue to evolve both with more efficient computational techniques and submodel capabilities. Some examples of strategies to reduce the computational effort and numerical inaccuracies are outlined by Shi *et al.* [21]. Many other references may be cited for examples of multidimensional engine simulation work (e.g., References 22 and 23). As this is not the main thrust of the current work, no further comments will be provided regarding multi-dimensional simulations.

3.6   Commercial Products

A listing of commercial products is problematic since after this book is published the companies listed may cease to exist and new ones may be formed. In spite of this concern, this subsection will provide a brief listing of the products that are available as of 2015. The following list is provided in two parts: (i) thermodynamic/quasi-dimensional simulations, and (ii) multi-dimensional simulations. Related software that is not focused on reciprocating engines is not included.

3.6.1   Thermodynamic Simulations

The following is a list of these types of engine cycle simulations—listed in alphabetical order.

a. AVL Boost (AVL List GmbH): Boost was created within AVL's department for applied thermodynamics. The task of engine analysis using thermodynamic measurements and calculations is a special focus of this code [24].

b. GT-Power (Gamma Technologies, Inc.): Formed in 1994, Gamma Technologies provides a suite of engine and vehicle simulations that include the base engine program, GT-Power [25].

c. Lotus Engine Simulation (Lotus Cars, Ltd.): This program appears to be based on the engine simulation software that the company uses in their own engine design and development activities [26].

d. Virtual Engine (Optimum Power Technology): Originally developed by Professor Gordon Blair [27,28], this engine simulation software is available in two versions: for two-stroke cycle and for four-stroke cycle engines. The company has other related software products [29].

e. Wave (Ricardo, plc): Originally formed in 1983 as Integral Technologies, Inc., Ricardo purchased the company in 1988. The existing engine cycle simulation was renamed *Wave*, and has continued to evolve [30].

3.6.2 Multi-dimensional Simulations

Multi-dimensional simulations have a close connection to other types of computational tools in the computational fluid dynamics (CFD) category. As described above, the first developments were supported by the United States government and originated from work at Los Alamos Laboratory. Commercial products were introduced beginning in the 1980s and new products continue to become available. The following is a list of these types of engine cycle simulations—listed in alphabetical order.

a. AVL FIRE (AVL List GmbH): Fire is a multi-purpose computational package with an emphasis on fluid flows for IC engines [31].

b. CFX (Ansys, Inc.): This software has been used for engine CFD computations, but has many other applications [32].

c. CONVERGE (Convergent Sciences, Inc.): An engine simulation code which claims robust spray and combustion models using adaptive mesh refinement (AMR) technology [33].

d. FLUENT (ANSYS, Inc.): A general CFD package that includes special models that give the software the ability to model in-cylinder combustion [34].

e. FORTÉ (Reaction Design): Simulation package for combustion engines that incorporates CHEMKIN-PRO solver technology [35].

f. KIVA (Los Alamos National Laboratory): As described above, this development began in the 1980s, and is available as an open source code [36].

g. OpenFOAM (OpenCFD, Ltd.): An open source CFD software package, licensed and distributed by the OpenFOAM Foundation and developed by OpenCFD Ltd. [37]. This software is being developed by several groups for engine applications (e.g., Reference 38).

h. STAR-CD (CD-Adapco): The original software for the current package was one of the first commercial multi-dimensional engine codes. The original software was produced by Computational Dynamics, London, England—a company that was started in 1987 [39].

i. VECTIS (Ricardo, plc): A three-dimensional fluid dynamics program that has been developed specifically to address fluid flow simulations in the vehicle and engine industries. This software includes an automatic mesh generator [40].

3.7 Summary

The three classifications of engine simulations provide three different sets of tools for understanding engine operation. Each type of simulation has its own strengths and weaknesses. Obviously, the basic thermodynamic engine cycle simulation may provide the most robust, rapid results. These simulations are ideal for routine "scoping" and fundamental thermodynamic studies. In addition, these simulations are ideal for intensive, parametric studies where a wide range of values of many parameters need to be investigated. The quasi-dimensional engine cycle simulation is a good choice where information is needed relative to different combustion geometry over a fairly wide range of values and variables. Although the quasi-dimensional simulations are not based on fundamental governing equations for the spatial dependence, these simulations have proved to be useful. Finally, the multi-dimensional cycle simulations may be used to study details at one or a few specific engine operating conditions. These multi-dimensional simulations are the only way to capture spatial information based on the governing equations.

As a summary of the above, Figure 3.1 shows a simplified chart of the evolution of engine cycle simulations. Two categories of features are used to illustrate this evolution. Along

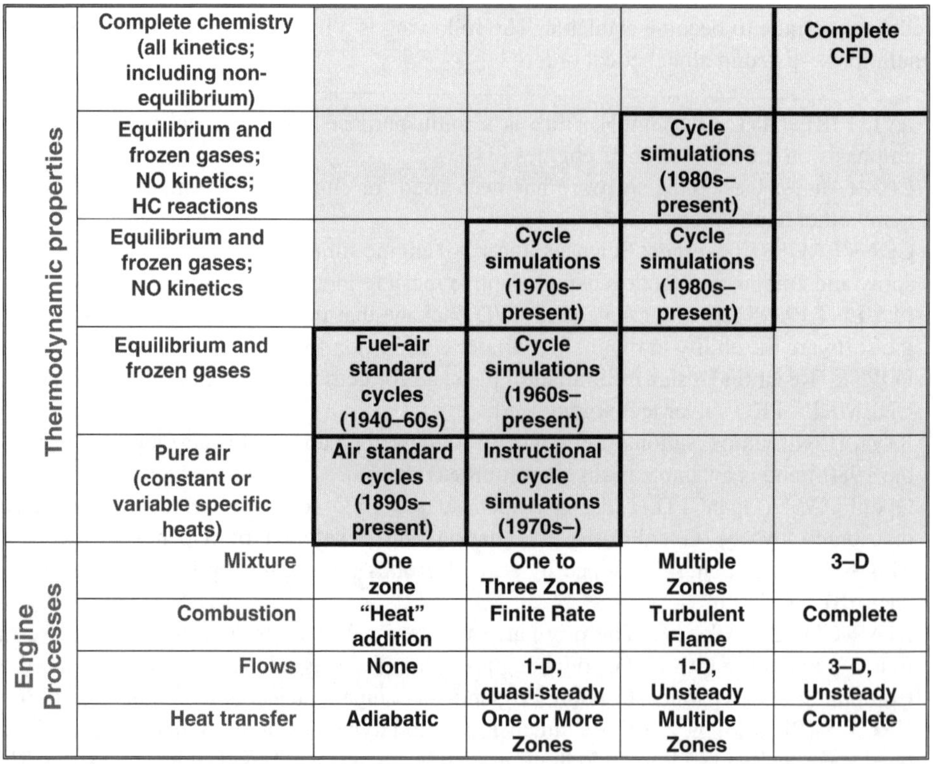

Thermodynamic properties	Complete chemistry (all kinetics; including non-equilibrium)				Complete CFD
	Equilibrium and frozen gases; NO kinetics; HC reactions			Cycle simulations (1980s–present)	
	Equilibrium and frozen gases; NO kinetics		Cycle simulations (1970s–present)	Cycle simulations (1980s–present)	
	Equilibrium and frozen gases	Fuel-air standard cycles (1940–60s)	Cycle simulations (1960s–present)		
	Pure air (constant or variable specific heats)	Air standard cycles (1890s–present)	Instructional cycle simulations (1970s–)		
Engine Processes	Mixture	One zone	One to Three Zones	Multiple Zones	3–D
	Combustion	"Heat" addition	Finite Rate	Turbulent Flame	Complete
	Flows	None	1-D, quasi-steady	1-D, Unsteady	3–D, Unsteady
	Heat transfer	Adiabatic	One or More Zones	Multiple Zones	Complete

Figure 3.1 A simplified description of the evolution of thermodynamic engine cycle simulations

the x-axis, the engine processes of mixture formation, combustion, flows, and heat transfer are listed. As the simulations improve from left to right, the description of these processes improves. For example, for combustion, the simplest sub-model is heat addition as used in the ideal (air standard) analyses. The next improvement for the combustion sub-model is a finite rate expression that is a simple input. A further improvement would involve some form of modeling the turbulent flame. Finally, at the furthest right location, the combustion would be modeled with the complete partial differential governing equations with proper sub-models for the turbulence, combustion reactions, and local heat transfer. The other engine processes improve from left to right in a similar manner.

Along the y-axis, the algorithms used to find the thermodynamic properties are listed. Starting at the bottom and proceeding to the top, the algorithms become progressively more thorough and representative of the actual working fluid. At the bottom, the properties are determined for pure air with constant specific heats. This would be the approach used in the ideal (air standard) cycle analyses. As the simulations improve from bottom to the top, the properties are more completely described. These improvements include using chemical equilibrium descriptions for the combustion products, using chemical kinetics to predict emissions, and at the most complete level, using complete chemistry at all times to have the most accurate depiction of the species concentrations and the thermodynamic properties.

The intersection of an engine process description and a thermodynamic property approach provides a specific type of engine simulation. The least accurate of both the process description and the property approach intersects at the "air standard cycles" while some of the intermediate intersections are for versions of the thermodynamic cycle simulation. The final intersection at the top, right-hand corner represents a complete computational fluid dynamic solution. This does not exist, but is a goal of current development.

References

1. Clerk, D. (1882). The theory of the gas engine, *Minutes of the Proceedings of the Institution of Civil Engineers*, **69**, 220–250.
2. Hershey, R. L., Eberhardt, J. E., and Hottel, H. C. (1936). Thermodynamic properties of the working fluid in internal combustion engines, *SAE Journal*, **39**, 409–419, October.
3. Heywood, J. B. (1988). *Internal Combustion Engine Fundamentals*, McGraw-Hill Book Company, New York.
4. Shyani, R. G., Jacobs, T. J., and Caton, J. A. (2011). Quantitative reasons that ideal air-standard engine cycles are deficient, *International Journal of Mechanical Engineering Education*, **39** (3) 232–248.
5. Patterson, D. J. and van Wylen, G. (1964). A digital computer simulation for spark-ignited engine cycles, in *Digital Calculations of Engine Cycles*, SAE Progress in Technology, 7, Society of Automotive Engineers, New York.
6. McAulay, K. J., Wu, T., Chen, S. K., Borman, G. L., Myers, P. S., and Uyehara, O. A. (1965). Development and evaluation of the simulation of the compression-ignition engine, Society of Automotive Engineers, SAE paper no. 650451.
7. Krieger, R. B. and Borman, G. L. (1966). The computation of apparent heat release for internal combustion engines, American Society of Mechanical Engineers, ASME paper no. 66WA/DGP-4.
8. Bracco, F. V. (1974). Introducing a new generation of more detailed and informative combustion models, Society of Automotive Engineers, SAE paper no. 741174, November.
9. Heywood, J. B., Higgins, J. M., Watts, P. A., and Tabaczynski, R. J. (1979). Development and use of a cycle simulation to predict SI engine efficiency and NO_x emissions, Society of Automotive Engineers, SAE Paper No. 790291.

10. Blumberg, P. N., Lavoie, G. A., and Tabaczynski, R. J. (1979). Phenomenological models for reciprocating internal combustion engines, *Progress in Energy and Combustion Science*, **5**, 123–167.
11. James, E. H. (1982). Errors in NO emission prediction from spark ignition engines, Society of Automotive Engineers, SAE Paper No. 790291.
12. Raine, R. R., Stone, C. R., and Gould, J. (1995). Modeling of nitric oxide formation in spark ignition engines with a multizone burned gas, *Combustion and Flame*, **102**, 241–255.
13. Mattavi, J. N. and Amann, C. A. (eds.) (1980). *Combustion Modeling in Reciprocating Engines*, Plenum Press, New York.
14. Ramos, J. I. (1989). *Internal Combustion Engine Modelling*, Hemisphere Press, Inc.
15. Primus, R. J. (2014). Reflections on the evolution of engine performance prediction, in proceedings of the 2014 ASME-ICED Fall Technical Conference, Columbus, IN, October.
16. Rakopoulos, C. D. and Giakoumis, E. G. (2006). Review of thermodynamic diesel engine simulations under transient operating conditions, Society of Automotive Engineers, paper no. 2006-01-0884.
17. Heywood, J. B. (1994). Combustion and its modeling in spark-ignition engines, International Symposium COMODIA 94.
18. Verhelst, S. and Sheppard, C. G. W. (2009). Multi-zone thermodynamic modelling of spark-ignition engine combustion – an overview, *Energy Conversion and Management*, **50**, 1326–35.
19. Shi, Y. and Reitz, R. D. (2010). Multi-dimensional modelling of diesel combustion: review, ch. 15 in *Modelling Diesel Combustion*, Lakshmi Narayanan, P. A. and Aghav, Y. V. (eds.), Springer Science+Business Media.
20. Amsden, D. C. and Amsden, A. A. (1993). The KIVA story: a paradigm of technology transfer, *IEEE Transactions on Professional Communication*, **36** (4) 190–195.
21. Shi, Y., Ge, H-W and Reitz, R. D. (2011). *Computational Optimization of Internal Combustion Engines*, Springer-Verlag, London.
22. Johnson, N. L. (1996). The legacy and future of CFD at Los Alamos, 1996 Canadian CFD Conference, Ottawa, Canada.
23. Reitz, R. D. and Rutland, C. J. (1995). Development and testing of diesel engine CFD models, *Progress in Energy and Combustion Science*, **21**, 173–196.
24. AVL Boost (2014). https://www.avl.com/boost (accessed June 27, 2015).
25. GT-Power (2014). http://www.gtisoft.com/ (accessed June 27, 2015).
26. Lotus Engine Simulation (2014). http://www.lotuscars.com/us/engineering/engineering-software, (accessed June 27, 2015).
27. Blair, G. P. (1999). *Design and Simulation of Four-Stroke Engines*, Society of Automotive Engineers, Inc., Warrendale, PA.
28. Blair, G. P. (1996). *Design and Simulation of Two-Stroke Engines*, Society of Automotive Engineers, Inc., Warrendale, PA.
29. Virtual Engine (2014). http://www.optimum-power.com/ (accessed June 27, 2015).
30. Wave (2014). http://www.ricardo.com/en-GB/What-we-do/Software/Products/WAVE/ (accessed June 27, 2015).
31. AVL Fire (2014). https://www.avl.com/web/ast/fire (accessed June 27, 2015).
32. CFX (2014). http://www.ansys.com/Products/Simulation+Technology/Fluid+Dynamics/Fluid+Dynamics+Products/ANSYS+CFX (accessed June 27, 2015).
33. Converge (2014). http://www.convergecfd.com/ (accessed June 27, 2015).
34. Fluent (2014). http://www.ansys.com/Products/Simulation+Technology/Fluid+Dynamics/Fluid+Dynamics+Products/ANSYS+Fluent (assessed 2014).
35. Forte (2014). http://www.reactiondesign.com/products/forte/ (accessed June 27, 2015).
36. KIVA (2014). https://www.lanl.gov/orgs/t/t3/codes/kiva.shtml (accessed June 27, 2015).
37. Open-FOAM (2014). http://www.openfoam.org/ (accessed June 27, 2015).
38. Contino, F., Jeanmart, H., Lucchini, T., and D'Errico, G. (2011). Coupling of *in situ* adaptive tabulation and dynamic adaptive chemistry: an effective method for solving combustion in engine simulations, *Proceedings of the Combustion Institute*, **33**, 3057–64.
39. Star-CD (2014). http://www.cd-adapco.com/products/star-cd%C2%AE (accessed June 27, 2015).
40. VECTIS (2014). http://www.ricardo.com/en-GB/What-we-do/Software/Products/VECTIS/ (accessed June 27, 2015).
41. Caton, J. A. (2014). On the importance of specific heats as regards efficiency increases for highly dilute IC engines, *Energy Conversion and Management*, **79**, 146–160.

Appendix 3.A: A Brief Summary of the Thermodynamics of the "Otto" Cycle Analysis[3]

This appendix contains a brief summary of the thermodynamics associated with the simple "Otto" cycle analysis. As described in the main text, the "Otto" cycle analysis is based on several simplifying assumptions and approximations. The analysis assumes adiabatic and reversible compression and expansion, and constant volume heat addition at TDC. Since the analysis does not include any gas exchange strokes, the geometric compression ratio is the effective compression ratio. Other simplifications may be found in complete descriptions of the analysis [3].

The results of the "Otto" cycle analysis are well known, and include the increase of thermal efficiency with increases of compression ratio and with increases of the ratio of specific heats. Examples of these results are provided below. Outside of these two parameters, however, the "Otto" cycle analysis provides little useful information. But the importance of compression ratio and the ratio of specific heats is significant, and the following will elaborate on these features of the "Otto" cycle results.

Of particular interest to the work presented in this book is the role of the ratio of specific heats. The ratio of specific heats is often given the symbol gamma (γ), and for simplicity "gamma" will be used in the following discussion as a replacement for the phrase "ratio of specific heats."

Although gamma is the property used throughout this discussion, the specific heats are the parameters of direct interest. Gamma and the specific heats are related:

$$\text{gamma} = \gamma = \frac{C_p}{C_v} \qquad C_p = C_v + R \qquad \gamma = 1 + \frac{R}{C_v}$$

Note that as the specific heats decrease, gamma increases.

For a simple "Otto" cycle [3], the effects of gamma (or the specific heats) can be illustrated for three processes. First, the compression stroke, the heat addition, and expansion stroke will be discussed. Then, the overall cycle will be examined. This analysis is based on "constant properties" and follows standard procedures [3].

For the (adiabatic) compression work,

$$W_{comp} = mC_v(T_2 - T_1)$$

$$W_{comp} = mC_vT_1(CR^{\gamma-1} - 1)$$

$$W_{comp} = m\frac{R}{\gamma - 1}T_1(CR^{\gamma-1} - 1)$$

$$\frac{W_{comp}}{mRT_1} = \frac{(CR^{\gamma-1} - 1)}{\gamma - 1}$$

---

[3] Another version of the material in this appendix is available in Reference 41.

Figure 3.A.1 shows the cylinder pressure (relative to the initial pressure, p_1) as functions of the cylinder volume (relative to the minimum volume, V_{min} at TDC) for three values of gamma for a compression ratio of 16. The compression work is proportional to the area under the curves. As the above expressions indicate, the compression work increases as gamma increases. The differences are modest for these conditions. For increases of gamma, however, this increase of compression work has a slightly negative effect for the overall cycle.

The heat addition (that simulates combustion) results in temperature and pressure increases, and these increases depend on the specific heats

$$\Delta T = \frac{Q_{in}}{C_v}; \quad \Delta p)_V = \frac{mR}{V} \Delta T = \frac{mR}{V} \frac{Q_{in}}{C_v}$$

As the specific heats decrease (gamma increases), the temperatures and pressures after the heat addition increase. These higher pressures provide the potential for higher expansion work. This is the major effect of the specific heats on engine performance, and these effects are clearly illustrated elsewhere in this book for actual engines with progressive combustion processes.

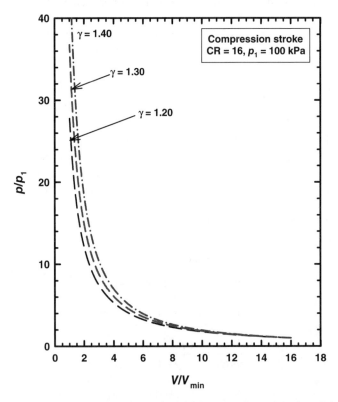

Figure 3.A.1 The cylinder pressure (relative to the initial pressure) as a function of the cylinder volume (relative to the minimum volume) for the compression stroke for three values of gamma for a compression ratio of 16. *Source:* Caton 2014b. Reproduced with permission from Elsevier

For the "Otto" cycle, Figure 3.A.2 shows the cylinder pressure (relative to the initial pressure) as functions of the cylinder volume (relative to the minimum volume) for three values of gamma for the heat addition portion of the cycle. "State 2" (end of compression) is assumed to have a p_2/p_1 of 50 for consistency. For this evaluation, a heat addition of 2600 kJ/kg is assumed. The resulting cylinder pressure after the heat addition is highest for the highest gamma value. This is the dominant characteristic that leads to higher efficiencies as gamma increases.

For the (adiabatic) expansion work,

$$W_{exp} = mC_v(T_3 - T_4)$$

$$W_{exp} = mC_vT_3(1 - \frac{1}{CR^{\gamma-1}})$$

$$W_{exp} = m\frac{R}{\gamma-1}T_3(1 - \frac{1}{CR^{\gamma-1}})$$

$$\frac{W_{exp}}{mRT_3} = \frac{(1 - \frac{1}{CR^{\gamma-1}})}{\gamma-1}$$

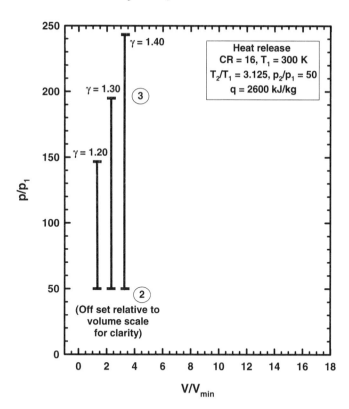

Figure 3.A.2 The cylinder pressure (relative to the initial pressure) as a function of the cylinder volume (relative to the minimum volume) for the heat addition for three values of gamma for a compression ratio of 16. All three results should be located at V/V_{min} of 1.0, but for clarity they are shown slightly offset from this value. *Source:* Caton 2014b. Reproduced with permission from Elsevier

As the specific heats decrease (gamma increases), the expansion work decreases. This is a slightly negative effect for the overall cycle. Figure 3.A.3 shows the cylinder pressure (relative to the initial pressure) as functions of the cylinder volume (relative to the minimum volume) for three values of gamma for the expansion stroke. For consistency, p_3/p_1 was the same for all three cases with a value of 200. The same starting point is important for a fair comparison. The differences between the three curves are modest, but the increase of gamma does result in a slightly negative effect for the overall cycle.

Figure 3.A.4 shows the non-dimensional work (described in the above equations) as functions of gamma for the compression and expansion strokes for a compression ratio of 16. The compression work increases and the expansion work decreases as gamma increases. Both of these trends are slightly negative relative to the overall efficiency.

In summary, the dominant effect of higher gamma (lower specific heats) are higher temperatures and pressures during the heat release portion of the cycle, and this increases the potential for greater expansion work. For the expansion and compression portions of the cycle, however, the lower specific heats (higher gamma) have a slightly negative impact on the net work delivered. The results presented elsewhere in this book support these conclusions for actual engine cycles and provide quantitative values for these increases.

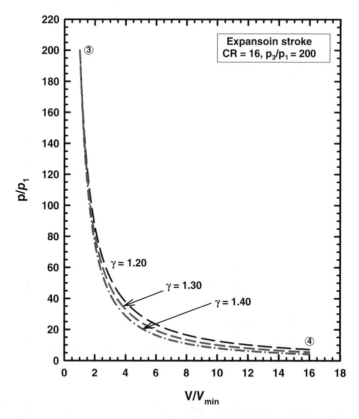

Figure 3.A.3   The cylinder pressure (relative to the initial pressure) as a function of the cylinder volume (relative to the minimum volume) for the expansion stroke for three values of gamma for a compression ratio of 16. *Source:* Caton 2014b. Reproduced with permission from Elsevier

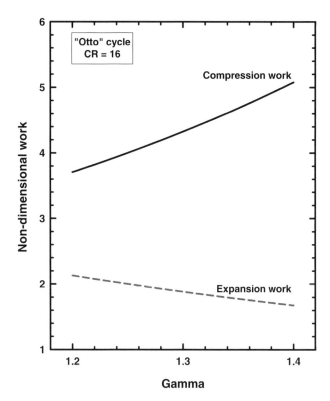

Figure 3.A.4   The non-dimensional work for the compression and expansion strokes as functions of gamma for a compression ratio of 16 based on the "Otto" cycle analysis. *Source:* Caton 2014b. Reproduced with permission from Elsevier

The results described above are combined for a complete cycle. Figure 3.A.5 shows the cylinder pressure (relative to the initial pressure) as functions of the cylinder volume (relative to the minimum volume, TDC) for two values of gamma for the complete cycle. Only two values of gamma are used for clarity. For the highest gamma ($\gamma = 1.40$), the state points (1, 2, 3, and 4) are denoted. This case is the solid line. The case with the lower gamma ($\gamma = 1.20$) is denoted with the dashed line. Each of the portions of this cycle may now be considered.

Since state 1 is the same for both cases, the compression stroke starts at the same pressure. For the higher gamma case, the final pressure (state 2) is higher and this is consistent with the higher work associated with the higher gamma.

For the heat addition portion of the cycle (state 2 to state 3), the higher gamma case results in a higher final pressure. This case begins at a slightly higher pressure (state 2), but as shown in Figure 3.A.2, the higher gamma case will produce a higher final pressure.

For the expansion stroke, the interpretation is not as direct. Since the starting point (state 3) is different for the two cases, the actual effect of gamma on expansion work is not obvious. From Figure 3.A.4, however, the expansion work would be less for the higher gamma case if state 3 was the same. In the actual cycle, the expansion work is less largely due to the lower pressure (state 3) associated with the lower gamma.

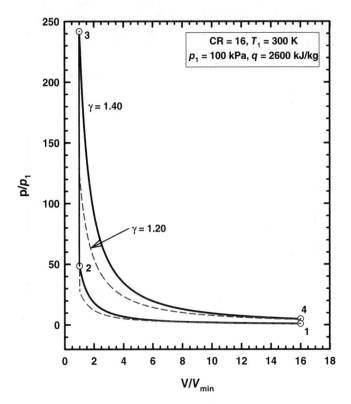

Figure 3.A.5 The cylinder pressure (relative to the initial pressure) as a function of the cylinder volume (relative to the minimum volume) for the "Otto" cycle for two values of gamma for a compression ratio of 16. *Source:* Caton 2014b. Reproduced with permission from Elsevier

Figure 3.A.6 shows the "Otto" cycle thermal efficiency as functions of gamma for two compression ratios. Increasing gamma increases the thermal efficiency in a major way. Further, this illustrates that the net effect of increasing gamma on thermal efficiency is due to the positive effect during the heat addition. And these results confirm the statement that higher gamma during compression and expansion produces relatively minor negative effects.

As Figure 3.A.6 demonstrates, small changes of gamma result in significant increase of the efficiency. For example, for a compression ratio of 16, the change from a gamma of 1.2 to 1.3 (about an 8% increase) results in an increase of the thermal efficiency of 13.9%, absolute (about a 33% relative increase). As described elsewhere in this book, an important contribution to the increased efficiencies of advanced, high efficiency engines can be shown to be the increase of gamma for these types of engines (see related case study on high efficiency engines).

In summary, the ideal cycle analysis produces results which are much too optimistic, and the analysis does not include many of the engine operating and design parameters. This is an

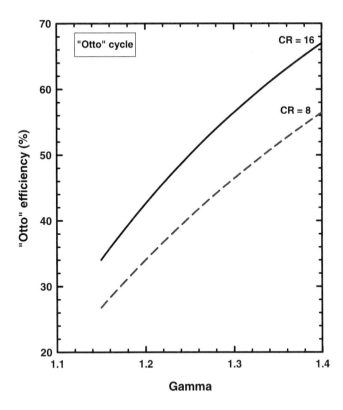

Figure 3.A.6 The "Otto" cycle thermal efficiency as functions of gamma for compression ratios of 8 and 16. *Source:* Caton 2014b. Reproduced with permission from Elsevier

elementary approach to understanding engine operation, but does provide the correct trends with respect to compression ratio and the ratio of specific heats. As shown in this appendix, modest increases of the ratio of specific heats (gamma) result in significant increases of the thermal efficiencies.

4

Properties of the Working Fluids

4.1 Introduction

This chapter is concerned with the required algorithms to compute the properties of the working fluids. The thermodynamic properties needed for solving the first law relations include the internal energy, enthalpy, gas constant, molecular mass, and a number of property derivatives. In addition, entropy is needed to complete any second-law analysis. These properties are needed as functions of time (crank angle) for the cylinder contents and for any material entering the cylinder. Most of the following description of the determination of composition and property values for the engine cycle simulation closely follows the procedures and nomenclature used by Heywood [1].

Depending on the specific processes during a cycle, the thermodynamic properties may be for mixtures of air, fuel vapor, and combustion products. The concentrations of the combustion products may be "frozen" for the lower temperatures, or these concentrations may be based on an instantaneous determination of chemical equilibrium for higher temperatures. In general, the determination of the equilibrium composition of the products is somewhat complex and one of the computationally intensive aspects of cycle simulations.

This chapter will include descriptions of the algorithms used for determining the thermodynamic properties for the (i) unburned mixtures, (ii) burned mixtures, and (iii) equilibrium compositions. This chapter will conclude with examples of the thermodynamic properties of the working fluids as functions of temperature, pressure and equivalence ratio.

4.2 Unburned Mixture Composition

The unburned mixture during the compression stroke prior to combustion consists of the inlet mixture of air and fuel, and the residual cylinder contents of combustion products from the preceding cycle. The residual fraction is defined as

$$x_r = \frac{m_r}{m_c} \tag{4.1}$$

An Introduction to Thermodynamic Cycle Simulations for Internal Combustion Engines, First Edition.
Jerald A. Caton © 2016 John Wiley & Sons, Ltd. Published 2016 by John Wiley & Sons, Ltd.

where m_r is the residual mass and m_c is the trapped charge mass. If exhaust gas recirculation (EGR) needs to be considered, this may be incorporated as follows:

$$\text{EGR}(\%) = \left(\frac{m_{\text{EGR}}}{m_i}\right) \times 100 \tag{4.2}$$

where m_{EGR} is the recycled exhaust gas mass and m_i is the inducted mass. The burned gas fraction[1] is therefore defined as

$$x_b = \frac{m_{\text{EGR}} + m_r}{m_c} = \left(\frac{\text{EGR}}{100}\right)(1 - x_r) + x_r \tag{4.3}$$

In general, the determination of the species in the burned gas fraction is a major part of determining the unburned mixture properties. For a simple hydrocarbon fuel (CH_y), the reaction may be represented (for *one mole of O_2 reactant*) as

$$\varepsilon\phi C + 2(1-\varepsilon)\phi H_2 + O_2 + \psi N_2 \rightarrow$$
$$n_{CO_2}CO_2 + n_{H_2O}H_2O + n_{CO}CO + n_{H_2}H_2 + n_{O_2}O_2 + n_{N_2}N_2 \tag{4.4}$$

where ψ is the molar N/O ratio (= 3.773[2] for pure air), y is the molar H/C ratio of the fuel, ϕ is the fuel–air equivalence ratio, n_i is the moles of species i per mole of O_2 reactant, and $\varepsilon = 4/(4+y)$. In the next subsection, the n_i values are determined.

The unburned mixture (*per mole of O_2 reactant*) contains the unreacted fuel and air ($1-x_b$), and the burned gas fraction (x_b). This mixture is represented by

$$(1-x_b)\left[\frac{4}{M_f}(1+2\varepsilon)\phi\left(CH_y\right)_\alpha + O_2 + \psi N_2\right]$$
$$+ x_b\left(n_{CO_2} + n_{H_2O} + n_{CO} + n_{H_2} + n_{O_2} + n_{N_2}\right) \tag{4.5}$$

where M_f is the molecular mass of the fuel. Table 4.1 summarizes the moles of each species *per one mole of reactant oxygen* for both (a) the stoichiometric and lean cases and (b) the rich cases. The total unburned moles for both cases are:

$$n_u = (1-x_b)\left[\frac{4(1+2\varepsilon)\phi}{M_f} + 1 + \psi\right] + x_b n_b \tag{4.6}$$

The fuel molecular mass, M_f, for the assumed fuel molecule $\left(CH_y\right)_\alpha$ is

$$M_f = \alpha(12 + y) \tag{4.7}$$

---

[1] The symbol, x_b, is also defined later for the burned mass fraction during combustion. These two definitions are common in the literature, are related, and are easily distinguished by considering the context in which they are used.
[2] The "3.773" for the molar N/O ratio is based on the use of an "apparent nitrogen species" which is weighted to include argon [1].

Table 4.1 Composition for the unburned mixtures (the moles of each species per one mole of reactant oxygen)

For $\phi \leq 1.0$ (stoichiometric and lean):	For $\phi > 1.0$ (rich):
$n_{\text{fuel}} = 4(1 - x_b)(1 + 2\varepsilon)\phi / M_f$	$n_{\text{fuel}} = 4(1 - x_b)(1 + 2\varepsilon)\phi / M_f$
$n_{CO_2} = x_b \varepsilon \phi$	$n_{CO_2} = x_b(\varepsilon\phi - c)$
$n_{H_2O} = 2x_b(1 - \varepsilon)\phi$	$n_{H_2O} = x_b[2(1 - \varepsilon\phi) + c]$
$n_{CO} = 0$	$n_{CO} = x_b c$
$n_{H_2} = 0$	$n_{H_2} = x_b[2(\phi - 1) - c]$
$n_{O_2} = 1 - x_b\phi$	$n_{O_2} = 1 - x_b$
$n_{N_2} = \psi$	$n_{N_2} = \psi$

$$n_u = (1 - x_b)\left[\frac{4(1 + 2\varepsilon)\phi}{M_f} + 1 + \psi\right] + x_b n_b$$

and the mass of mixture (burned or unburned) per mole O_2 in the mixture is given by

$$m_{RP} = 32 + 4\phi(1 + 2\alpha) + 28.013\,\psi \tag{4.8}$$

The molecular mass of the burned and unburned mixture, respectively, are given by

$$M_b = \frac{m_{RP}}{n_b} \tag{4.9}$$

$$M_u = \frac{m_{RP}}{n_u} \tag{4.10}$$

The values in Table 4.1 may be converted to mole fractions. For example, for oxygen in the burned gases of a lean or stoichiometric mixture,

$$y_{O2} = \text{mole fraction} = \frac{n_{O2}}{n_{\text{tot}}} \tag{4.11}$$

$$y_{O2} = \frac{(1 - x_b)\phi}{(1 - x_b)\left[\dfrac{4(1 + 2\varepsilon)\phi}{M_f} + 1 + \psi + \dfrac{x_b n_b}{(1 - x_b)}\right]} \tag{4.12}$$

or simplifying,

$$y_{O2} = \frac{\phi}{\left[\dfrac{4(1 + 2\varepsilon)\phi}{M_f} + 1 + \psi + \dfrac{x_b n_b}{(1 - x_b)}\right]} \tag{4.13}$$

and (from below),

$$n_b = (1-\varepsilon)\phi + 1 + \psi \tag{4.14}$$

Therefore, combining,

$$y_{O2} = \frac{\phi}{\left[\dfrac{4(1+2\varepsilon)\phi}{M_f} + 1 + \psi + \dfrac{x_b(1-\varepsilon)\phi + 1 + \psi}{(1-x_b)}\right]} \tag{4.15}$$

Each of the species may be converted to a mole fraction by this procedure.

4.2.1   Oxygen-containing Fuels

The above may be extended to include alcohol or other oxygen-containing fuels. In this case, the fuel is represented by CH_yO_z, and the balanced equation is

$$
\begin{aligned}
&\varsigma\,\varepsilon\phi C + 2\varsigma\,(1-\varepsilon)\phi H_2 + O_2 + \left(1-\frac{\varepsilon z}{2}\right)\varsigma\,\psi N_2 \rightarrow \\
&n_{CO_2}CO_2 + n_{H_2O}H_2O + n_{CO}CO + n_{H_2}H_2 + n_{O_2}O_2 + n_{N_2}N_2
\end{aligned}
\tag{4.16}
$$

where the following is defined

$$\varsigma = \frac{2}{2-\varepsilon z(1-\phi)} \tag{4.17}$$

Note the following:

$$
\begin{aligned}
y &= \frac{C}{H} \\
z &= \frac{C}{O} \\
\varepsilon &= \frac{4}{4+y}
\end{aligned}
\tag{4.18}
$$

To use Table 4.1 for oxygen-containing fuels, the following definitions are needed:

$$
\begin{aligned}
\phi^* &= \varsigma\phi \\
\psi^* &= \left(1-\frac{\varepsilon z}{2}\right)
\end{aligned}
\tag{4.19}
$$

The reactant expression is now

$$\phi^*\varepsilon C + 2\phi^*(1-\varepsilon)H_2 + O_2 + \psi^* N_2 \tag{4.20}$$

This reactant expression for oxygen-containing fuels (excluding the "*") is identical to the expression for the simple hydrocarbon. This allows Table 4.1 to be used for the composition for oxygen-containing fuels using the following replacement:

- replace ϕ with $\phi*$
- replace ψ with $\psi*$

4.2.2   Oxidizers

Conventional air ($21\%_v$ O_2) is the dominant oxidizer used by engines and in the following computations. Other oxidizers may be used, however. These other oxidizers may include pure oxygen, oxygen-enriched air or oxygen-depleted air. Examples of the use of oxygen-enriched air are provided in one of the case studies.

4.2.3   Fuels

The majority of the following computations have been completed for isooctane as the fuel. Any pure fuel (or mixtures of pure fuels), however, may be used. As described above, the current formulations are designed for fuels of the form: CH_yO_z. Other forms of the fuel could be accommodated with some slight modifications of the above formulations. In any case, for use with the cycle simulation, the fuel properties need to be available in a form similar to that described above.

Table 4.2 lists eight fuels that have been used in previous computations using the cycle simulation described here. The list includes the fuel's molecular mass, stoichiometric mass air–fuel ratio, lower heating value, exergy value, and the ratio of the exergy value and the lower heating value. Although the fuel properties span a significant range of values, engine performance is similar for all fuels for the conditions examined [2]. The greatest differences were the values of the exergy destruction during combustion [2]. More details on engine results for different fuels are provided in one of the case studies.

Table 4.2   Fuel specifications [1]

Fuel	Molecular mass (kg/kmol)	AF (stoich)	LHV (MJ/kg$_f$)	Fuel exergy (MJ/kg$_f$)	Fuel exergy/ LHV
Hydrogen, H_2	2.016	34.15	120.0	113.5	0.946
Methane, CH_4	16.04	17.17	50.0	49.9	0.998
Propane, C_3H_8	44.10	15.61	46.4	47.1	1.015
Hexane, C_6H_{14}	86.18	15.16	45.1	46.1	1.022
Isooctane, C_8H_{18}	114.2	15.07	44.4	45.5	1.025
Methanol, CH_3OH	32.04	6.45	20.0	21.1	1.055
Ethanol, C_2H_5OH	46.07	8.97	26.9	28.4	1.056
Carbon monoxide	28.01	2.46	10.1	9.18	0.909

With the above information, the unburned mixture composition is specified. Results for the mole fractions of the species and thermodynamic properties for the unburned mixtures for isooctane-air mixtures will be provided later in this chapter.

4.3   Burned Mixture ("Frozen" Composition)

As mentioned above, for low enough temperatures, the chemical reactions are too slow to significantly alter the composition. For these cases, the composition is essentially frozen, and does not change until the next cycle. The frozen composition described next, therefore, is sufficient for the residual gases, exhaust gas recirculation, and the low temperature product gases during expansion (below some temperature value).

For determining the frozen composition of the burned mixture, six major product species are considered. The general combustion equation for a fuel molecule of the form CH_y with standard air for these six product species is

$$\varepsilon \phi C + 2(1-\varepsilon)\phi H_2 + O_2 + \psi N_2 \rightarrow$$
$$n_{CO_2} CO_2 + n_{H_2O} H_2O + n_{CO} CO + n_{H_2} H_2 + n_{O_2} O_2 + n_{N_2} N_2 \tag{4.21}$$

where ψ is the molar N/O ratio (= 3.773 for pure air), y is the molar H/C ratio of the fuel, ϕ is the fuel–air equivalence ratio, n_i is the moles of species i per mole of O_2 reactant, and $\varepsilon = 4/(4+y)$. The next step is to determine the composition (i.e., values for the n_i). The following assumptions are used:

1. For lean and stoichiometric mixtures, n_{CO} and n_{H_2} are assumed zero.
2. For rich and stoichiometric mixtures, n_{O_2} is assumed zero.
3. For rich mixtures, the "water gas" reaction is assumed to be sufficiently accurate to determine the n_{CO} and n_{H_2}. This case is explained in more detail next.

The water gas reaction is:

$$CO_2 + H_2 \rightarrow CO + H_2O \tag{4.22}$$

Typically, this reaction is assumed to be in equilibrium. An equilibrium constant may be defined as

$$K(T) = \frac{n_{H_2O} n_{CO}}{n_{CO_2} n_{H_2}} \tag{4.23}$$

For this work, K(T) is assumed constant with a value of 3.5. Other approaches have assumed a polynomial expression for "K" in terms of temperature. The information leads to a quadratic equation, which can be solved yielding expressions for the individual moles [1]. This is described next.

Using the above three assumptions and atom balances, expressions for the various n_i for the combustion products may be determined. The total burned moles for the stoichiometric and lean cases are $n_b = (1-\varepsilon)\phi + 1 + \psi$. The total burned moles for the rich cases are $n_b = (2-\varepsilon)\phi + \psi$. Table 4.3 provides a summary of the expressions for the moles of each product species for (a) the stoichiometric and lean cases and (b) the rich cases.

Table 4.3 Composition for the burned mixtures based on the frozen species assumption (the moles of each species per *one mole of reactant oxygen*)

For $\phi \leq 1.0$ (Stoichiometric and lean):	For $\phi > 1.0$(rich):
(moles of species *i per mole of O_2 reactant*)	
$n_{CO_2} = \varepsilon\phi$	$n_{CO_2} = \varepsilon\phi - c$
$n_{H_2O} = 2(1-\varepsilon)\phi$	$n_{H_2O} = 2(1-\varepsilon\phi) + c$
$n_{CO} = 0$	$n_{CO} = c$
$n_{H_2} = 0$	$n_{H_2} = 2(\phi-1) - c$
$n_{O_2} = 1 - \phi$	$n_{O_2} = 0$
$n_{N_2} = \psi$	$n_{N_2} = \psi$
$n_b = (1-\varepsilon)\phi + 1 + \psi$	$n_b = (2-\varepsilon)\phi + \psi$

The value of "*c*" in the expressions in Table 4.3 is found from the solution of the following quadratic expression from the water–gas reaction:

$$(K-1)c^2 - \left\{K[2(\phi-1) + \varepsilon\phi] + 2(1-\varepsilon\phi)\right\}c + 2K\varepsilon\phi(\phi-1) = 0 \tag{4.24}$$

So, "*c*" may be found from,

$$c = \frac{-b \pm \sqrt{b^2 - 4ad}}{2a} \tag{4.25}$$

where

$a = K - 1$

$b = -\left(K\left(2(\phi-1) + \varepsilon\phi\right) + 2\left(1 - \varepsilon\phi\right)\right)$

$d = 2K\varepsilon\phi(\phi-1)$

4.4 Equilibrium Composition

For high temperature portions of the cycle, descriptions of the products of combustion are needed. These descriptions need to identify the major species and their concentrations. One approach that has been successful is to model the products of combustion as possessing the chemical equilibrium composition by using the instantaneous temperature and pressure for a given fuel and equivalence ratio. This is sometimes referred to as a "shifting equilibrium" and is based on the assumption that the species have enough time (reactions are fast enough) to continually adjust to the new conditions. For a speed of 5000 rpm, one crank angle of revolution is equivalent to about 0.03 ms. Therefore, the "relaxation time" of the chemical reactions must be less than this for the assumption of a shifting equilibrium to be valid. This evaluation depends on the kinetic rates of the relevant reactions at the temperature and pressure. In general, for the major species during the high temperature portions of the cycle, these rates are fast enough such that this is a valid assumption.

The use of equilibrium compositions for the purpose of engine cycle simulations was recognized by the earliest modelers of engine cycles. For example, users of the air standard cycle models improved the models by using equilibrium compositions for the products. With the incorporation of equilibrium composition, the models have been known as fuel–air cycles to better reflect the fact that the air, fuel and products of combustion are considered. This early use of equilibrium compositions used charts. With the advent of computers, the use of charts was abandoned. By the early 1970s, the use of equilibrium compositions to describe the combustion products was fairly routine [1].

Solution procedures for the complete equilibrium relations (described below) are complex and nonlinear. One reason for this complexity is that the concentrations of the various species can range over many orders of magnitude [1]. A number of comprehensive equilibrium programs are available [3–7] for the complete determination of equilibrium compositions for a wide range of temperatures, pressures, and equivalence ratios. Comprehensive equilibrium programs are available from individuals, universities, and commercial entities. A number of these programs are available via the internet on various sites.

The comprehensive equilibrium programs described above are generally too extensive for the routine, repetitive use in an engine cycle simulation. For the purpose of this work, a limited version of these more complete programs is not only adequate, but preferred due to the more rapid execution associated with a more concise program. The following development is based on the work of Olikara and Borman [8].

For the purposes of this work, the equilibrium determinations are limited to a relatively small subset of possible gaseous species (12 product species) formed from the atoms: C, H, O, Ar and N. For consistency, the internal energy and enthalpy are defined to consist of both the sensible and chemical components. The equilibrium thermodynamic properties are functions of the composition (equivalence ratio) and two state variables (temperature and pressure). In general, the thermodynamic properties may be expressed as

$$\beta = \beta(T, p, \phi) \qquad (4.26)$$

where β is any property, T is temperature, p is pressure, and ϕ is the fuel–air equivalence ratio. The fuel–air equivalence ratio is defined as

$$\phi = \frac{(AF)_{stoich}}{(AF)_{actual}} \qquad (4.27)$$

An appropriate fuel–air equivalence ratio must be used to obtain the accurate properties. Typically, for the inlet mixture, this is the actual equivalence ratio. For products, this is the equivalence ratio of a reactant mixture which would result in the required product composition. For products mixed with air or other species, the appropriate equivalence ratio is the one for an imaginary reactant mixture which would result in the desired composition.

In general, the fuel and air may be assumed to react as follows:

$$x_{13}\left[C_nH_mO_lN_k+(1\ \phi)(n+m/4-l/2)\{O_2+\psi N_2+\xi Ar\}\right]\rightarrow$$
$$x_1H+x_2O+x_3N+x_4H+x_5OH+x_6CO+x_7NO \qquad (4.28)$$
$$+x_8O_2+x_9H_2O+x_{10}CO_2+x_{11}N_2+x_{12}Ar$$

where $C_nH_mO_lN_k$ is the general fuel molecule (n and m must be non-zero, but l and k are optional), ψ is the molar N/O ratio (3.773^3 for air), x_1 through x_{12} are the mole fractions of the respective product species, and x_{13} is the moles of fuel to produce one (1.0) mole of total products. The moles of argon in the air is zero (since the nitrogen is based on "apparent" nitrogen with a weighted molecular mass of 28.16 to account for the argon); and, therefore, since $\xi = 0$ then $x_{12} = 0$. To solve for the remaining 12 unknown mole fractions, seven (independent) equilibrium expressions are available, four atom balances (for C, H, O, and N), and the fact that the mole fractions must sum to one:

$$\sum_{i=1}^{11} x_i = 1.0 \qquad (4.29)$$

The equilibrium expressions follow general formulations [3–8] and are listed here for completeness:

$$\frac{1}{2}H_2 \leftrightarrow H$$

$$\frac{1}{2}O_2 \leftrightarrow O$$

$$\frac{1}{2}N_2 \leftrightarrow N$$

$$\frac{1}{2}H_2 +\frac{1}{2}O_2 \leftrightarrow OH \qquad (4.30)$$

$$\frac{1}{2}N_2 +\frac{1}{2}O_2 \leftrightarrow NO$$

$$H_2 +\frac{1}{2}O_2 \leftrightarrow H_2O$$

$$CO+\frac{1}{2}O_2 \leftrightarrow CO_2$$

The equilibrium constants are curve-fits to the JANAF Thermochemical Tables [9]. The solution techniques for the above equation set are described by Olikara and Borman [8].

---

[3] As mentioned above, the "3.773" for the molar N/O ratio is based on the use of an "apparent nitrogen species" which is weighted to include argon. This, of course, is consistent with the statement that the number of moles of pure argon is zero.

4.5   Determinations of the Thermodynamic Properties

Once the compositions of the gas mixtures are known, the individual thermodynamic properties may be determined. The temperatures and pressures during engine cycles are such that the species may be approximated as ideal gases with a high level of accuracy. For ideal gas mixtures, the properties of the mixture are the mole- or mass-weighted sums of the properties of the individual species. For example, for internal energy and enthalpy,

$$u = \sum x_i u_i$$
$$h = \sum x_i h_i \tag{4.31}$$

where x_i are the mole fractions, and the properties are "per mole" values. The other thermodynamic properties are obtained in similar ways.

The final aspect of defining the thermodynamic properties for the gas mixtures is to determine the individual properties for each species. These properties, for ideal gases, are only functions of temperature. The most successful approach for use in engine cycle simulations is to use polynomial curve-fits to the property data for the species. This approach is used for many applications, not just for engine cycle simulations. The following is a brief summary of this approach, but more details are available elsewhere [1].

For example, for enthalpy, the following expression may be used:

$$\frac{h_i}{RT} = a_{i1} + \frac{a_{i2}}{2} T + \frac{a_{i3}}{3} T^2 + \frac{a_{i4}}{4} T^3 + \frac{a_{i5}}{5} T^4 + \frac{a_{i6}}{T} \tag{4.32}$$

where each of the a_{ij} are fit coefficients which are obtained from regression analysis using the actual data. These coefficients may be obtained from a variety of sources (e.g., Reference [1]).

As another example, the specific entropy of a mixture is determined by

$$s_{mix} = \sum x_i s_i \tag{4.33}$$

The specific entropy values for each species, s_i, include two parts:

$$s_i = s_i^\circ - R_u \ln\left(\frac{p_i}{p_o}\right) \tag{4.34}$$

The first part, s_i°, depends only on temperature and is the specific entropy of species " i " when p_i equals p_o. For all properties used here (and in the majority of combustion applications), the standard pressure, p_o, is 1.0 atm (101.325 kPa). Combining the above equations,

$$s_{mix} = \sum \left(s_i^\circ - R_u \ln\left(\frac{p_i}{p_o}\right) \right) \tag{4.35}$$

This may be rearranged to use the mole fractions of each species,

$$s_{mix} = \sum x_i \left(s_i^\circ - R_u \ln\left(\frac{p_i}{p_{mix}}\right) \right) - R_u \ln\left(\frac{p_{mix}}{p_o}\right) \tag{4.36}$$

Table 4.4 Coefficients for polynomial fit for isooctane vapor [1]

A_{f1}	A_{f2}	A_{f3}	A_{f4}	A_{f5}	A_{f6}	A_{f7}
−0.55313	181.62	−97.787	20.402	−0.03095	−60.751	51.007

$$s_{\text{mix}} = \sum x_i \left(s_i^\circ - R_u \ln x_i \right) - R_u \ln \left(\frac{p_{\text{mix}}}{p_o} \right) \tag{4.37}$$

The standard specific entropy for each species may be determined from polynomial expressions much like presented above for enthalpy:

$$\frac{s_i^\circ}{R_u} = a_{i1} \ln T + a_{i2} T + \frac{a_{i3}}{2} T^2 + \frac{a_{i4}}{3} T^3 + \frac{a_{i5}}{4} T^4 + a_{i7} \tag{4.38}$$

The coefficient, a_{i7}, is selected for a certain datum temperature. This datum temperature is typically 298.15 K or 0 K. The current work uses 298.15 K.

Fuels: Heywood [1] uses a slightly different polynomial expression[4] for the properties of the fuels. For enthalpy, he has used a dimensional form:

$$h_f = A_{f1} t + \frac{A_{f2}}{2} t^2 + \frac{A_{f3}}{3} t^3 + \frac{A_{f4}}{4} t^4 - \frac{A_{f5}}{t} + A_{f6} \tag{4.39}$$

where $t = T(\text{K})/1000$ and h_f has units of cal/gmol-K. The coefficients for different fuels are available from Heywood [1].

A similar approach is used for the entropy of fuels:

$$s_f \left(\frac{\text{cal}}{\text{gmol K}} \right) = A_{f1} \ln t + A_{f2} t + \frac{A_{f3}}{2} t^2 + \frac{A_{f4}}{3} t^3 - \frac{A_{f5}}{2t^2} + A_{f7} \tag{4.40}$$

where t is T(K)/1000. Since isooctane vapor is the fuel used the most in the current work, the coefficients for isooctane are listed in Table 4.4.

4.6 Results for the Thermodynamic Properties

With the above algorithms, a number of thermodynamic properties may be examined for a range of variables. The results presented in this section are not relative to specific engine operating conditions. Rather, the following are fundamental results as functions of temperature, pressure, and equivalence ratio. Property results as functions of engine operating conditions are presented in Chapter 7 and in some of the other chapters. For the following, the fuel is isooctane and the oxidizer is standard air. The properties of the unburned mixture will be discussed first, and then the properties for the burned mixtures will be discussed.

---

[4] Note the negative sign before A_{f5}. This is consistent with the fit coefficients supplied here and in Reference 1.

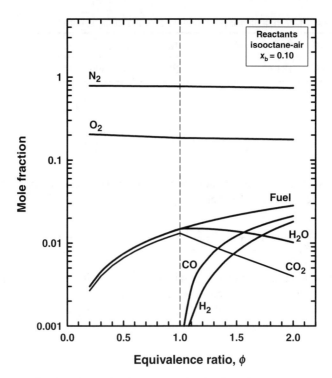

Figure 4.1   The mole fraction (log scale) for each species for the unburned charge as a function of equivalence ratio for a burned gas fraction of 0.10 for isooctane and air mixtures

Figures 4.1 and 4.2 show the mole fractions of the species of the unburned mixture for isooctane-air mixtures with a burned gas fraction of 10% ($x_b = 0.10$) as functions of equivalence ratio. Figure 4.1 shows the mole fractions (logarithmic scale) of seven species, and Figure 4.2 shows the mole fractions (linear scale) of five species for this case. Note that the mole fractions for nitrogen and oxygen are about an order of magnitude higher concentration than the other species. For this range of conditions, the nitrogen and oxygen account for about 92% of the moles of the unburned mixture. For lean mixtures, the species are only nitrogen, oxygen, fuel, water, and carbon dioxide. For the rich mixtures, the additional species are carbon monoxide and hydrogen. As the mixture becomes richer from stoichiometric, the mole fractions of fuel, carbon dioxide and hydrogen increase, and the mole fractions of water and carbon dioxide decrease.

The mole fraction of the fuel increases linearly with equivalence ratio. Carbon dioxide and water vapor increase nearly linearly until an equivalence ratio of 1.0. Both of these species then decrease as the mixture becomes richer. Carbon monoxide and hydrogen are zero until the equivalence ratio is greater than stoichiometric. For mixtures richer than stoichiometric, carbon monoxide and hydrogen increase. At an equivalence ratio of 2.0, the mole fractions of carbon monoxide and hydrogen are about 2.1 and 1.8%, respectively. For these conditions, the total of these five species are always less than about 8% of the total unburned moles. For cases with higher burned gas fractions, the concentrations of the water, carbon dioxide, carbon monoxide and hydrogen would be higher.

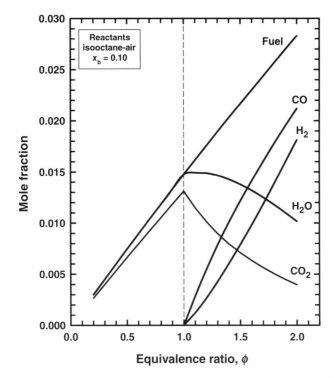

Figure 4.2   The mole fraction for five of the species for the unburned charge as a function of
equivalence ratio for a burned gas fraction of 0.10 for isooctane and air mixtures

Next, the molecular mass of the unburned mixtures will be examined as a function of
equivalence ratio and burned fraction (x_b) for a range of temperatures. Figure 4.3 shows the
molecular mass for the unburned mixture as functions of equivalence ratio for three values of
the burned gas fraction[5] (x_b). For reference, Figure 4.3 also shows the value of the molecular
mass for pure air. For these conditions, the unburned molecular mass ranges between 28 and
32, and for cases with lower burned gas fraction, the values are greater than for pure air (due
to the higher molecular mass of the fuel). Also, note that the higher the burned gas fraction,
the lower the molecular mass due to the contribution of the products of combustion (see the
following).

Figures 4.4 and 4.5 show the specific heat at constant pressure and the ratio of the specific
heats, respectively, for unburned mixtures as a function of temperature for three constant
equivalence ratios for a burned gas fraction of 0.1. The specific heat increases with temperature, and with equivalence ratio. For these conditions, the specific heat at constant pressure ranges from about 1.0 to 1.4 kJ/kg K. The ratio of specific heats for unburned mixtures
decreases as temperature and equivalence ratio increases. For these conditions, the ratio of
specific heats for unburned mixtures ranges from about 1.24 to 1.40.

---

[5] Note that for 0% EGR, the burned gas fraction is identical to the residual fraction.

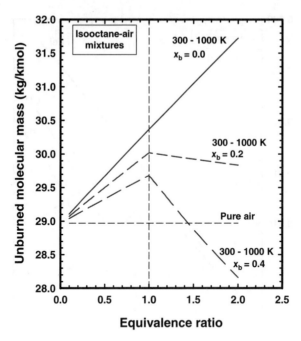

Figure 4.3   The molecular mass (M_u) for the unburned mixture as a function of equivalence ratio for isooctane and air mixtures

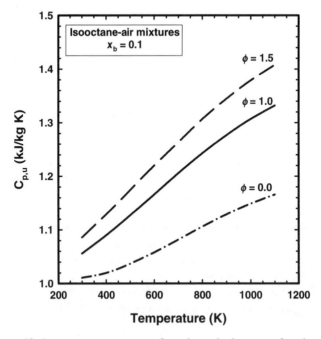

Figure 4.4   The specific heat at constant pressure for unburned mixtures as functions of temperature for three equivalence ratios for a burned gas fraction of 0.1 for isooctane and air mixtures

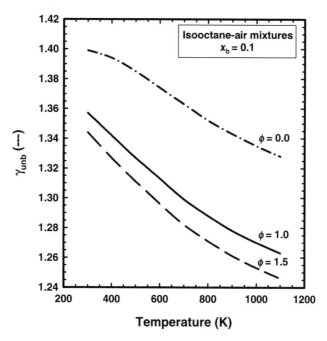

Figure 4.5 The ratio of specific heats for unburned mixtures as functions of temperature for three equivalence ratios for a burned gas fraction of 0.1 for isooctane and air mixtures

Next, results for the burned mixtures will be discussed. For these mixtures, the composition determined by equilibrium considerations is required for the higher temperatures. Results will be presented based on both the equilibrium compositions and the frozen mixtures.

Figures 4.6, 4.7, and 4.8 show the mole fractions of the species of the burned gases for isooctane-air mixtures as functions of equivalence ratio for three temperatures (2000, 2500, and 3000 K, respectively) for 3000 kPa based on equilibrium and frozen assumptions. First, considering Figure 4.6, the oxygen mole fraction decreases from its highest values for the leanest mixtures to a value of zero at an equivalence ratio of 1.0. Water vapor and carbon dioxide increase from the leanest mixtures to their highest values near an equivalence ratio of 1.0. For water and carbon dioxide, the maximum values are 14.0 and 13.4%, respectively, for these conditions. As the mixture continues to become richer, the mole fractions of water and carbon dioxide decrease. Carbon monoxide and hydrogen are zero until the equivalence ratio is greater than stoichiometric, then the mole fractions for these two species increase. The mole fractions are about 0.17 and 0.135 for carbon monoxide and hydrogen, respectively, at an equivalence ratio of 2.0.

Figures 4.6 and 4.7 show both the equilibrium and frozen values for the mole fractions. For 2000 K, the frozen approach provides similar values to the equilibrium approach; especially for lean mixtures. For 2500 K, the frozen approach is in less agreement with the equilibrium approach. For accurate engine cycle computations, the frozen approach is only used for temperatures less than about 1500 K. For the computations reported in this book, the frozen assumption was used only for temperatures below 1200 K—so accuracy was assured.

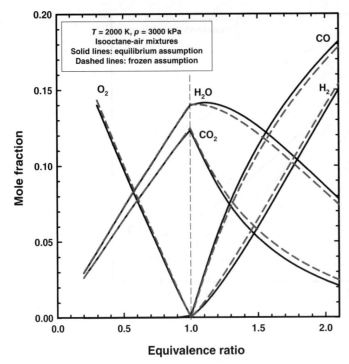

Figure 4.6 The mole fractions of the burned gases as functions of equivalence ratio for a temperature of 2000 K and a pressure of 3000 kPa based on equilibrium (solid lines) and frozen (dashed lines) assumptions for isooctane and air mixtures

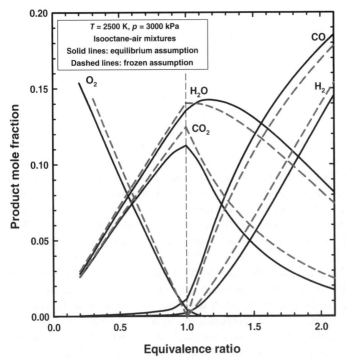

Figure 4.7 The mole fractions of the burned gases as functions of equivalence ratio for a temperature of 2500 K and a pressure of 3000 kPa based on equilibrium (solid lines) and frozen (dashed lines) assumptions for isooctane and air mixtures

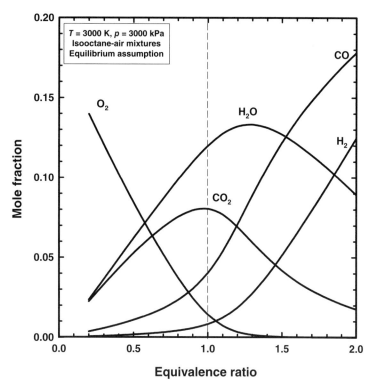

Figure 4.8   The mole fractions of the burned gases as functions of equivalence ratio for a temperature of 3000 K and a pressure of 3000 kPa based on equilibrium assumptions for isooctane and air mixtures

Figure 4.8 shows the mole fractions of the major species for equilibrium mixtures as a function of equivalence ratio for a constant temperature of 3000 K at a constant pressure of 3000 kPa. As shown, the mole fractions of CO_2 and H_2O are maximum at stoichiometric or slightly rich conditions. The mole fraction of O_2 continually decreases from the leanest condition to the rich conditions. The mole fractions of CO and H_2 continually increase from the leanest conditions to the richest conditions. Due to this relatively high temperature (3000 K), dissociation is significant. Due to the dissociation, the mole fraction of oxygen is non-zero for slightly rich mixtures, and the mole fractions of carbon monoxide and hydrogen are non-zero for lean mixtures and are significantly high for most of the rich mixtures.

Figure 4.9 shows the molecular mass (M_b) for burned mixtures as a function of equivalence ratio for three constant temperatures at a constant pressure of 3000 kPa based on equilibrium and frozen assumptions. For the conditions of Figure 4.9, the molecular masses for the burned mixtures range from about 26 to 29 kg/kmol. The molecular mass decreases slightly from low equivalence ratios to stoichiometric conditions. From stoichiometric to rich conditions, the molecular mass decreases more rapidly. This is due to the increasing concentrations of the lighter molecules (such as, CO, H_2, and H_2O) and the decreasing concentration of the heavier molecules (such as O_2).

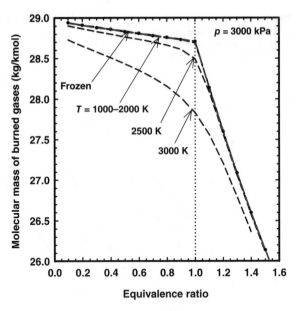

Figure 4.9 The molecular mass of the burned gases as functions of equivalence ratio for temperatures of 1000–3000 K and a pressure of 3000 kPa based on equilibrium and frozen assumptions for isooctane and air mixtures

As the gas temperature increases, the molecular mass decreases due to the greater extent of dissociation of the heavier molecules such as CO_2 and O_2. As the temperature decreases (say below about 2000 K), the molecular mass does not significantly change for a specific equivalence ratio. This is due to the decreasing importance of dissociation—the mixture is "frozen" in composition.

As noted above, the molecular mass decreases as the equivalence ratio increases due to the increasing concentrations of the lighter molecules (such as CO, H_2, and H_2O) and the decreasing concentration of the heavier molecules (such as O_2). For the lower temperatures, the molecular mass has a "sharp" change in value for an equivalence ratio of 1.0. This is largely a result of the discontinuous change in composition from the lean to rich conditions (see the derivation in Section 4.3). For the higher temperatures (say, above 2000 K), the dependence of the molecular mass on equivalence ratio is more continuous since the composition changes in a continuous fashion as dictated by the equilibrium determinations.

Figure 4.10 shows the specific heat at constant pressure for burned mixtures as a function of equivalence ratio for three temperatures for a pressure of 3000 kPa based on the equilibrium assumption. The specific heats increase as temperature increases, and they increase as the mixture increases from the leanest mixtures to near stoichiometric. As the mixtures become richer than stoichiometric, the specific heats first decrease and then increase slightly. These characteristics are generally a result of the species concentrations. As the concentrations of carbon dioxide and water increase, the specific heats increase. This is the major reason for the sharp increase near stoichiometric for the 2000 K case.

Figure 4.11 shows the ratio of specific heats (γ_b) for equilibrium mixtures as functions of equivalence ratio for three constant temperatures at a constant pressure of 3000 kPa. These

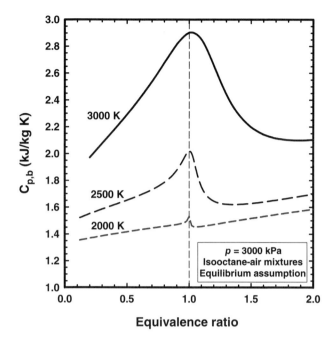

Figure 4.10 The specific heat at constant pressure of the burned gases as functions of equivalence ratio for three temperatures for a pressure of 3000 kPa based on equilibrium assumptions for isooctane and air mixtures

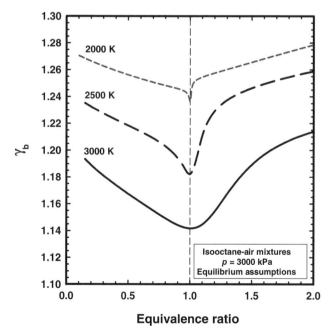

Figure 4.11 The ratio of specific heats of the burned gases as functions of equivalence ratio for three temperatures for a pressure of 3000 kPa based on equilibrium and frozen assumptions for isooctane and air mixtures

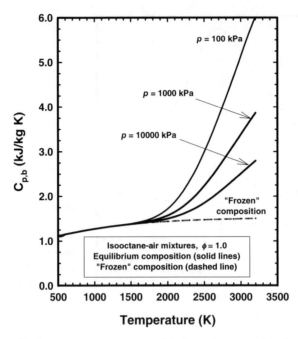

Figure 4.12 The specific heat at constant pressure of the burned gases as functions of temperatures for three pressures for an equivalence ratio of 1.0 based on equilibrium and frozen assumptions for isooctane and air mixtures

curves are roughly the inverse of the curves in Figure 4.10 for the specific heats. The ratio of specific heats decreases as the equivalence ratio increases from low values to stoichiometric. A minimum value is attained at stoichiometric, and then the values increase as the equivalence ratio becomes richer. For these conditions, the ratio of specific heats ranges between about 1.14 and 1.28. As explained above, the sharp change at stoichiometric is the result of the increase of the concentrations of carbon dioxide and water.

Figure 4.12 shows the specific heat at constant pressure for burned mixtures for the equilibrium compositions and for the frozen mixtures as functions of temperature for three pressures for an equivalence ratio of 1.0. The specific heats for the frozen mixtures are in good agreement with the values for the equilibrium mixtures for temperatures below about 1800 K. For higher temperatures, the specific heats for the "frozen" composition under predict the specific heats. Further, for the equilibrium mixtures, the specific heats are also a function of the pressure due to the importance of chemical dissociation. The high pressure cases result in some suppression of the dissociation.

Next, the functional characteristics of the specific internal energy for burned gases (u_b) based on the equilibrium and frozen composition will be examined. These thermodynamic properties will be examined in a fashion which will help interpret the partial derivatives used in the energy equation (Chapter 5). These partial derivatives are based on the characteristics of the specific internal energy and the specific gas constant.

Figure 4.13 shows the specific internal energy as functions of temperature for a constant equivalence ratio of 1.0 for three pressures. As shown, the specific internal energy increases

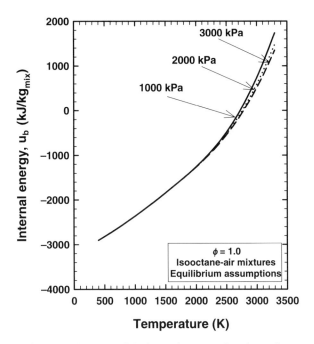

Figure 4.13 The specific internal energy of the burned gases as functions of temperatures for three pressures for an equivalence ratio of 1.0 based on equilibrium assumptions for isooctane and air mixtures

with increasing temperature. The higher pressures inhibit dissociation which slightly lowers the internal energy for the same temperature. The slopes of these curves are the partial derivative, $\frac{\partial u}{\partial T} = C_{v,b}$. This will be discussed later in the section on engine results.

Although not shown directly, the specific internal energy has only a modest dependence on pressure, and only for the highest temperatures. This modest dependence is due to chemical dissociation at the higher temperatures which is somewhat suppressed by high pressures. The partial derivative of significance is $\frac{\partial u}{\partial p}$ and this derivative is small for all temperatures, and near zero for the lower temperatures.

Figure 4.14 shows the specific internal energy as a function of equivalence ratio for a constant pressure of 2000 kPa for three temperatures. The specific internal energy decreases as the mixture goes from the most lean condition to stoichiometric, and increases as the mixture continues from stoichiometric to the rich conditions. Of particular importance is the slope of these curves which represents the partial derivative $\frac{\partial u}{\partial \phi}$. As can be seen, the partial derivative is different on the lean side and rich side of stoichiometric for the low temperature cases. On the other hand, for the high temperature cases, the derivative is continuous.

Although not shown, the gas constant is nearly constant at low temperatures (say, below about 2000 K) for all pressures. For the higher temperatures, the gas constant increases with

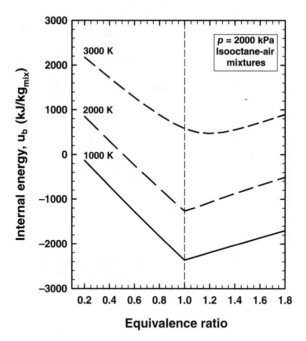

Figure 4.14   The specific internal energy of the burned gases as functions of equivalence ratio for three temperatures for 2000 kPa based on equilibrium assumptions for isooctane and air mixtures

increasing temperature, and the higher pressures slightly inhibit this increase. The partial derivative of significance is $\dfrac{\partial R_b}{\partial T}$. This derivative is near zero for the low temperatures, and increases slightly for the higher temperatures. At 2000 K, this derivative is about 2.0×10^{-6} kJ/kg-K^2. The gas constant is only a modest function of pressure and only significantly for the higher temperatures. The partial derivative, $\dfrac{\partial R_b}{\partial p}$, is near zero for the lower temperatures and higher pressures (since dissociation is inhibited for these conditions).

The gas constant increases with increasing equivalence ratio for all temperatures. For the higher temperatures, the partial derivative, $\dfrac{\partial R_b}{\partial \phi}$, is continuous with equivalence ratio. For the lower temperatures, the partial derivative, $\dfrac{\partial R_b}{\partial \phi}$, changes at an equivalence ratio of 1.0. For example, for temperatures below 2000 K, for equivalence ratios less than 1.0, the partial derivative is approximately equal to 0.0027 kJ/kg-K. Again, information about the partial derivatives of some of the properties described here is relevant to the solution of the thermodynamic relationships presented in Chapter 5.

Figure 4.15 shows the specific enthalpy of the reactants and products as functions of temperature for an equivalence ratio of 1.0 for three pressures. This figure includes values from the unburned routines (including frozen composition for the residuals and low temperature products) as well as values from the equilibrium computations for these conditions. As shown

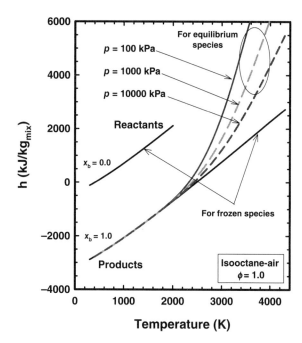

Figure 4.15 The specific enthalpy of the reactants and burned gases as functions of temperatures for three pressures for an equivalence ratio of 1.0 based on equilibrium and frozen assumptions for isooctane and air mixtures

for the low temperature products, the specific enthalpy is well represented by the routines using fixed or frozen composition. For temperatures above 2000 K, however, the specific enthalpy for the frozen assumption becomes increasingly too low compared to the values obtained from the equilibrium computations. Further, the effect of pressure is important for the temperatures above 2000 K. This is a result of the increasing importance of dissociation at the higher temperatures, and of course, chemical dissociation will be affected by the level of pressure. Also note that as the pressure increases, the results tend toward those of the frozen composition due to the inhibiting effect of higher pressures on dissociation.

Figure 4.16 shows the specific entropy for reactant mixtures as functions of temperature for constant pressures of 100, 1000, and 10,000 kPa for an equivalence ratio of 1.0. As shown, the specific entropy increases with increases in temperature and pressure. Recall that the specific entropy has two components: a temperature component and a pressure component. This latter component exists even though the reactant mixture obeys the ideal gas relation.

Figure 4.17 shows the specific entropy for product mixtures as functions of temperature for constant pressures of 100, 1000, and 10,000 kPa for an equivalence ratio of 1.0. As was shown for reactant mixtures, the specific entropy increases with increases in temperature. Results in the figure are for both equilibrium compositions and for "frozen" compositions. As shown, the "frozen" composition provides excellent agreement with the equilibrium composition for temperatures below about 2000 K for these conditions. Above about 2000 K, the frozen composition is not accurate and the specific entropy for the equilibrium composition must be used.

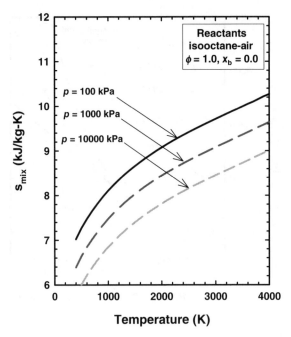

Figure 4.16   The specific entropy of the reactants as functions of temperatures for three pressures for an equivalence ratio of 1.0 for a burned gas fraction of 0.0 for isooctane and air mixtures

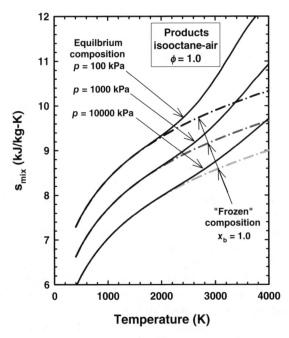

Figure 4.17   The specific entropy of the products as functions of temperatures for three pressures for an equivalence ratio of 1.0 based on equilibrium and frozen assumptions for isooctane and air mixtures

4.7   **Summary**

This chapter has presented the algorithms that are used to compute the thermodynamic properties. The working fluid in the engine may be mixtures of air, fuel, and combustion products. The properties needed for these mixtures include the internal energy, enthalpy, gas constant, molecular mass, and a number of property derivatives. For combustion products, two approaches were described: equilibrium and frozen compositions. For low enough product gas temperatures, the frozen composition was shown to be a good approximation to the equilibrium composition.

The thermodynamic properties have been presented in this chapter as functions of temperature, pressure, and composition. In the section on the engine results, these properties will be revisited, but they will be examined for conditions of variable temperature, pressure, and composition that are consistent with specific engine operating conditions.

References

1. Heywood, J. B. (1988). *Internal Combustion Engine Fundamentals*, McGraw-Hill Company, New York.
2. Caton, J. A. (2010). Implications of fuel selection for an SI engine: results from the first and second laws of thermodynamics, *Fuel*, **89**, 3157–3166.
3. Gordon, S. and McBride, B. J. (1971). Computer program for calculation of complex chemical equilibrium compositions, rocket performance, incident and reflected shocks, and Chapman-Jouquet detonations, NASA SP-273.
4. Svehla, R. A. and McBride, B. J. (1973). Fortran IV computer program for calculation of thermodynamic and transport properties of complex chemical systems, NASA Technical Note TN D-7056, NASA Lewis Research Center.
5. McBride, B. J., Reno, M. A., and Gordon, S. (1994). CET93 and CETPC: An interim updated version of the NASA Lewis computer program for calculating complex chemical equilibria with applications, NASA TM 4557, March.
6. Gordon, S. and McBride, B. J. (1994). Computer program for calculation of complex chemical equilibrium compositions and applications I: analysis, NASA RP 1311, October.
7. McBride, B. J. and Gordon, S. (1996). Computer program for calculation of complex chemical equilibrium compositions and applications II: users' manual and program description, NASA RP 1311, June.
8. Olikara, C. and Borman G. L. (1975). A computer program for calculating properties of equilibrium combustion products with some applications to IC engines, Society of Automotive Engineers, SAE paper no. 750468.
9. *JANAF Thermochemical Tables* (1971). National Bureau of Standards publication NSRDS-NBS37.

5

Thermodynamic Formulations

5.1 Introduction

This chapter describes the development of the formulations of the thermodynamic relations for the engine cycle simulation. The overall process includes identifying the thermodynamic system, applying the first law of thermodynamics, identifying the heat and work interactions, and developing the resulting ordinary differential equations. These ordinary differential equations are most convenient in terms of pressure and temperatures. Many of the features of the overall cycle simulation have been documented elsewhere (e.g., see Reference 1).

As described in this chapter, the first law is applied to an open engine cylinder with volume changes according to the engine kinematics. Cylinder heat transfer is assumed positive into the system, and boundary work (due to the volume change) is assumed positive out of the system. The energy values (internal energy and enthalpy) are all consistent with the same datum, and hence include the appropriate chemical energy. In other words, during combustion, the "energy release" is automatically provided by the difference of the thermodynamic states between reactants and products. This approach is in contrast to a simple use of the heating value of the fuel and the fuel mass burned. The procedure which uses the total energy values is a more accurate approach, and captures any chemical dissociation and other such effects. As part of the case study on combustion, the relation of these two approaches is quantified.

This chapter will include descriptions of the approximations and assumptions used in the development, and the details of the thermodynamic formulations for the various models. Items (such as heat transfer correlations) needed for solving the resulting differential equations are presented in the following chapter, and results are included in their own chapters. In particular, results for items such as internal energy as functions of temperature are presented in Chapter 7. In all cases, energy conservation is achieved. Results illustrating the partition of the fuel energy among items such as work, cylinder heat transfer and exhaust are included in Chapter 7 (and in some of the other chapters).

An Introduction to Thermodynamic Cycle Simulations for Internal Combustion Engines, First Edition.
Jerald A. Caton © 2016 John Wiley & Sons, Ltd. Published 2016 by John Wiley & Sons, Ltd.

5.2   Approximations and Assumptions

The major assumptions and approximations used in the development include the following:

1. The thermodynamic system is the cylinder contents (see Figure 5.1).
2. The engine is in steady-state such that the thermodynamic state at the beginning of each cycle (two crank-shaft revolutions) is equivalent to the state at the end of the cycle.
3. For the compression, expansion, and exhaust processes, the cylinder contents are spatially homogeneous and occupy one zone.
4. For the intake process, two zones (each spatially homogeneous) are used. One zone consists of the fresh charge and the other zone consists of the residual gases.
5. For the combustion processes, one-zone, two-zone and three-zone approaches are outlined.
6. The thermodynamic properties (including pressure and temperature) vary only with time (crank angle) and are spatially uniform in each zone.
7. The instantaneous composition is obtained from generally accepted algorithms [1] and the species obey the ideal gas equation of state (see Chapter 4).
8. The instantaneous thermodynamic properties are computed from established formulations [1] based on the appropriate compositions (see Chapter 4).
9. The flow rates are determined from quasi-steady, one-dimensional flow equations, and the intake and exhaust manifolds are infinite plenums containing gases at constant temperature and pressure (see Chapter 6).
10. The fuel is completely vaporized and mixed with the incoming air.
11. The combustion efficiency is 100% (i.e., no unburned fuel[1]).
12. The blow-by is zero.

The majority of the above assumptions and approximations have been validated and used in a number of previous simulations (e.g., References 1–3). Note that many other sub-models may be developed for use with the simulation. These sub-models could include, for example, models for knock, exhaust gas recirculation (EGR), oxygen enriched inlet air, and over-expanded engines. The use of some of these sub-models is illustrated in the case studies at the end of this book.

The current form of the simulation is limited to in-cylinder processes. For simplicity, the combustion chamber was assumed to be a cylindrical shape. For this work, all cylinders of a multiple-cylinder engine are assumed to be identical, to possess the same thermodynamics, and to operate with identical conditions. This means, therefore, that overall results for a multiple-cylinder engine are obtained by multiplying the results from the single-cylinder analysis by the number of cylinders.

---

[1] The use of the Wiebe function for the mass fraction burned (described below) results in a small amount of unburned fuel at the end of the combustion period. For example, for values of the constants $a = 5.0$ and $m = 2.0$, this small amount is 0.67% of the original fuel mass. This results in a fuel consumption of 99.33%. Further details of this are provided in Chapter 6.

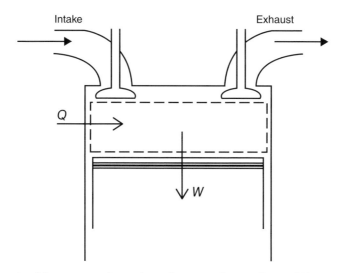

Figure 5.1 Schematic of the one-zone thermodynamic system. *Source:* Caton 1999. Reproduced with permission from ASME

5.3   Formulations

The primary feature used in the development of this cycle simulation is the first law of thermodynamics which is utilized to derive expressions for the time (crank-angle) derivative of the pressure, gas temperatures, and volumes in terms of engine design variables, operating conditions, and sub-model parameters.

The following description is divided into three parts: (i) the one-zone consideration (inlet, compression, expansion, and exhaust processes), (ii) the two-zone consideration for the combustion process (burned and unburned zones), and (iii) the three-zone consideration for the combustion process (adiabatic, boundary layer, and unburned zones).

5.3.1   One-Zone Formulation

Figure 5.1 is a schematic of the engine cylinder which shows cylinder heat transfer, work, and intake and exhaust flows. The first law of thermodynamics for the one-zone formulation for this system is

$$\frac{dE}{dt} = \frac{dQ}{dt} - \frac{dW}{dt} + \dot{m}_{in} h_{in} - \dot{m}_{out} h_{out} \tag{5.1}$$

The only significant energy of the system is internal energy (U), and the only significant work term is due to the piston motion (system boundary motion). So eq. (5.1) becomes

$$\frac{d(mu)}{dt} = \dot{Q} - p\dot{V} + \dot{m}_{in} h_{in} - \dot{m}_{out} h_{out} \tag{5.2}$$

Expanding,

$$m\dot{u} + u\dot{m} = \dot{Q} - p\dot{V} + \dot{m}_{in} h_{in} - \dot{m}_{out} h_{out} \tag{5.3}$$

Since, in general, the thermodynamic properties are functions of temperature, pressure and equivalence ratio (composition), the derivative of u may be written as

$$\dot{u} = \frac{\partial u}{\partial T}\dot{T} + \frac{\partial u}{\partial p}\dot{p} + \frac{\partial u}{\partial \phi}\dot{\phi} \tag{5.4}$$

(For the case of a premixed fuel–air inlet mixture, the time derivative of the equivalence ratio, $\dot{\phi}$, is zero.)

To obtain a derivative of the gas temperature which is independent of the pressure, an expression for the derivative of the pressure, $\dot{p}$, is needed. This may be obtained from the derivative of the ideal gas equation of state:

$$\dot{p} = \left(mR\dot{T} + m\dot{R}T + \dot{m}RT - p\dot{V}\right)/V \tag{5.5}$$

The derivative of R may be given by

$$\dot{R} = \frac{\partial R}{\partial T}\dot{T} + \frac{\partial R}{\partial p}\dot{p} + \frac{\partial R}{\partial \phi}\dot{\phi} \tag{5.6}$$

Combining eqs. (5.3) through (5.6), the following expression is obtained for $\dot{T}$:

$$\dot{T} = \frac{G(1+P_1)}{mC_v(1+P_2)} \tag{5.7}$$

where,

$$G = \left(\dot{Q} - p\dot{V} + \dot{m}_{in}h_{in} - \dot{m}_{out}h_{out} - u\dot{m}\right) \tag{5.8}$$

$$P_1 = \alpha_1/G \text{ and } P_2 = \alpha_2/C_v \tag{5.9}$$

$$\alpha_1 = m\left\{\frac{\partial u}{\partial \phi}\dot{\phi} + \frac{\partial u/\partial p}{V\beta}\left(-mT\frac{\partial R}{\partial \phi}\dot{\phi} - \dot{m}RT + p\dot{V}\right)\right\} \tag{5.10}$$

$$\alpha_2 = \frac{\partial u/\partial p}{\beta}p\left(\frac{1}{T} + \frac{1}{R}\frac{\partial R}{\partial T}\right) \tag{5.11}$$

$$\beta = \left(1 - \frac{p}{R}\frac{\partial R}{\partial p}\right) \tag{5.12}$$

where the term $\partial u/\partial T$ is the specific heat at constant volume, C_v.

Integration of eq. (5.7), therefore, provides a value for the average, one-zone gas temperature. The corresponding cylinder pressure is then found from the ideal gas equation,

$$p = \frac{mRT}{V} \tag{5.13}$$

For the solution of eq. (5.7), several partial derivatives of the thermodynamic properties are needed (e.g., $\dfrac{\partial u}{\partial p}, \dfrac{\partial R}{\partial p}, \dfrac{\partial R}{\partial T}, ...$). These partial derivatives were discussed in Chapter 4.

For the inlet process, an expression similar to eq. (5.7) is needed, but this expression must account for the two different substances that exist during intake (fresh charge and residual gases). This expression is somewhat more complicated, but in general, has the same form. For this situation, a mass rate of mixing is required, and for the inlet process, this is given by the inlet flow rate.

5.3.2   Two-Zone Formulation

For a spark-ignition engine, the flame propagation proceeds from the spark into the unburned gases. Behind the flame, a burned gas zone is created. To capture this flame propagation more precisely, multiple zones are used. First, the formulation is presented for two zones: an unburned zone and a burned zone. Then, the formulation is extended such that the burned zone may be divided into an adiabatic core zone and a boundary layer zone (three-zone formulation).

First, the two-zone formulation will be presented. Figure 5.2 is a schematic of the two zones. As shown, the combustion gases are divided into a burned and an unburned portion. The total heat transfer is divided into portions for each zone. Figure 5.3 shows the details of the mass transfer between the two zones. The burned zone entrains mass from the unburned zone as the flame front proceeds.

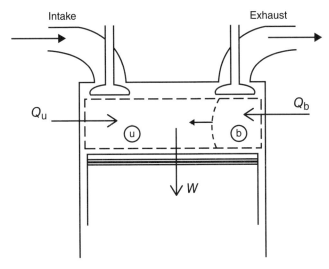

Figure 5.2   Schematic of the two-zone thermodynamic system

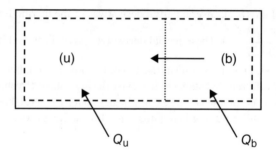

Figure 5.3   Schematic of the expansion of the burned zone toward the unburned zone

The energy equations for the two zones are[2]

$$\frac{d(mu)_b}{d\theta} = \dot{Q}_b - p\dot{V}_b + \dot{m}_b h_u \tag{5.14}$$

$$\frac{d(mu)_u}{d\theta} = \dot{Q}_u - p\dot{V}_u + \dot{m}_u h_u \tag{5.15}$$

As before, the derivatives may be expanded,

$$m_b \dot{u}_b + \dot{m}_b u_b = \dot{Q}_b - p\dot{V}_b + \dot{m}_b h_u \tag{5.16}$$

$$m_u \dot{u}_u + \dot{m}_u u_u = \dot{Q}_u - p\dot{V}_u + \dot{m}_u h_u \tag{5.17}$$

The derivative of the internal energy may then be found for each zone,

$$\dot{u}_u = \frac{\dot{Q}_u - p\dot{V}_u + \dot{m}_u(h_u - u_u)}{m_u} \tag{5.18}$$

$$\dot{u}_b = \frac{\dot{Q}_b - p\dot{V}_b + \dot{m}_b(h_u - u_b)}{m_b} \tag{5.19}$$

Of course, for the unburned zone for an ideal gas,

$$(h_u - u_u) = R_u T_u \tag{5.20}$$

Therefore, the two equations are,

$$\dot{u}_u = \frac{\dot{Q}_u - p\dot{V}_u + \dot{m}_u R_u T_u}{m_u} \tag{5.21}$$

$$\dot{u}_b = \frac{\dot{Q}_b - p\dot{V}_b + \dot{m}_b(h_u - u_b)}{m_b} \tag{5.22}$$

---

[2] The subscript "*u*" denotes the unburned zone, and the subscript "*b*" denotes the burned zone.

Since, in general, the thermodynamic properties are functions of temperature, pressure and equivalence ratio (composition), the derivative of u_b may be written as

$$\dot{u}_b = \frac{\partial u_b}{\partial T_b}\dot{T}_b + \frac{\partial u_b}{\partial p}\dot{p} + \frac{\partial u_b}{\partial \phi}\dot{\phi} \tag{5.23}$$

where the specific heat at constant pressure for the burned gases has been identified

$$C_{v,b} = \frac{\partial u_b}{\partial T_b} \tag{5.24}$$

For the unburned component, the derivatives of the internal energy with respect to pressure and equivalence ratio are zero. Therefore,

$$\dot{u}_u = \frac{\partial u_u}{\partial T_u}\dot{T}_u = C_{v,u}\dot{T}_u \tag{5.25}$$

Combining,

$$\dot{T}_u = \frac{\dot{Q}_u - p\dot{V}_u + \dot{m}_u R_u T_u}{C_{v,u}m_u} \tag{5.26}$$

$$C_{v,b}\dot{T}_b + \frac{\partial u_b}{\partial p}\dot{p} + \frac{\partial u_b}{\partial \phi}\dot{\phi} = \frac{\dot{Q}_b - p\dot{V}_b + \dot{m}_b(h_u - u_b)}{m_b} \tag{5.27}$$

$$\dot{T}_b = \frac{\dot{Q}_b - p\dot{V}_b + \dot{m}_b(h_u - u_b) - m_b\dfrac{\partial u_b}{\partial p}\dot{p} - m_b\dfrac{\partial u_b}{\partial \phi}\dot{\phi}}{m_b C_{v,b}} \tag{5.28}$$

Now, an expression for the derivative of pressure, p, is needed. This may be obtained from the derivative of the ideal gas equations of state (one for the burned component and one for the unburned component):

$$pV_b = m_b R_b T_b \tag{5.29}$$

$$pV_u = m_u R_u T_u \tag{5.30}$$

and adding eqs. (5.29) and (5.30):

$$pV = (m_b R_b T_b + m_u R_u T_u) \tag{5.31}$$

And the derivative of eq. (5.31) is

$$\dot{p}V + p\dot{V} = m_b R_b \dot{T}_b + m_b \dot{R}_b T_b + \dot{m}_b T_b R_b + \\ m_u R_u \dot{T}_u + m_u \dot{R}_u T_u + \dot{m}_u T_u R_u \tag{5.32}$$

Since the unburned mixture is fixed in composition:

$$\dot{R}_{\mathrm{u}} = 0 \tag{5.33}$$

And since,

$$\dot{m}_{\mathrm{b}} = -\dot{m}_{\mathrm{u}} \tag{5.34}$$

Now, eq. (5.32) provides

$$\dot{p} = (m_{\mathrm{b}} R_{\mathrm{b}} \dot{T}_{\mathrm{b}} + m_{\mathrm{b}} \dot{R}_{\mathrm{b}} T_{\mathrm{b}} + \dot{m}_{\mathrm{b}} (T_{\mathrm{b}} R_{\mathrm{b}} - T_{\mathrm{u}} R_{\mathrm{u}}) + \\ m_{\mathrm{u}} R_{\mathrm{u}} \dot{T}_{\mathrm{u}} - p\dot{V}) / V \tag{5.35}$$

The derivative of R_{b} may be given by

$$\dot{R}_{\mathrm{b}} = \frac{\partial R_{\mathrm{b}}}{\partial T_{\mathrm{b}}} \dot{T}_{\mathrm{b}} + \frac{\partial R_{\mathrm{b}}}{\partial p} \dot{p} + \frac{\partial R_{\mathrm{b}}}{\partial \phi} \dot{\phi} \tag{5.36}$$

Combining,

$$\dot{p} = (m_{\mathrm{b}} R_{\mathrm{b}} \dot{T}_{\mathrm{b}} + m_{\mathrm{b}} (\frac{\partial R_{\mathrm{b}}}{\partial T_{\mathrm{b}}} \dot{T}_{\mathrm{b}} + \frac{\partial R_{\mathrm{b}}}{\partial p} \dot{p} + \frac{\partial R_{\mathrm{b}}}{\partial \phi} \dot{\phi}) T_{\mathrm{b}} + \\ \dot{m}_{\mathrm{b}} (T_{\mathrm{b}} R_{\mathrm{b}} - T_{\mathrm{u}} R_{\mathrm{u}}) + m_{\mathrm{u}} R_{\mathrm{u}} \dot{T}_{\mathrm{u}} - p\dot{V}) / V \tag{5.37}$$

Solving for $\dot{P}$,

$$\dot{p}(1 - \frac{m_{\mathrm{b}} \dfrac{\partial R_{\mathrm{b}}}{\partial p} T_{\mathrm{b}}}{V}) = (m_{\mathrm{b}} R_{\mathrm{b}} \dot{T}_{\mathrm{b}} + m_{\mathrm{b}} (\frac{\partial R_{\mathrm{b}}}{\partial T_{\mathrm{b}}} \dot{T}_{\mathrm{b}} + \frac{\partial R_{\mathrm{b}}}{\partial \phi} \dot{\phi}) T_{\mathrm{b}} + \\ \dot{m}_{\mathrm{b}} (T_{\mathrm{b}} R_{\mathrm{b}} - T_{\mathrm{u}} R_{\mathrm{u}}) + m_{\mathrm{u}} R_{\mathrm{u}} \dot{T}_{\mathrm{u}} - p\dot{V}) / V \tag{5.38}$$

$$\dot{p} = \frac{(m_{\mathrm{b}} R_{\mathrm{b}} \dot{T}_{\mathrm{b}} + m_{\mathrm{b}} (\dfrac{\partial R_{\mathrm{b}}}{\partial T_{\mathrm{b}}} \dot{T}_{\mathrm{b}} + \dfrac{\partial R_{\mathrm{b}}}{\partial \phi} \dot{\phi}) T_{\mathrm{b}}}{V(1 - \dfrac{m_{\mathrm{b}} \dfrac{\partial R_{\mathrm{b}}}{\partial p} T_{\mathrm{b}}}{V})} \\ + \frac{\dot{m}_{\mathrm{b}} (T_{\mathrm{b}} R_{\mathrm{b}} - T_{\mathrm{u}} R_{\mathrm{u}}) + m_{\mathrm{u}} R_{\mathrm{u}} \dot{T}_{\mathrm{u}} - p\dot{V}}{V(1 - \dfrac{m_{\mathrm{b}} \dfrac{\partial R_{\mathrm{b}}}{\partial p} T_{\mathrm{b}}}{V})} \tag{5.39}$$

In addition to the above expressions, expressions are needed for the unburned and burned volumes. From volume continuity,

$$\dot{V}_{\mathrm{b}} = \dot{V} - \dot{V}_{\mathrm{u}} \tag{5.40}$$

By obtaining the derivative of eq. (5.30), the following may be found for $\dot{V}_u$:

$$\dot{V}_u = \frac{m_u R_u \dot{T}_u + \dot{m}_u T_u R_u - \dot{p} V_u}{p} \tag{5.41}$$

or,

$$\dot{V}_u = V_u \left(\frac{\dot{m}_u}{m_u} + \frac{\dot{T}_u}{T_u} - \frac{\dot{p}}{p} \right) \tag{5.42}$$

To summarize, the five main ordinary derivative equations are

$$\dot{T}_b = \frac{\dot{Q}_b - p\dot{V}_b + \dot{m}_b(h_u - u_b) - m_b \dfrac{\partial u_b}{\partial p} \dot{p} - m_b \dfrac{\partial u_b}{\partial \phi} \dot{\phi}}{m_b C_{v,b}}$$

$$\dot{T}_u = \frac{\dot{Q}_u - p\dot{V}_u + \dot{m}_u R_u T_u}{C_{v,u} m_u}$$

$$\dot{V}_b = V_b \left(\frac{\dot{T}_b}{T_b} + \frac{\dot{m}_b}{m_b} + \frac{\dot{R}_b}{R_b} - \frac{\dot{p}}{p} \right)$$

$$\dot{V}_u = \dot{V} - \dot{V}_b \tag{5.43}$$

$$\dot{p} = \frac{(m_b R_b \dot{T}_b + m_b(\dfrac{\partial R_b}{\partial T_b} \dot{T}_b + \dfrac{\partial R_b}{\partial \phi} \dot{\phi}) T_b}{V(1 - \dfrac{m_b \dfrac{\partial R_b}{\partial p} T_b}{V})}$$

$$+ \frac{\dot{m}_b(T_b R_b - T_u R_u) + m_u R_u \dot{T}_u - p\dot{V})}{V(1 - \dfrac{m_b \dfrac{\partial R_b}{\partial p} T_b}{V})}$$

The solution of the above set of ordinary differential equations (ODEs) is described below. Once the unburned and burned temperatures (T_u and T_b, respectively) are known, an overall, average cylinder gas temperature may be determined from the overall energy balance:

$$m_{cyl} u_{cyl} = m_u u_u + m_b u_b \tag{5.44}$$

Since the overall cylinder internal energy (u_{cyl}) is not known, an estimate is obtained for the overall cylinder temperature:

$$T_{cyl}^{est} = \frac{m_u C_{v,u} T_u - m_b C_{v,b} T_b}{m_{cyl} C_{v,cyl}} \tag{5.45}$$

Using the ideal gas equation, the corresponding estimated cylinder pressure may be found:

$$p^{est} = \frac{m_{cyl} R_{cyl} T_{cyl}}{V_{cyl}} \tag{5.46}$$

Now the internal energy may be obtained,

$$u_{cyl}^{est} = function\,(T_{cyl}^{est}, p^{est}, ...) \tag{5.47}$$

The overall cylinder temperature (and pressure) is (are) adjusted at each time step until the energy balance (eq. 5.44) is satisfied.

5.3.3   Three-Zone Formulation

The next step is to divide the burned zone into the boundary layer and adiabatic core zones. As mentioned elsewhere in this book, the use of a three-zone formulation is particularly advantageous for capturing the high temperatures associated with combustion. This will be shown to be especially important for determining nitric oxide emissions which are a strong function of the gas temperature. Figure 5.4 is a schematic of the engine cylinder with the three zones during combustion, and Figure 5.5 is a schematic of the details of the boundary layer and adiabatic zones which shows the mass exchange between the various zones.

Initially, the boundary layer has zero mass, and the adiabatic core is equal to the burned mass. As shown, the boundary layer increases in mass by entraining mass from the adiabatic core. The rate of mass entrainment of the boundary layer is dictated by the thermodynamics and temperature relations (described below). All the burned gas heat transfer is assigned to the boundary layer.

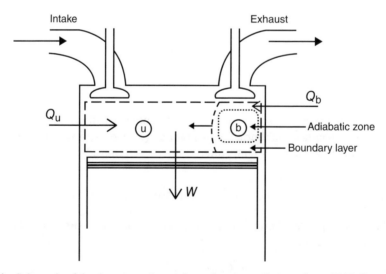

Figure 5.4   Schematic of the three-zone thermodynamic system. *Source:* Caton 2003. Reproduced with permission from ASME

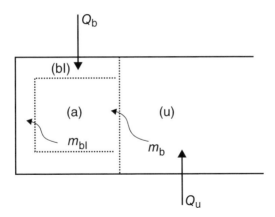

Figure 5.5 Schematic of the adiabatic and boundary layer zone. *Source:* Caton 2007. Reproduced with permission from ASME

For these additional two zones, the formulation of the energy equation is similar to the above for the overall system. The first law, then, for the two additional zones[3]:

$$\frac{d(mu)_a}{d\theta} = -p\dot{V}_a + \dot{m}_b h_u - \dot{m}_{bl} h_a \tag{5.48}$$

$$\frac{d(mu)_{bl}}{d\theta} = \dot{Q}_b - p\dot{V}_{bl} + \dot{m}_{bl} h_a \tag{5.49}$$

Expanding these equations yields:

$$m_a \dot{u}_a + \dot{m}_a u_a = -p\dot{V}_a + \dot{m}_b h_u - \dot{m}_{bl} h_a \tag{5.50}$$

$$m_{bl} \dot{u}_{bl} + \dot{m}_{bl} u_{bl} = \dot{Q}_b - p\dot{V}_{bl} + \dot{m}_{bl} h_a \tag{5.51}$$

But mass conservation for the burned zone:

$$\dot{m}_a = \dot{m}_b - \dot{m}_{bl} \tag{5.52}$$

Combining,

$$\dot{u}_a = \frac{-p\dot{V}_a + \dot{m}_b h_u - \dot{m}_{bl} h_a - (\dot{m}_b - \dot{m}_{bl}) u_a}{m_a} \tag{5.53}$$

$$\dot{u}_a = \frac{-p\dot{V}_a + \dot{m}_b (h_u - u_a) - \dot{m}_{bl}(h_a - u_a)}{m_a} \tag{5.54}$$

and since,

$$h_a - u_a = R_a T_a \tag{5.55}$$

---

[3] The subscript "a" denotes the adiabatic zone, and the subscript "bl" denotes the boundary layer.

then,

$$\dot{u}_a = \frac{-p\dot{V}_a + \dot{m}_b(h_u - u_a) - \dot{m}_{bl}(R_a T_a)}{m_a} \tag{5.56}$$

$$\dot{u}_{bl} = \frac{\dot{Q}_b - p\dot{V}_{bl} + \dot{m}_{bl}(h_a - u_{bl})}{m_{bl}} \tag{5.57}$$

Also, as before, the time derivatives of the internal energies may be expressed as,

$$u_a = fnc(T_a, p, \phi) \tag{5.58}$$

Therefore,

$$\dot{u}_a = \frac{\partial u_a}{\partial T_a}\dot{T}_a + \frac{\partial u_a}{\partial p}\dot{p} + \frac{\partial u_a}{\partial \phi}\dot{\phi} \tag{5.59}$$

Similarly,

$$u_{bl} = fnc(T_{bl}, p, \phi) \tag{5.60}$$

and therefore,

$$\dot{u}_{bl} = \frac{\partial u_{bl}}{\partial T_{bl}}\dot{T}_{bl} + \frac{\partial u_{bl}}{\partial p}\dot{p} + \frac{\partial u_{bl}}{\partial \phi}\dot{\phi} \tag{5.61}$$

Combining,

$$C_{v,a}\dot{T}_a + \frac{\partial u_a}{\partial p}\dot{p} + \frac{\partial u_a}{\partial \phi}\dot{\phi} =$$
$$\frac{-p\dot{V}_a + \dot{m}_b(h_u - u_a) - \dot{m}_{bl}(R_a T_a)}{m_a} \tag{5.62}$$

$$\dot{T}_a = \frac{-p\dot{V}_a + \dot{m}_b(h_u - u_a) - \dot{m}_{bl}(R_a T_a)}{C_{v,a}m_a}$$
$$+ \frac{-m_a\dfrac{\partial u_a}{\partial p}\dot{p} - m_a\dfrac{\partial u_a}{\partial \phi}\dot{\phi}}{C_{v,a}m_a} \tag{5.63}$$

$$C_{v,bl}\dot{T}_{bl} + \frac{\partial u_{bl}}{\partial p}\dot{p} + \frac{\partial u_{bl}}{\partial \phi}\dot{\phi} =$$
$$\frac{\dot{Q}_b - p\dot{V}_{bl} + \dot{m}_{bl}(h_a - u_{bl})}{m_{bl}} \tag{5.64}$$

$$\dot{T}_{\text{bl}} = \frac{\dot{Q}_{\text{b}} - p\dot{V}_{\text{bl}} + \dot{m}_{\text{bl}}(h_{\text{a}} - u_{\text{bl}})}{C_{\text{v,bl}}m_{\text{bl}}}$$
$$+ \frac{-m_{\text{bl}}\dfrac{\partial u_{\text{bl}}}{\partial p}\dot{p} - m_{\text{bl}}\dfrac{\partial u_{\text{bl}}}{\partial \phi}\dot{\phi}}{C_{\text{v,bl}}m_{\text{bl}}} \tag{5.65}$$

This relation may be rearranged,

$$\dot{m}_{\text{bl}} = \frac{1}{u_{\text{bl}} - h_{\text{a}}}(\dot{Q}_{\text{b}} - p\dot{V}_{\text{bl}}$$
$$- m_{\text{bl}}(\frac{\partial u_{\text{bl}}}{\partial p}\dot{p} - m_{\text{bl}}\frac{\partial u_{\text{bl}}}{\partial \phi}\dot{\phi} - \dot{T}_{\text{bl}}C_{\text{v,bl}})) \tag{5.66}$$

Next, from the ideal gas equations of state,

$$pV_{\text{a}} = m_{\text{a}}R_{\text{a}}T_{\text{a}} \tag{5.67}$$

$$pV_{\text{bl}} = m_{\text{bl}}R_{\text{bl}}T_{\text{bl}} \tag{5.68}$$

And the derivative of these expressions,

$$\dot{p}V_{\text{a}} + p\dot{V}_{\text{a}} = m_{\text{a}}R_{\text{a}}\dot{T}_{\text{a}} + m_{\text{a}}\dot{R}_{\text{a}}T_{\text{a}} + \dot{m}_{\text{a}}T_{\text{a}}R_{\text{a}} \tag{5.69}$$

$$\dot{p}V_{\text{bl}} + p\dot{V}_{\text{bl}} = m_{\text{bl}}R_{\text{bl}}\dot{T}_{\text{bl}} + m_{\text{bl}}\dot{R}_{\text{bl}}T_{\text{bl}} + \dot{m}_{\text{bl}}T_{\text{bl}}R_{\text{bl}} \tag{5.70}$$

where,

$$\dot{R}_{\text{a}} = \frac{\partial R_{\text{a}}}{\partial T_{\text{a}}}\dot{T}_{\text{a}} + \frac{\partial R_{\text{a}}}{\partial p}\dot{p} + \frac{\partial R_{\text{a}}}{\partial \phi}\dot{\phi} \tag{5.71}$$

and,

$$\dot{R}_{\text{bl}} = \frac{\partial R_{\text{bl}}}{\partial T_{\text{bl}}}\dot{T}_{\text{bl}} + \frac{\partial R_{\text{bl}}}{\partial p}\dot{p} + \frac{\partial R_{\text{bl}}}{\partial \phi}\dot{\phi} \tag{5.72}$$

The above derivatives of the ideal gas relations for the two zones can be solved for $\dot{V}_{\text{a}}$ and $\dot{V}_{\text{bl}}$

$$\dot{V}_{\text{a}} = \frac{m_{\text{a}}R_{\text{a}}\dot{T}_{\text{a}} + m_{\text{a}}\dot{R}_{\text{a}}T_{\text{a}} + \dot{m}_{\text{a}}T_{\text{a}}R_{\text{a}} - \dot{p}V_{\text{a}}}{p} \tag{5.73}$$

or,

$$\dot{V}_{\text{a}} = V_{\text{a}}(\frac{\dot{m}_{\text{a}}}{m_{\text{a}}} + \frac{\dot{R}_{\text{a}}}{R_{\text{a}}} + \frac{\dot{T}_{\text{a}}}{T_{\text{a}}} - \frac{\dot{p}}{p}) \tag{5.74}$$

Similarly,

$$V_{bl} = \frac{m_{bl} R_{bl} \dot{T}_{bl} + m_{bl} \dot{R}_{bl} T_{bl} + \dot{m}_{bl} T_{bl} R_{bl} - \dot{p} V_{bl}}{p} \tag{5.75}$$

or,

$$\dot{V}_{bl} = V_{bl} \left(\frac{\dot{m}_{bl}}{m_{bl}} + \frac{\dot{R}_{bl}}{R_{bl}} + \frac{\dot{T}_{bl}}{T_{bl}} - \frac{\dot{p}}{p} \right) \tag{5.76}$$

And, finally, volume continuity requires,

$$\dot{V}_{bl} = \dot{V}_{b} - \dot{V}_{a} \tag{5.77}$$

The growth of the boundary layer is dictated by the mass and energy conservation relations for the burned zone. A choice at this point is how the average boundary layer temperature is defined. One classical definition for the average boundary layer temperature is related to the wall temperature and the adiabatic zone temperature.

$$T_{bl} = \frac{T_a - T_{wall}}{\ln(T_a / T_{wall})} \tag{5.78}$$

In addition, the derivative of this expression may be derived,

$$\dot{T}_{bl} = \left[\frac{1}{\ln(T_a / T_{wall})} - \frac{1 - T_{wall}/T_a}{(\ln(T_a / T_{wall}))^2} \right] \dot{T}_a \tag{5.79}$$

To summarize, the new unknowns for the boundary layer and adiabatic core zones are T_a, T_{bl}, V_a, V_{bl}, m_a and m_{bl}. A number of differential relations may be derived for these variables. Because these relations are highly nonlinear and exhibit complex interrelations, not all combinations of the above equations were successful. After extensive trials, the following six ordinary differential equations for these variables were found to be the most robust in terms of obtaining stable solutions.

$$\dot{T}_a = \frac{-p\dot{V}_a + \dot{m}_b(h_u - u_a) - \dot{m}_{bl}(R_a T_a)}{C_{v,a} m_a} + \frac{-m_a \dfrac{\partial u_a}{\partial p} \dot{p} - m_a \dfrac{\partial u_a}{\partial \phi} \dot{\phi}}{C_{v,a} m_a}$$

$$\dot{T}_{bl} = \left[\frac{1}{\ln(T_a / T_{wall})} - \frac{1 - T_{wall}/T_a}{(\ln(T_a / T_{wall}))^2} \right] \dot{T}_a$$

$$\dot{V}_{bl} = V_{bl} \left(\frac{\dot{m}_{bl}}{m_{bl}} + \frac{\dot{R}_{bl}}{R_{bl}} + \frac{\dot{T}_{bl}}{T_{bl}} - \frac{\dot{p}}{p} \right) \tag{5.80}$$

$$\dot{V}_a = \dot{V}_b - \dot{V}_{bl}$$

$$\dot{m}_{bl} = \frac{1}{u_{bl} - h_a} \left(\dot{Q}_b - p\dot{V}_{bl} - m_{bl} \left(\frac{\partial u_{bl}}{\partial p} \dot{p} - m_{bl} \frac{\partial u_{bl}}{\partial \phi} \dot{\phi} - \dot{T}_{bl} C_{v,bl} \right) \right)$$

$$\dot{m}_a = \dot{m}_b - \dot{m}_{bl}$$

5.4 Comments on the Three Formulations

The three different formulations (based on one, two, or three zones) presented in this chapter provide increasing resolution on the details of the combustion process. The overall thermodynamics, however, are the same for each formulation. For example, the overall instantaneous cylinder pressure and the performance parameters are the same regardless of the number of zones selected to depict the combustion process. The two- and three-zone formulations provide items such as the temperatures of the unburned and burned gases which are not available from the one-zone formulation.

5.5 Summary

This chapter has presented the detail thermodynamic formulations for the engine cylinder contents which results in a set of differential equations for temperatures, pressures, volumes, and masses as functions of time (crank angle). For the combustion process, these formulations were derived for zones one, two, and three in the cylinder. The rest of the material in this book is based on the three-zone formulations. To solve the differential equations obtained from the formulations a number of items are needed such as the mass fraction burn, cylinder heat transfer, flow rates, and property data. Chapter 6 describes these items.

References

1. Heywood, J. B. (1988). *Internal Combustion Engine Fundamentals*, McGraw-Hill Book Company, New York.
2. Heywood, J. B., Higgins, J. M., Watts, P. A., and Tabaczynski, R. J. (1979). Development and use of a cycle simulation to predict SI engine efficiency and NO_x emissions, Society of Automotive Engineers, SAE Paper No. 790291.
3. Heywood, J. B. and Watts, P. (1979). Parametric studies of fuel consumption and NO emissions of dilute spark-ignition engine operation using a cycle simulation, *Institute of Mechanical Engineers Conference Publication C*, **98**, 117–127.

6

Items and Procedures for Solutions

6.1 Introduction

This chapter will include descriptions of the items needed to solve the ordinary differential equations described in Chapter 5. These items include the thermodynamic properties (Chapter 4), cylinder volume, rate of change of the cylinder volume, surface areas, expressions for the combustion process (mass fraction burned), cylinder heat transfer, mass flow rates, and algorithms for the friction. The algorithms for the friction will be illustrated with values as functions of engine parameters. In addition, this chapter will include a brief overview of the way nitric oxide emissions could be computed (a case study will supply much more detail). The chapter will end with a brief discussion of the solution procedures.

6.2 Items Needed to Solve the Energy Equations

6.2.1 Thermodynamic Properties

The thermodynamic properties needed for solving the first law expressions in Chapter 5 include the instantaneous values of the specific internal energy, specific enthalpy, specific gas constant, molecular mass, entropy, exergy, and six property derivatives. The algorithms used for determining the properties were described in Chapter 4. Chapter 9 describes the property relations needed to quantify the second law considerations. All the properties are needed as a function of time (crank angle) for the cylinder contents for the three zones, and for any matter entering the cylinder.

Depending on the specific processes during a cycle, the thermodynamic properties may be for different mixtures of air, fuel vapor, and combustion products. The concentrations of the combustion products may be "frozen" for the lower temperatures, or these concentrations may be based on an instantaneous determination of chemical equilibrium (e.g., see References 1,2) for higher temperatures. Complete descriptions of the algorithms used for determining the compositions are presented in Chapter 4.

Once the composition is known, the individual thermodynamic properties may be determined. The properties of each species in the mixture are first determined for the given temperature and pressure by the use of polynomial curve-fits (e.g., Reference 2) to the thermodynamic data. The overall mixture properties are then determined by suitable expressions. For example, for the specific internal energy,

$$u_k = \sum_i x_i \tilde{u}_i \tag{6.1}$$

where u_k is the mixture specific internal energy for unburned ($k = u$) or burned mixtures ($k = b$), x_i is the mole fraction of species i and $\tilde{u}_i$ is the specific internal energy for species i on a per mole basis. For entropy, a similar expression is used,

$$s_k = -R_u \ln\left(\frac{p}{p_o}\right) + \sum_i x_i (\tilde{s}_i^o - R_u \ln x_i) \tag{6.2}$$

where s_k is the mixture specific entropy for the unburned or burned mixture, R_u is the universal gas constant, p is the mixture pressure, p_o is the reference pressure, and $\tilde{s}_i^o$ is the temperature dependent portion of the specific entropy for species i on a per mole basis.

Entropy and exergy are more fully discussed in Chapter 9 on the second law analyses.

6.2.2 Kinematics

The kinematics and geometry of reciprocating engines were briefly mentioned in Chapter 2 (engine overview). Some of the details are provided here for completeness.

Figure 6.1 is a simple sketch of the key geometrical items of a conventional reciprocating engine. The bore and stroke are denoted as "B" and "S." The piston travels between "top dead center" (TDC) and "bottom dead center" (BDC). The volume above the TDC position is the clearance volume (V_c), the volume between BDC and TDC is the displaced volume (V_d), and the sum of V_c and V_d is the total cylinder volume (V_{tot}). The compression ratio (CR) is an important geometric parameter.

$$CR = \frac{V_{tot}}{V_c} = \frac{V_c + V_d}{V_c} = 1 - \frac{V_d}{V_c} \tag{6.3}$$

Note that this is the geometric compression ratio and is determined solely by the geometry. In practice, due to valve timings, the effective or actual compression ratio can be different. The effective compression ratio is used by the cycle simulation in this work.

The crank radius, a, is related to the stroke

$$S = 2a \tag{6.4}$$

The connecting rod length is given by the symbol ℓ.

The final geometrical measure is the distance from the crank center to the end of the connecting rod—this is given the symbol "s" in the following and is a function of crank angle position.

For the simulation, an important item is the instantaneous cylinder volume. This may be determined from simple trigonometry. The cylinder volume at any crank angle position, θ, is given by

$$V = V_c + \frac{\pi B^2}{4}(\ell + a - s) \tag{6.5}$$

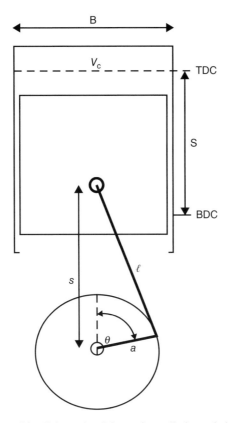

Figure 6.1   Schematic of the engine cylinder and piston

From the triangular relationship, an expression for "s" is

$$s = a\cos\theta + (\ell^2 - a^2 \sin^2\theta)^{1/2} \tag{6.6}$$

Combining the above two relations provides an expression for the instantaneous volume,

$$V = V_c + \frac{\pi B^2}{4}(\ell + a - a\cos\theta + (\ell^2 - a^2 \sin^2\theta)^{1/2}) \tag{6.7}$$

Also needed is the rate of change of cylinder volume:

$$\dot{V} = \frac{\pi B^2}{4} a\sin\theta \left[1 + \frac{a\cos\theta}{\left(\ell^2 - a^2 \sin^2\theta\right)^{1/2}} \right] \tag{6.8}$$

where, if B and a have units of meters, then $\dot{V}$ has units of "m^3/rad."

In addition to the instantaneous cylinder volume and rate of change of cylinder volume, the instantaneous surface area is needed for the heat transfer

$$A = A_{\text{top}} + A_{\text{side}} + A_{\text{bottom}} \tag{6.9}$$

For a combustion chamber with a right circular cylinder shape,

$$A_{top} = A_{bottom} = \frac{\pi B^2}{4} \tag{6.10}$$

And the surface area on the sides (cylinder walls) at any crank angle position is the product of the perimeter and the distance from the top to the piston (x_{side}),

$$A_{side} = \pi B x_{side} \tag{6.11}$$

$$x_{side} = (\ell + a - s) \tag{6.12}$$

The total instantaneous surface area is then given by,

$$A = 2\frac{\pi B^2}{4} + \pi B(\ell + a - s) \tag{6.13}$$

6.2.3   Combustion Process (Mass Fraction Burned)

The mass fraction of fuel (or fuel mixture) burned (x_b) may be determined from a first-law analysis of the measured cylinder pressure [2]. The result of this analysis is almost always an "S" shape curve. The mass fraction burned is defined as

$$x_b = \frac{(m_f)_{burned}}{(m_f)_{total}} \tag{6.14}$$

The current work has used the Wiebe function [3] to represent the mass fraction burned

$$x_b = 1 - \exp\left\{-ay^{m+1}\right\} \tag{6.15}$$

where y is a non-dimensional time variable,

$$y = \frac{\theta - \theta_o}{\theta_b} \tag{6.16}$$

and where "a" and "m" are parameters that are selected to provide a match with experimental information. The values that provide a good match with the data depend on complex functions of the turbulence, chemistry, fuel, equivalence ratio, chamber geometry, and a number of other features. The values used here are based on the recommendation of Heywood [2]:

$$a = 5.0 \text{ and } m = 2.0 \tag{6.17}$$

In eq. 6.16, θ is the current crank angle, θ_o is the crank angle for the start of combustion, and θ_b is the burn duration (crank angle duration of combustion). In general, the combustion duration is a complex function of temperature, pressure, turbulence, composition, chemistry and other factors. For the work reported here, the combustion duration is specified. Other work has used empirical relationships to predict the combustion duration based on engine speed, load, equivalence ratio and other engine parameters.

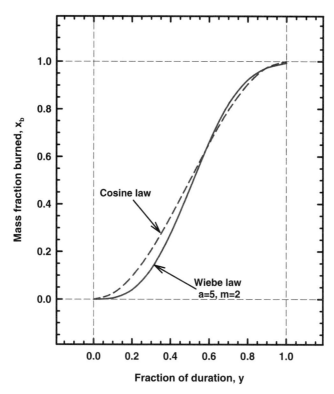

Figure 6.2   Mass fraction burned as function of the fraction of the burn duration using either the Wiebe or the cosine function

Figure 6.2 shows the mass fraction burned as a function of the fraction of the combustion duration variable, y, using the Weibe function ($a = 5.0$; $m = 2.0$) and a cosine function. The cosine function is a good alternative to the Weibe function and captures the major features of the mass burn curve.

$$x_b = \frac{1}{2}\{1 - \cos(\pi y)\} \tag{6.18}$$

For slightly more precision, however, the rest of the work reported here will use the Wiebe function.

For most of the work reported here, the combustion duration is based on 0–100% fuel burned. Other definitions of the combustion duration exist. One popular definition is 10–90% fuel burned. This is a particularly useful definition for use with experimental work. Since the start of combustion is gradual, detecting the exact 0% point is difficult if not impossible. The 10% point is much easier to detect from experimental data.

The combustion duration definitions are related. Figure 6.3 shows the mass fraction burned as a function of the fraction of the combustion duration. Denoted on the figure are the 10% and 90% fuel burned locations. As this demonstrates, the 10–90% period is much shorter than the 0–100% period for typical burn profiles. For the Wiebe constants used here, the relation

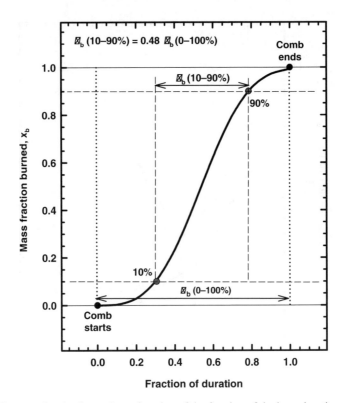

Figure 6.3 The mass fraction burned as a function of the fraction of the burn duration

is a factor of 0.48. That is, the if the 0–100% burn duration is 60°CA, then the 10–90% burn duration is 28.8°CA.

The Wiebe function and the cosine function capture the main features of the combustion process. The burning process begins slowly due to the low initial temperatures and turbulence. As the process develops, the burning becomes more rapid and the mass fraction burned increases more rapidly. Then, as the process begins to reach completion, the rate of the mass fraction burned becomes slower due to less reactant, dilution by combustion products and cooling as the flame front reaches the far walls.

Any expression which is selected to represent the mass fraction burned must satisfy the end conditions. These end conditions include

$$x_b = 0 \text{ at } y = 0$$
$$x_b = \left(x_b\right)_{\text{final}} \text{ at } y = 1 \tag{6.19}$$

where the final mass fraction burned, $(x_b)_{\text{final}}$, will have a value near 1.0, but for some expressions may be less than 1.0. In addition to eq. 6.19, the expression may satisfy the derivative conditions at the end points. Namely,

$$\dot{x}_b = 0 \text{ at } y = 0$$
$$\dot{x}_b = 0 \text{ as } y \rightarrow 1 \tag{6.20}$$

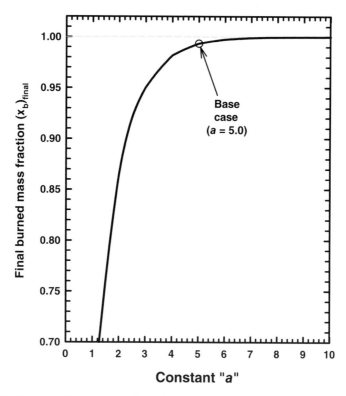

Figure 6.4   The final mass fraction burned as a function of the constant "*a*" using the Wiebe function

For the Wiebe function (eq. 6.15), the final mass fraction burned depends on the value of the constant "*a*." Figure 6.4 shows the final mass fraction burned as a function of "*a*." For values of the constant "*a*" greater than 7.0 the final value is nearly 1.0. For the value of "*a*" used in this work (*a* = 5.0), the final mass fraction burned is 0.9933. This means that for all cases using this value of "*a*," at the end of combustion, 0.67% of the original fuel remains unburned. This is essentially insignificant and, if anything, represents a portion of the unburned fuel that is probable for most engines.

6.2.4   Cylinder Heat Transfer

The cylinder heat transfer is a complex phenomenon that remains an active area of engine research. A vast literature is available that discusses various aspects of this topic. In the case study concerning cylinder heat transfer, actual quantitative evaluations are provided. Additional comments on cylinder heat transfer correlations may be found in [4].

At this point, however, only the basics will be described. The overall convective heat transfer to the cylinder gases is given by

$$\dot{Q} = h_c \, A(\theta) \, (T_{wall} - T) \tag{6.21}$$

where h_c is the instantaneous convective heat transfer coefficient, A is the instantaneous surface area, T_{wall} is the cylinder wall temperature, and T is the instantaneous one-zone gas temperature. A number of correlations for the instantaneous convective heat transfer coefficient (h_c) have been proposed and used throughout the years. Further discussion of the choice of the heat transfer correlation is provided as part of the case study on cylinder heat transfer.

During the combustion process, the total heat transfer is allocated to the various zones (but the total is still given by eq. 6.21). This allocation may be completed in a number of ways. For this work, the heat transfer is assumed to be proportional to the surface area of each zone. The appropriate surface area is assumed to be proportional to the volume raised to the 2/3 power. In addition, the heat transfer is proportional to the temperature difference. For example, since the unburned zone temperature is lower, the proportion of the heat transfer allocated to the unburned zone should be lower as well. This results in the following for the individual allocated heat transfer values:

$$\dot{Q}_u = \left(\frac{V_u}{V_{total}}\right)^{2/3}\left(\frac{T_{wall} - T_u}{T_{wall} - T}\right)\dot{Q}_{total} \tag{6.22}$$

$$\dot{Q}_b = \dot{Q}_{total} - \dot{Q}_u \tag{6.23}$$

Since the burned zone is divided into an adiabatic core and a boundary layer, the burned zone heat transfer is assigned in total to the boundary layer.

6.2.5 Mass Flow Rates

The next item is the mass flow rates for use in the energy equation. These may be found from basic considerations outlined elsewhere (e.g., see Reference 2). These relationships are based on the following assumptions and approximations:

- The flow is quasi-steady
- The flow is one-dimensional
- The flow is reversible
- The flow is adiabatic
- The flow discharge coefficients are assumed constant[1]

Since a real flow would not conform to the above assumptions and approximations, these considerations are corrected by use of an empirical discharge coefficient (C_D),

$$C_D = \frac{\text{actual mass flow rate}}{\text{ideal mass flow rate}} \tag{6.24}$$

Discharge coefficients are functions of the flow geometry, Reynolds number, Mach number, and the gas properties [2]. For purposes of this study, the discharge coefficient has been assumed constant with a value of 0.7, which is a typical average value [2].

---

[1] Note that within the scope of this work, the flow discharge coefficients easily could be a function of instantaneous valve lift.

The flow rates may be determined from the following relationships. The flow rates may be either subsonic or sonic. The subsonic flow rates are given by

$$\dot{m} = A_t P_u \sqrt{\frac{2}{RT_u}\left(\frac{P_t}{P_u}\right)^{2/\gamma} \frac{\gamma}{(\gamma - 1)}\left[1 - \left(\frac{P_t}{P_u}\right)^{(\gamma-1)/\gamma}\right]} \tag{6.25}$$

where A_t is the actual open flow area, P_u is the upstream pressure, R is the specific gas constant, T_u is the upstream gas temperature, P_t is the throat (or downstream) pressure, and γ is the ratio of specific heats.

On the other hand, the flows may be sonic (also known as choked or critical) if the upstream pressure is sufficiently high relative to the downstream pressure such that

$$\frac{P_u}{P_t} \geq \left(\frac{\gamma + 1}{2}\right)^{\gamma/(\gamma-1)} \tag{6.26}$$

If the flows are sonic, then the flow rate is given by

$$\dot{m} = A_t P_u \sqrt{\frac{\gamma}{RT_u}\left(\frac{2}{\gamma + 1}\right)^{(\gamma+1)/(\gamma-1)}} \tag{6.27}$$

The actual open flow area, A_t, is based on a "curtain" area definition at the throat (valve opening),

$$A_t = C_D \pi D_v L(\theta) \tag{6.28}$$

where D_v is the valve diameter, and $L(\theta)$ is the instantaneous valve lift. Since the valve lift profiles are not necessarily known, this work has used a standard assumption for the valve lift based solely on the maximum lift and the valve open duration, and using a sinusoidal shape as recommended by Sherman and Blumberg [5]:

$$L(\theta) = L_{max} \sin\left(\pi \frac{\theta - \theta_{vo}}{\theta_{dur}}\right) \tag{6.29}$$

where L_{max} is the maximum valve lift, θ_{vo} is the crank angle of the valve opening, and θ_{dur} is the valve open duration.

Figure 6.5 shows the instantaneous valve lift as functions of crank angle for both the exhaust and for the intake for the base engine considered here. The valve lift is a sinusoidal function of crank angle. The valve overlap period is denoted.

Figure 6.6 shows the instantaneous ideal open flow area as functions of crank angle for both the exhaust and for the intake for the base engine considered here. The ideal open flow area is,

$$A_{ideal} = \pi D_v L(\theta) \tag{6.30}$$

This area is proportional to the valve lift, and remains a sinusoidal function of crank angle. As with the previous figure, the valve overlap period also is denoted in Figure 6.6.

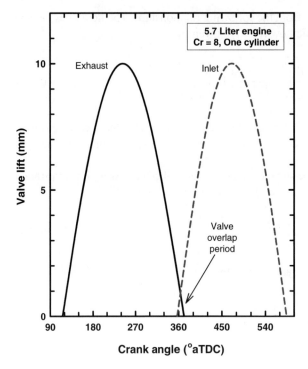

Figure 6.5 The instantaneous valve lift as functions of crank angle

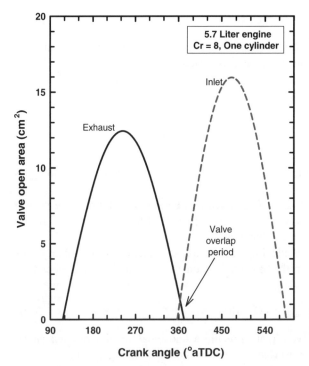

Figure 6.6 The valve open area as functions of crank angle

The formulation used in the cycle simulation includes the capability for "reverse" flows. These are flows that are counter to the normal direction. For example, when the inlet valves are open and the inlet manifold pressure is lower than the cylinder pressure, flows will be from the cylinder into the manifold. Similarly, when the exhaust valves are open and the exhaust manifold pressure is higher than the cylinder pressure, flows will be from the exhaust manifold into the cylinder. By comparing the manifold and cylinder pressure at each instant, the direction of the flow can be determined. Results in Chapter 7 will illustrate this feature.

6.2.6   Mass Conservation

In addition to the energy equation described above, mass conservation must be satisfied throughout the computation. The instantaneous change in the cylinder mass is equal to the net result of the mass flow rates.

$$\frac{dm}{dt} = \dot{m}_{in} - \dot{m}_{out} \tag{6.31}$$

6.2.7   Friction

Information concerning engine friction is needed for a variety of reasons. One of these reasons is to use realistic frictional losses to determine "brake" quantities. Engine friction is estimated with a series of algorithms which have been developed for current (1990s) spark-ignition engines. The friction algorithms used here are based on the algorithms proposed by Sandoval and Heywood [6] which were updated correlations originally published by Patton *et al.* [7]. Original work by Bishop [8] has been used as a starting point for a number of these algorithms. For the current work, the algorithms are used exactly as presented.

The total friction will be given by

$$fmep_{total} = \sum fmep_{various} \tag{6.32}$$

The overall friction is divided into the following components:

Rubbing:	1. Crankshaft: Main bearings Seals for main bearings Work of moving lubricant (turbulent dissipation)
	2. Reciprocating: Connecting rod bearings Pistons and piston skirts Piston rings (no gas pressure) Increase of piston ring friction due to gas pressure loading
	3. Valvetrain: Camshaft bearings Cam followers Valve actuation mechanisms

(continued)

Pumping: 1. Intake system
                                  2. Intake valves
                                  3. Exhaust system
                                  4. Exhaust valves
Auxiliaries: 1. Oil pump
                                  2. Water pump
                                  3. Alternator

Introducing these components:

$$fmep_{total} = fmep_{rub} + fmep_{pump} + fmep_{aux} \qquad (6.33)$$

Next, the $fmep_{rub}$ and $fmep_{aux}$ may be expressed in terms of their components:

$$fmep_{rub} = fmep_{crank} + fmep_{recip} + fmep_{valve} \qquad (6.34)$$

$$fmep_{aux} = fmep_{oil\text{-}pump} + fmep_{water\text{-}pump} + fmep_{alternator} \qquad (6.35)$$

And the components of $fmep_{crank}$, $fmep_{recip}$, and $fmep_{valve}$ may be expressed in terms of even smaller aspects:

$$fmep_{crank} = fmep_{bearings} + fmep_{seals} + fmep_{dissip} \qquad (6.36)$$

$$fmep_{recip} = fmep_{conrod} + fmep_{piston} + fmep_{rings} + fmep_{gas\text{-}loading} \qquad (6.37)$$

$$fmep_{valve} = fmep_{cam} + fmep_{followers} + fmep_{actuation} \qquad (6.38)$$

Algorithms have been developed for each of these components [6].

To illustrate the results of the friction algorithms, the algorithms were applied for a 5.7 liter V-8, spark-ignition engine. The friction algorithms require much detail about the engine. For example, required information includes the number and size of bearings, the number of valves, the type of valve mechanism, and the type and location of the camshaft. Sandoval and Heywood [6] may be consulted for specific examples of this information.

For the 5.7 liter V-8, spark-ignition engine, Figure 6.7 shows the various components of the friction as functions of engine speed for a part load condition with an inlet manifold pressure of 50 kPa (which results in a $bmep$ of about 325 kPa). For this constant load condition, the pumping friction increases slightly with increases of engine speed. The mechanical friction due to the auxiliaries, reciprocating components, and (to a lesser extent) crank components increases as engine speed increases. The mechanical friction of the valve train decreases slightly with increases of speed.[2]

---

[2] At relatively low engine speeds, the valve train friction is dominated by valve spring compression which explains the higher relative friction. At high speeds, however, inertia plays a major role as the acceleration of the tappet has a negative value (and this acceleration becomes more negative as engine speed increases) resulting in a slight decrease of the valve train friction.

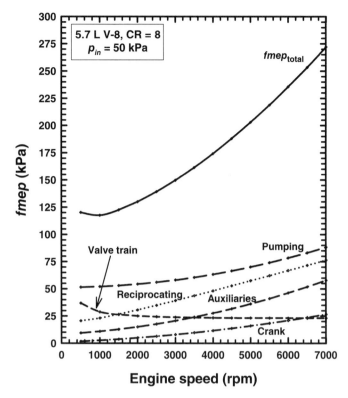

Figure 6.7   The individual *fmep* components as functions of the engine speed for an inlet pressure of 50 kPa for the base engine

Figure 6.8 shows the same data as presented in Figure 6.7, but the data are shown in an accumulative fashion. For these conditions, the reciprocating component is the largest contributor to the mechanical component of the total friction. The pumping friction is a major portion of the total friction due to the part load condition. The figure shows the total mechanical component for the base case (1400 rpm).

Figure 6.9 shows data similar to that presented in Figure 6.3, but for an inlet pressure of 95 kPa (which represents WOT). Due to higher cylinder pressures, the total friction is higher than the previous case—especially for the higher speeds. The pumping friction is much less at lower speeds, but does increase for higher speeds.

Figure 6.10 shows the total friction as functions of engine speed for an inlet pressure of 50 kPa for two compression ratios. The higher compression ratio results in higher friction due to higher cylinder pressures—especially for lower speeds.

Figure 6.11 shows the various friction components as functions of inlet manifold pressure (load) for a constant engine speed of 1400 rpm. The mechanical components increase nearly linearly with increasing inlet manifold pressure. The pumping component decreases as inlet manifold pressure increases (reflecting less throttle losses). The net effect for these conditions is that the total friction decreases as inlet manifold pressure increases.

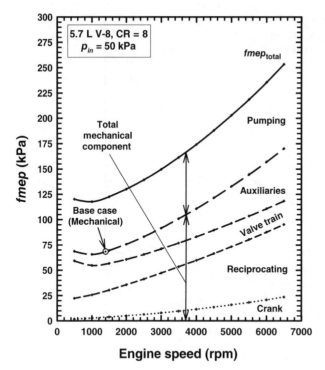

Figure 6.8   The accumulative *fmep* components as functions of the engine speed for an inlet pressure of 50 kPa for the base engine

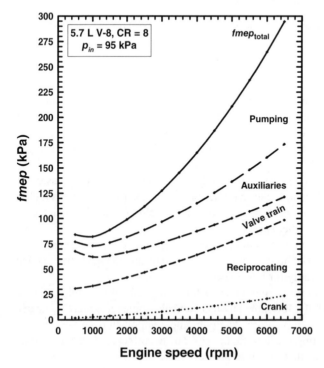

Figure 6.9   The individual *fmep* components as functions of the engine speed for an inlet pressure of 95 kPa for the base engine

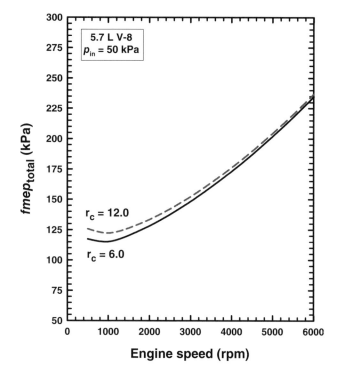

Figure 6.10 The total *fmep* as functions of the engine speed for an inlet pressure of 50 kPa for two compression ratios for the base engine

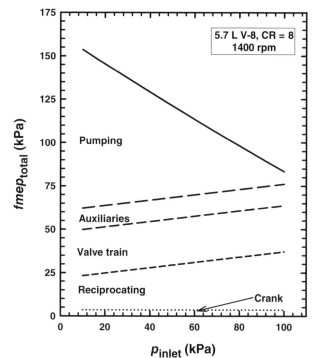

Figure 6.11 The *fmep* as functions of the inlet pressure (load) for an engine speed of 1400 rpm for the base engine

6.2.8 Pollutant Calculations

As mentioned above, thermodynamic simulations may be of some use for estimating pollutant emissions. Of the regulated emissions, thermodynamic simulations have had the most success in estimating nitric oxides emissions for homogeneous charge engines (approximated by SI engines). This success is largely due to the fact that the nitric oxide formation and destruction processes are primarily functions of temperature, pressure, and residence time. In addition, the nitric oxide processes are not strongly affected by surfaces, complex chemistry, or physical process descriptions (such as mixing or diffusion). Hydrocarbon, carbon monoxide, and soot emissions may be estimated, but due to the complex physical and chemical processes, these estimates have not been as successful. More information on these topics may be found in Heywood [2]. Calculations of nitric oxide emissions using the thermodynamic cycle simulation are reported in one of the case studies in this book.

6.2.9 Other Sub-models

The above items (for the solution of the mathematical formulations presented in Chapter 5) provide all the necessary components to obtain engine performance output values such as work, power, torque, and mean effective pressure. Other sub-models, however, can be used to enhance the simulation or to provide additional features. For example, the prediction of knock in the end gases may be added by providing a relationship for autoignition of the unburned gases. This is at best an approximation since the knock phenomena is a complex chemical process. Further comments on modeling the knock phenomena are available (e.g., References 9,10).

Other examples of additional sub-models include items such as exhaust gas recirculation (EGR) and oxygen-enriched reactants. These are described later in this book in the chapters on results and case studies. Finally, the simulation may be modified in many ways to consider other engine configurations or applications. As an example, an over-expanded engine is described and evaluated in one of the following case studies.

6.3 Numerical Solution

The instantaneous cylinder conditions (temperatures, pressure, volumes, masses, and thermodynamic properties) as a function of crank angle are obtained by the simultaneous numerical integration of the various differential equations. Since a number of the differential equations are themselves dependent on the other variables, an iterative approach was chosen. That is, the set of differential equations was evaluated, and then the results from this evaluation were used to determine the final values of the derivatives. For the two-zone portion, two iterations were sufficient for convergence. For the three-zone portion, four iterations were necessary.

A number of numerical methods are available to solve a set of differential equations like the ones outlined in Chapter 5. Both the Runge–Kutta technique and a simple Euler technique were evaluated. The two methods provided surprisingly similar results. Since the application of the Euler technique was more straightforward, it was selected for this work. A description of the sensitivity to the time step interval for the calculations is presented next.

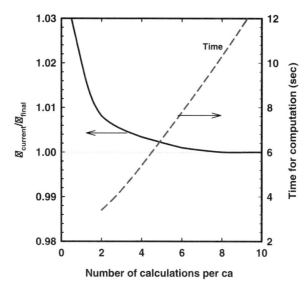

Figure 6.12   The ratio of the thermal efficiency from the current and final calculations and the time to complete the computation as functions of the number of calculations per crank angle

The number of calculations per crank angle relates to both the time to complete the computation and the accuracy of the final result. Figure 6.12 shows the ratio of the thermal efficiency from different numbers of calculations per crank angle and the thermal efficiency value as determined for eight calculations per crank angle (η_{final}). The result is highly nonlinear, and acceptable results (less than 0.3% difference) are obtained for four or more calculations per crank angle. The time to complete the computation, on the other hand, is fairly linear with the number of calculations per crank angle. That means that doubling the number of calculations doubles the amount of time.

6.3.1   Initial and Boundary Conditions

To complete the required input information, the boundary conditions for the inlet (temperature and pressure) and for the exhaust (pressure) are specified. For most of the work reported here, these three quantities were specified. In other work, these values could be found as functions of the operating conditions via empirical relationships.

To begin a particular engine cycle calculation, several parameters are not known. The initial amount of exhaust gases (residual), and the initial cylinder gas temperature and pressure must be assumed. The complete calculation is repeated until the final values agree (within a specified tolerance) with the initial values. Depending on the specified tolerances and the first values selected, this procedure usually finds convergence within about three complete cycles. Figure 6.13 shows the ratio of the residual mass fraction ("f") for the current cycle and the final value as functions of the calculation cycle for two different starting values of the residual fraction. The two starting values were selected to be significantly different. The solid line represents the progression from a very poor first value (off the scale of the figure), but by the

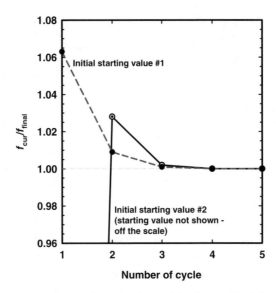

Figure 6.13   The ratio of the current residual fraction and the final residual fraction as functions of the number of the cycle of calculation for two different starting values

third cycle, the residual fraction is essentially at its final value. The dashed line represents the progression from a better estimate of the initial value (about 6% too high), but in this case as well, the residual fraction is essentially at its final value by the third cycle. For results reported here, the calculations are continued until at least cycle number "5" to insure convergence.

6.3.2   Internal Consistency Checks

A number of calculations may be completed to serve as internal consistency checks. These types of calculations do not validate the overall computations from a fundamental perspective, but do serve to at least verify that the computed results are self-consistent. For example, the mass and the energy of the burned zone must equal the masses and energy in the adiabatic and boundary layer zones. Other such mass and energy balances can be used throughout the simulation.

6.4   Summary

This chapter has provided descriptions of the items needed to solve the ordinary differential equations described in Chapter 5. These items included the thermodynamic properties (Chapter 4), the cylinder volume, the rate of change of the cylinder volume, the surface areas, expressions for the combustion process (mass fraction burned), cylinder heat transfer, mass flow rates, and algorithms for the friction. The algorithms for the friction were illustrated with values as functions of engine parameters. In addition, this chapter included a brief overview of the way nitric oxide emissions could be computed (a case study at the end of this book includes much more detail). The chapter ended with a brief discussion of the solution procedures.

References

1. Olikara C. and Borman, G. L. (1975). A computer program for calculating properties of equilibrium combustion products with some applications to IC engines, Society of Automotive Engineers, Paper no. 750468.
2. Heywood, J. B. (1988). *Internal Combustion Engine Fundamentals*, McGraw-Hill Book Company, New York.
3. Wiebe, J. J. (1970). Brennverlauf und kreisprozess von verbrennungsmotoren, VEB Verlag Technik Berlin.
4. Caton, J. A. (2011). Comparisons of global heat transfer correlations for conventional and high efficiency reciprocating engines, in Proceedings of the ASME Internal Combustion Engine Division 2011 Fall Technical Conference, paper no. ICES2011-60017, Morgantown, WV, 02–05 October.
5. Sherman, R. H. and Blumberg, P. N. (1977). The influence of induction and exhaust processes on emissions and fuel consumption in the spark ignited engine, Society of Automotive Engineers, SAE Paper no. 770990.
6. Sandoval, D. and Heywood, J. B. (2003). An improved friction model for spark-ignition engines, Society of Automotive Engineers, SAE paper no. 2003–01–0725.
7. Patton, K. J., Nitschke, R. G., and Heywood, J. B., (1989). Development and evaluation of a friction model for spark-ignition engines, Society of Automotive Engineers, SAE paper no. 890836.
8. Bishop, I. N. (1965). Effect of design variables on friction and economy, Society of Automotive Engineers, SAE paper no. 812A (1964), *SAE Transactions*, **73,** 334–358.
9. Ho, S. Y. and Kuo, T.-W. (1997). A hydrocarbon autoignition model for knocking combustion in SI engines, Society of Automotive Engineers, SAE paper no. 971672.
10. Giglio, V., Police, G., Rispoli, N., Iorio B., and diGaeta, A. (2011). Experimental evaluation of reduced kinetic models for the simulation of knock in SI engines, Society of Automotive Engineers, SAE paper no. 2011–24–0033.

7

Basic Results

7.1 Introduction

This chapter provides some examples of the results from engine thermodynamic simulations with an emphasis on the first law of thermodynamics. These results will focus on the cylinder conditions for the base case. The following chapter (Chapter 8) will provide global, performance results for a range of operating and design parameters, and Chapter 9 will provide results based on the second law of thermodynamics. The current chapter includes sections on the engine specifications and operating conditions, detailed time-resolved results, and overall energy results. The presentation in this chapter includes highly detailed results. This is designed to not only highlight the basic results but also illustrate the range of information that is available from these types of simulations.

7.2 Engine Specifications and Operating Conditions

Almost all of the results presented in this book are for the same engine: a 5.7 liter, V–8 configuration with a bore and stroke of 101.6 and 88.4 mm, respectively. Table 7.1 lists the engine specifications. For the Wiebe combustion parameters, the following values were used as recommended by Heywood [1]: m = 2.0 and a = 5.0. The combustion duration was 60°CA. For the cylinder heat transfer, the correlation recommended by Woschni [2] has been used with an *htm* of 1.0.

A base case operating condition has been selected for the following results. This condition is a part load condition that is near frequent operating conditions for many light-duty driving cycles, and includes an engine speed of 1400 rpm and a *bmep* of 325 kPa. Table 7.2 lists some of the values of parameters (and "how obtained") which were needed in this work. The inlet pressure (50 kPa) was selected to obtain a *bmep* of 325 kPa, and the combustion start was selected to maximize the brake torque (MBT).

An Introduction to Thermodynamic Cycle Simulations for Internal Combustion Engines, First Edition.
Jerald A. Caton © 2016 John Wiley & Sons, Ltd. Published 2016 by John Wiley & Sons, Ltd.

Table 7.1   Engine specifications

Item	Value
Number of cylinders	8
Bore (mm)	101.6
Stroke (mm)	88.4
Crank rad/con rod ratio	0.305
Inlet valves:	
Diameter (mm)	50.8
Max Lift (mm)	10.0
Opens (°CA aTDC)	357
Closes (°CA aTDC)	−136
Exhaust valves:	
Diameter (mm)	39.6
Max lift (mm)	10.0
Opens (°CA aTDC)	116
Closes (°CA aTDC)	371
Valve overlap (degrees)	14°

Table 7.2   Some engine and fuel input parameters (base case: 325 kPa, 1400 rpm)

Item	Value used	How obtained
Displaced volume (dm^3)	5.733	Computed
Compression ratio	8.0	Input
AF$_{stoich}$	15.07	Computed
Equivalence ratio	1.0	Input
Inlet (air-fuel) temperature	319.3 K	Input
Inlet pressure (kPa)	50.0	Input
Exhaust pressure (kPa)	102.6	From algorithm [3]
Start of combustion (°bTDC)	22.0	Determined for MBT
Combustion duration (°CA)	60	Input
Cylinder wall temp (K)	450	Input
Mech frictional mep (kPa)	68.3	From algorithm [3]
Heat transfer multiplier (htm)	1.0	Input
Heat transfer correlation	Woschni	Input
Fuel LHV (kJ/kg)	44,400	For isooctane [1]
Fuel exergy (kJ/kg)	45,670	For isooctane [1]

7.3 Results and Discussion

Tables 7.3 and 7.4 list results from the simulation for this 325 kPa, 1400 rpm condition. The net indicated and brake thermal efficiencies were 31.5% and 26.0%, respectively. The associated net indicated and brake specific fuel consumption values were 257.2 and 311.5 g/kW-h, respectively. The energy distribution indicates that about 23.7% and 44.1% of the fuel energy was assigned to heat transfer and exhaust gas energy, respectively. Also listed are the peak cylinder pressure (1904 kPa at 16.0°aTDC) and the peak cylinder temperature (2393 K). The maximum gas temperature in the burn zone is 2554 K at 10.2°aTDC. Other results listed in Table 7.3 are described below.

Table 7.4 lists items such as total mass (air + fuel vapor + residuals) per cylinder, and fuel energy and exergy inputs per event for each cylinder. Also listed are the cylinder one-zone temperature (1265 K) at TDC and the cylinder pressure (1250 kPa) at TDC. In addition, the table lists the fuel *energy* (0.894 kJ/cyl) and the fuel *exergy* (0.916 kJ/cyl).

Table 7.3   Results for the base case[*]

	Value	
Item	Net Indicated[**]	Brake
mep (kPa)	393.2	324.6
sfc (g/kW·hr)	257.2	311.5
η (%)	31.5	26.0
Torque (N·m)	178.8	147.7
Power (kW)	26.2	21.6
p_{peak} (kPa)	1904	
CA of p_{peak}	16.0	
max T_b (K)	2554	
CA of max T_b	10.2	
T_{peak} (K)	2393	
T_{exh} (K)	1331	
$\dot{m}_{fuel}$ (g/s)	1.88	
$\dot{m}_{air}$ (g/s)	28.3	
Residual fraction	0.104	
Energy distribution:		
Brake work (%)	26.03	
Friction (%)	5.50	
Heat loss (%)	23.69	
Exhaust (%)	44.11	
Unused (%)	0.68	
Total (%)	100.0	

[*]Results for T_{switch} = 1200 K, and ΔCA = 0.25°

[**]Net indicated is for all four strokes.

Table 7.4 Additional results for the base case

Item	Value
Volume at TDC (dm³/cyl)	0.1024
Volume at BDC (dm³/cyl)	0.8191
Total mass, valves closed (g/cyl)	0.3609
Air mass (g/cyl)	0.3234
Fuel mass (g/cyl)	0.0201
Residual mass (g/cyl)	0.0374
Overall temperature at TDC (K)	1265
Cylinder pressure at TDC (kPa)	1250
Average gas constant at TDC (kJ/kg-K)	0.2790
Fuel energy for each event (kJ/cyl)	0.8940
Fuel exergy for each event (kJ/cyl)	0.9159

7.3.1   Cylinder Volumes, Pressures, and Temperatures

The presentation of the results will begin with some of the basic considerations such as cylinder volume, and then will progress to cylinder pressures and temperatures. Figure 7.1 shows the instantaneous cylinder volume as a function of crank angle for one cylinder of the base case engine with a compression ratio of 8. The cylinder volume as a function of crank angle is a sinusoidal type curve. The minimum and maximum volumes are the clearance and total volumes, respectively.

Figure 7.2 shows the rate of change of cylinder volume as a function of crank angle. This quantity is proportional to the piston velocity. The rate of change of the cylinder volume is zero at TDC and BDC, and will be maximum at some point in between TDC and BDC. The first maximum value after TDC for these conditions occurs at 75°aTDC.

Figure 7.3 shows cylinder pressures and temperatures as functions of the crank angle for the base case conditions. Both the motoring and firing pressures are included. The start and end of combustion are indicated for reference. The motoring pressure is nearly symmetrical about TDC (0°aTDC) with a maximum pressure of about 720 kPa. The firing pressure has a maximum of 1904 kPa at 16.0°aTDC. The one-zone (average of all zones) cylinder gas temperature increases rapidly during the combustion period and reaches a maximum of 2393 K at 27.8°aTDC.

Figure 7.4 shows the cylinder pressure as a function of cylinder volume for the base case conditions. The start and end of combustion are denoted, as are the valve events. For this part load condition, the "pumping loop" represents work done to the cylinder gases during the gas exchange. The area of the compression-expansion "loop" is proportional to the gross indicated work done by the gases.

Figure 7.5 shows the log of the cylinder pressure as a function of the log of the cylinder volume. This figure provides an alternative representation of the cylinder pressure and volume relationship. In particular, the "pumping loop" is shown in more detail, and the nearly

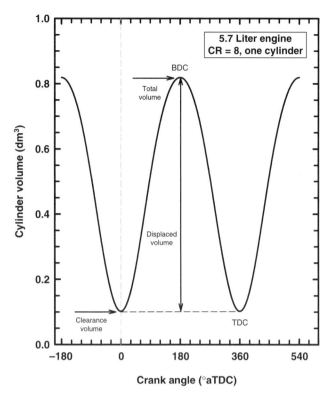

Figure 7.1   Instantaneous cylinder volume as a function of crank angle for a compression ratio of 8

linear "log p versus log V" characterizes a polytropic process[1] for portions of the compression and expansion processes. For these conditions, the compression process has a polytropic constant, n, of 1.313, and the expansion process has a polytropic constant of 1.288. This exponent reflects both the ratio of specific heats and heat transfer. Values for the exponent typically range from about 1.25 to 1.35 for most engines for a range of operating conditions. Due largely to the high gas temperatures and the significant cylinder heat transfer during the expansion stroke, the polytropic constant for the expansion stroke is always smaller than the value for the compression stroke.

Figure 7.6 shows the average overall (one-zone) gas temperature, and the temperatures associated with each zone as functions of crank angle for the base case conditions. At the bottom of the figure is the unburned zone temperature which remains below about 1000 K. The boundary layer temperatures range between about 1050 and 1250 K. The adiabatic zone has the highest temperatures, and for this case, a maximum of about 2625 K is attained. The burned zone temperatures represent the temperatures of the boundary layer and adiabatic zones. After combustion, the burned zone temperature and the average temperature are the same.

---

[1] A polytropic process is a process which obeys pV^n = constant.

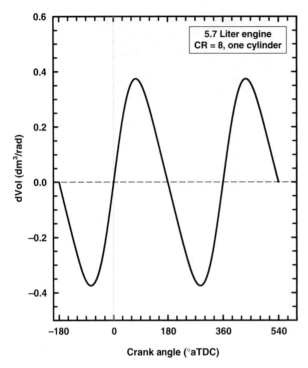

Figure 7.2   Instantaneous rate of change of cylinder volume as a function of crank angle for a compression ratio of 8

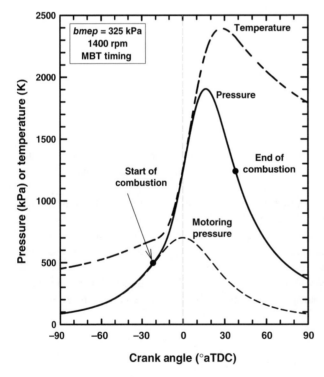

Figure 7.3   Cylinder pressures and temperatures as functions of crank angle for the base case conditions. *Source:* Caton 2015. Reproduced with permission from John Wiley & Sons, Inc

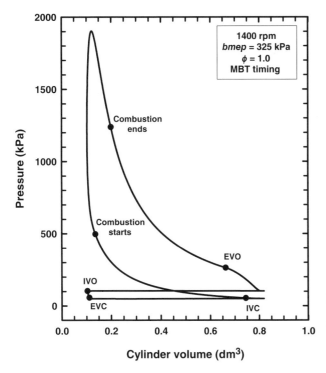

Figure 7.4 Cylinder pressure as a function of cylinder volume for the base case conditions

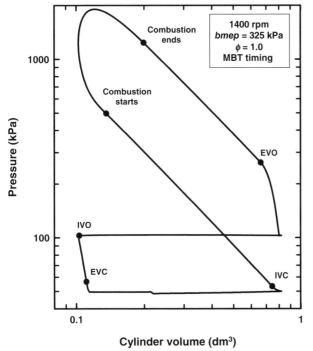

Figure 7.5 Log of the cylinder pressure as a function of log of the cylinder volume for the base case conditions. *Source:* Caton 2001. Reproduced with permission from ASME

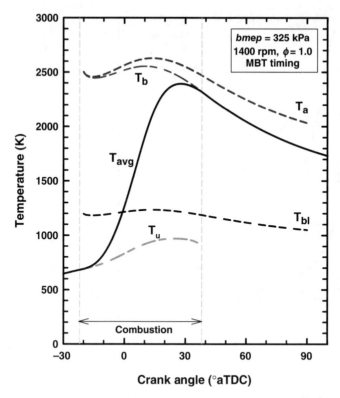

Figure 7.6 Multiple zone cylinder gas temperatures as functions of crank angle for the base case conditions

Figure 7.7 shows the cylinder pressure and the average, one-zone, gas temperature as functions of crank angle for the complete four strokes (720°CA) for the base case conditions. Also noted on the figure are the open and close crank angles for each of the valves. At exhaust valve open (EVO), both the pressure and temperature change slope as the mass flow out adds to the expansion decrease of pressure and temperature. At intake valve open (IVO), a further decrease occurs for both the pressure and temperature. After exhaust valve close (EVC), the temperature increases slightly during the portion of flow that is from the intake manifold back into the cylinder (see figure 7.9 with mass flow rates).

7.3.2 Cylinder Masses and Flow Rates

Figure 7.8 shows the cylinder mass for each zone as functions of crank angle for the base case conditions for a period near TDC. The unburned (m_u) and burned (m_b) zone masses are reciprocals of each other—one increases and one decreases such that the total mass (m_{total}) remains constant. The boundary layer mass (m_{bl}) begins to become significant at about TDC, and continues to increase until the computations are terminated at 90°aTDC. The mass (m_{ad}) of the adiabatic zone is equal to the mass of the burned zone minus the mass of the boundary layer.

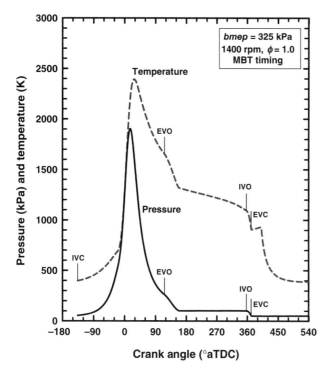

Figure 7.7 Cylinder pressures and one-zone gas temperatures as functions of crank angle for the base case conditions. *Source:* Caton 2012. Reproduced with permission from Inderscience Publishers

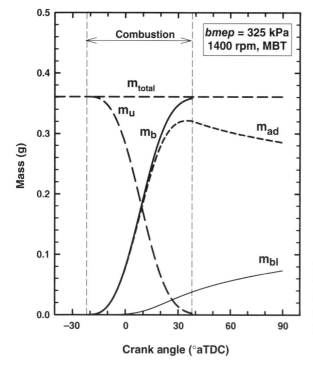

Figure 7.8 Cylinder mass in the various zones as functions of crank angle for the base case conditions. *Source:* Caton 2001. Reproduced with permission from ASME

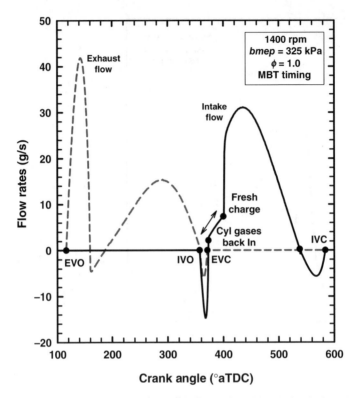

Figure 7.9   Exhaust and intake mass flow rates as functions of crank angle for the base case conditions. *Source:* Caton 2015. Reproduced with permission from John Wiley & Sons, Inc

Figure 7.9 shows the mass flow rates as functions of crank angle for both the intake and exhaust flows for the base case conditions. The positive flow rates are the flow rates in the natural directions—the intake flow is positive into the cylinder and the exhaust flow is positive out of the cylinder. The "blow-down" after the exhaust valve opens (EVO) attains a maximum flow of about 42.6 g/s, and then the exhaust flow decreases. The displacement phase of the exhaust flow results in a maximum of about 16 g/s. The flow is back into the cylinder during the valve overlap period—the time between the intake valve open and the exhaust valve close times. When the intake valve first opens (IVO), the flow is into the intake manifold. The flow reverses shortly after the exhaust valve closes. The first mass into the cylinder is the mass that previously flowed into the intake manifold. Once that mass has returned, then the fresh air and fuel vapor mixture flows into the cylinder. The intake flow attains a maximum flow rate of about 32 g/s and then decreases. At the end of the intake valve open period, the flow reverses back into the manifold. For this condition, at the end of the intake valve open period, the cylinder pressure is greater than the intake manifold pressure which causes the reverse flow.

7.3.3   Specific Enthalpy and Internal Energy

Figures 7.10 and 7.11 show the average specific enthalpy and internal energy, respectively, as functions of temperature for the base case conditions. These curves have similar shapes but the

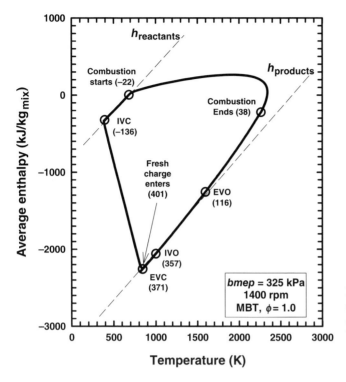

Figure 7.10 Average mixture enthalpy as a function of the one-zone temperature for the base case conditions

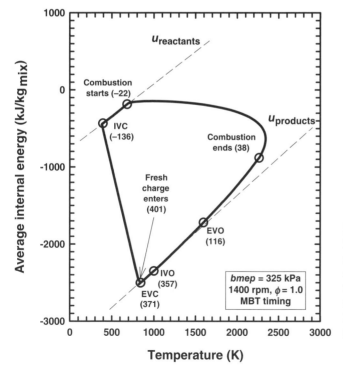

Figure 7.11 The specific internal energy for the mixtures as functions of the one-zone, overall cylinder gas temperature for the base case conditions. *Source:* Caton 2001. Reproduced with permission from ASME

absolute values are different. Also denoted on the figures are selected crank angle "times" such as the start of combustion, exhaust valve opening (EVO), and so forth. The specific enthalpy and internal energy increase during compression, attain a maximum, and then decrease during combustion and through expansion. After the reverse flow into the intake manifold, the fresh charge (at 401°aTDC) begins to enter the cylinder and the specific enthalpy and internal energy begin to increase until they attain their original values at intake valve close.

7.3.4   Molecular Masses, Gas Constants, and Mole Fractions

Figure 7.12 shows the molecular mass of the mixture as functions of crank angle for the base case conditions. Two curves are provided: a solid curve for the detail species (including equilibrium), and a dashed curve for the "frozen" species approximation. Again, denoted on the figure are selected crank angle "times." Beginning at intake valve close (IVC), the molecular mass remains constant during the compression process. During combustion, the molecular mass decreases until near the end of combustion. The molecular mass remains at low values until the fresh charge begins to enter the cylinder. At this point, the molecular mass increases until it has the value at intake valve close. For these conditions, the molecular mass of the mixture varies between 28.5 and 30.2.

The molecular mass of the mixture as determined by using the frozen species assumption is in general agreement with the values determined from the equilibrium composition assumption (Figure 7.12). At all times, the two values are well within 1% agreement. As an example of the detailed information provided, the results based on the equilibrium composition shows some variation of the molecular mass after combustion until the exhaust valve opens. During this portion, some dissociated species are recombining and the molecular mass decreases slightly.

For these conditions, the simulation "switches" from the detailed properties to the frozen properties at a temperature of 1200 K (crank angle of 290°aTDC). Although for the scale of Figure 7.12, this switch appears significant, the actual change of the molecular mass is from 28.51 to 28.61 (less than a 0.3% difference). This feature is described in more detail below.

Figure 7.13 shows the gas constants as functions of the temperature for the base case conditions. The figure shows the gas constants for the burned gas (R_b, upper dashed line), unburned gas (R_{unb}, lower dashed line), and complete mixture (R_{mix}, solid line). The gas constant for the unburned gas remains unchanged since the composition is fixed. The gas constant for the burned gas changes only slightly—mostly near the end of combustion due to dissociation and other reactions. The gas constant for the mixture increases from the start of combustion, attains a maximum value near the end of combustion, and then decreases during the intake process as fresh charge enters the cylinder.

Table 7.5 lists the results for the mole fractions of the species at two specific crank angles for the base case conditions. The two crank angles are 290.00°aTDC and 290.25°aTDC, and represent the point where the computation "switched" from the equilibrium to the "frozen" assumption. This allows a comparison between the estimated values from the "frozen" assumption and from the equilibrium computations. The "switch" temperature between the equilibrium and frozen assumption is 1200 K for this work. At a crank angle of 290.25°aTDC, the temperature decreases below 1200 K, and the composition is computed from the "frozen" assumption. The major species (oxygen, nitrogen, carbon dioxide, and water) are in excellent

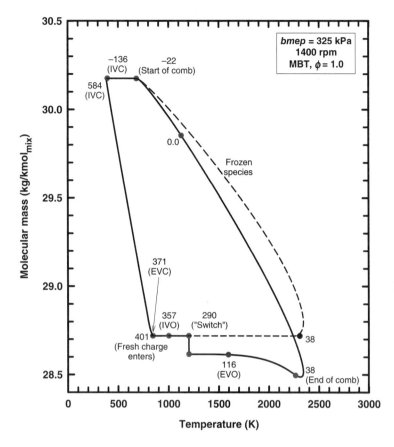

Figure 7.12 Average molecular mass for the mixture as functions of overall, one-zone temperature for the base case conditions— for detailed properties and for the "frozen" assumption

agreement, but the minor species are set to zero for the "frozen" assumption. As shown at the bottom of the list, the overall molecular mass is in good agreement between the two approaches (within about 0.3%).

Figure 7.14 shows the mole fractions of the major species and the one-zone average temperature as functions of crank angle for the base case conditions. During the combustion period, the fuel and oxygen mole fractions decrease to near zero, and the carbon dioxide and water mole fractions increase to near their maximum values. Once fresh charge begins to enter the cylinder, the oxygen and fuel mole fractions begin to increase, and the carbon dioxide and water mole fractions decrease to near their minimum values (as part of the residual mass). Note that at the crank angle (290°aTDC), where the switch from equilibrium composition to frozen composition occurs, the lines for the mole fractions are continuous and no change can be detected (at least for the scale of this plot).

Figure 7.15 shows the mole fractions of some of the other more minor species and the one-zone average temperature near TDC as functions of crank angle for the base case conditions. These species (CO, OH, H_2, NO, H, and O) are largely a result of dissociation reactions, and are highly dependent on temperature. As combustion starts, these species increase, attain their maximum values, and then during the expansion process, continue to decrease to near-zero concentration levels. Carbon monoxide, of all of these species, attained the highest

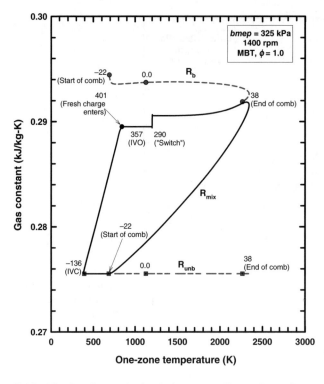

Figure 7.13 The mixture (R_{mix}), unburned (R_{unb}), and burned (R_b) gas constants as functions of the one-zone gas temperature for the base case conditions

Table 7.5 Species results for the base case: Comparison of concentrations for "Frozen" and equilibrium assumptions

Item	Equilibrium	Frozen
CA	290.00	290.25
Temp (K)	1200.1	1199.9
Mole fractions:		
O_2	0.00172	0.001177
N_2	0.7353	0.7353
CO_2	0.1239	0.1240
H_2O	0.1394	0.1395
Fuel vapor	0.000093	0.000093
H_2	1.798×10^{-07}	0.0
CO	2.184×10^{-07}	0.0
H	8.221×10^{-11}	0.0
N	3.555×10^{-18}	0.0
O	2.398×10^{-10}	0.0
NO	4.418×10^{-06}	0.0
OH	5.042×10^{-07}	0.0
MW (g/gmol)	28.61	28.71

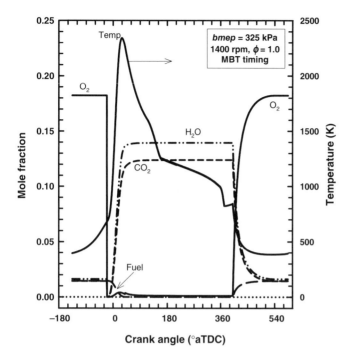

Figure 7.14 Mole fractions of major species and one-zone cylinder gas temperature as functions of crank angle for the base case conditions

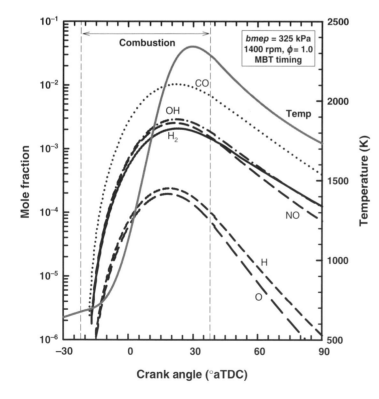

Figure 7.15 Mole fractions of minor species and the one-zone cylinder gas temperature as functions of crank angle for the base case conditions

concentration—about a mole fraction of 0.01 (1.0%). The nitric oxide (NO) reported here is the equilibrium concentration for the instantaneous conditions. The case study on nitric oxide emissions describes the relation of the equilibrium and actual nitric oxide exhaust concentrations.

7.3.5   Energy Distribution and Work

Table 7.6 is a summary of the energy and exergy values for the base case. The fuel *energy* is divided among work, friction, heat transfer, net exhaust, and unburned fuel. The fuel *exergy* is divided among the same items plus destruction during the combustion processes and intake mixing processes. These quantities are listed in both energy units (kJ) and as percentages (%) of the fuel energy or exergy. The quantities for the brake work and friction combine to provide the values for the net indicated work. Note that the pumping work is already a part of the work values and cannot be listed separately. Further comments on these types of energy and exergy distributions are provided next and in subsequent chapters.

Figure 7.16 shows a "pie" chart of the energy distribution for the base case conditions. Energy inputs into the defined thermodynamic system (the cylinder contents) are from the fuel and the pumping work. This approach is used so that the pumping work can be identified, but the comments provided above are still accurate. The energy outputs are the gross indicated work, heat transfer, net exhaust gas energy, and unburned fuel. The gross indicated work consists of the brake work, mechanical friction, and pumping work. For this operating condition, significant energy is associated with the heat transfer and exhaust gas flow. Again, subsequent chapters will provide much more discussion of the energy distribution as functions of engine operating and design parameters.

Figure 7.17 shows the energy items as a function of crank angle for the base case conditions. The energy units are set to equal 1.0 kJ at the start (IVC) for convenience. The dark line represents the cylinder total energy throughout the four strokes. The total energy starts and ends at 1.0 kJ. During the compression stroke, the total cylinder energy increases due to compression work. After TDC, the total energy continues to decrease due to work extraction, heat losses, and exhaust flow. Once the fresh charge enters (401°aTDC), the total energy increases back to the original value.

Table 7.6   Summary of energy values for the base case

Item	Energy		Exergy	
	Value (kJ)	Percent (%)	Value (kJ)	Percent (%)
Brake work	0.2326	26.02	0.2325	25.38
Friction	0.0490	5.50	0.0490	5.35
Indicated work	0.2816	31.52	0.2815	30.73
Heat loss	0.2118	23.70	0.1758	19.19
Net flow out	0.3943	44.11	0.2490	27.19
Destruction (comb)	n/a	n/a	0.1896	20.70
Destruction (intake)	n/a	n/a	0.0138	1.51
Fuel not used	0.0060	0.68	0.0062	0.68
Total	0.8939	100.0	0.9159	100.0

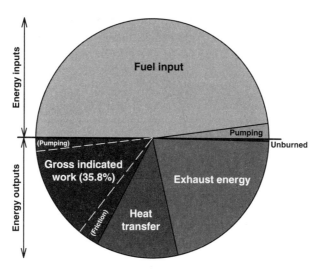

Figure 7.16   Distribution of the energy inputs and outputs for the base case conditions

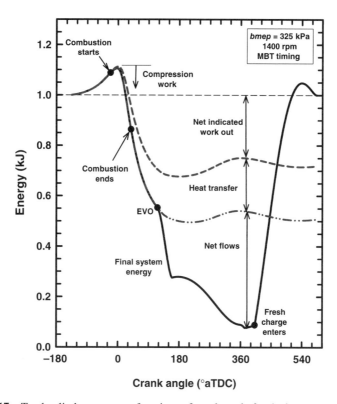

Figure 7.17   Total cylinder energy as functions of crank angle for the base case conditions

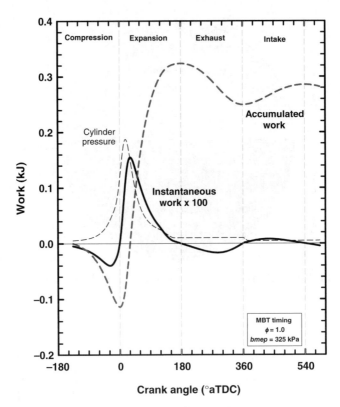

Figure 7.18 Instantaneous and accumulated work, and cylinder pressure as functions of crank angle for the base case conditions

After TDC at the end of the compression stroke (Figure 7.17), work continues to be extracted, and at the end of the cycle, the net indicated work is the value listed in Table 7.6, 0.282 kJ. The energy associated with the heat transfer is designated below the line for the work. The majority of the heat transfer is during the expansion process due to the high temperatures. At the end, the total heat transfer equals 0.212 kJ. A figure similar to Figure 7.17 will be used in Chapter 9 to illustrate a similar distribution of *exergy* as a function of crank angle for this case.

Finally, Figure 7.18 shows the instantaneous and accumulated work, and the cylinder pressure as functions of crank angle for the base case. Work produced by the system (output) is a positive quantity. During the compression process, the instantaneous work is negative (into the gases) until TDC. The instantaneous work reaches a peak slightly after TDC and then decreases during the remainder of the expansion process. The instantaneous work is slightly negative during the exhaust stroke and then slightly positive during the intake stroke. The final value of the *accumulated* work (0.282 kJ) represents the work for the cycle. Chapter 8 contains further discussion concerning these work values.

7.4   Summary and Conclusions

This chapter has provided detailed results from the engine cycle simulation for one operating condition (*bmep* = 325 kPa; 1400 rpm). The results are largely illustrated as functions of crank angle. Cylinder volume, the derivative of the cylinder volume, temperatures, pressures, cylinder mass, and flow rates were illustrated as functions of crank angle and cylinder volume. Some of the properties were also described. Enthalpy, internal energy, molecular mass, and gas constant were shown as functions of the one-zone gas temperature. In addition, the mole fractions of the major and some minor species were illustrated as functions of crank angle. Finally, the distributions of energy and exergy were described for this operating condition. For this one operating condition, some general findings and conclusions may be listed.

- The engine cycle simulation includes three zones during combustion. The maximum one-zone cylinder gas temperature and pressure were 2393 K and 1904 kPa, respectively. The maximum burned and adiabatic zone gas temperatures were 2554 K and 2625 K, respectively.
- Portions of the compression and expansion processes satisfied a polytropic process with exponents of 1.313 and 1.288, respectively.
- When the intake valve first opens, the mass flow is from the cylinder into the intake system and from the exhaust system into the cylinder. Starting at 401°aTDC, fresh charge enters the cylinder.
- Plots of specific enthalpy and internal energy illustrate the process from reactants to products.
- For the equilibrium composition assumption, species concentrations for CO, OH, H_2, NO, H, and O are provided.
- Energy and exergy distributions, and comments on the role of pumping work are provided.

References

1. Heywood, J. B. (1988). *Internal Combustion Engine Fundamentals*, McGraw-Hill Book Company, New York.
2. Woschni, G. (1968). A universally applicable equation for the instantaneous heat transfer coefficient in the internal combustion engine, *SAE Transactions*, SAE paper no. 670931, **76**, 3065–3083.
3. Sandoval D. and Heywood, J. B. (2003). An improved friction model for spark-ignition engines, Society of Automotive Engineers, SAE paper no. 2003–01–0725.

8

Performance Results

8.1 Introduction

This chapter provides examples of the overall engine performance results which include items such as power, torque, mean effective pressures, thermal efficiencies, specific fuel consumption, and volumetric efficiencies. Other results include the distribution of the fuel energy among work, friction, heat transfer, and exhaust as functions of load and speed. Also, the results include thermal efficiency as functions of engine design and operating variables. In addition, items such as the residual mass fraction, exhaust pressure, and exhaust temperatures are presented. Finally, a brief discussion on the thermodynamics of the use of exhaust gas recirculation (EGR), and of pumping work are provided.

8.2 Engine and Operating Conditions

The engine examined in this chapter is the same 5.7 liter, V–8 configuration described previously. Many of the following results are for one cylinder of the engine. The engine specifications and some of the input parameters are provided in Chapter 7. For the most part, the following results are based on the variation of one engine operating or design parameter while all other parameters remain at their base values. The major parameters for the base condition include a *bmep* of 325 kPa, a speed of 1400 rpm, an equivalence ratio of 1.0, MBT start of combustion timing, a burn duration of 60°CA, a cylinder wall temperature of 450 K, and a compression ratio of 8.0. The fuel is isooctane, C_8H_{18}, with a LHV of 44,400 kJ/kg, the heat transfer correlation is from Woschni [1], and *htm* is 1.0.

8.3 Performance Results (Part I)—Functions of Load and Speed

The first set of results are presented as functions of load and speed. This will be followed by a second set of results which are functions of engine operating and design parameters. Figure 8.1 shows the classic presentation of engine performance results—either obtained from experiments or from simulations. The figure shows total engine power and torque as functions

An Introduction to Thermodynamic Cycle Simulations for Internal Combustion Engines, First Edition.
Jerald A. Caton © 2016 John Wiley & Sons, Ltd. Published 2016 by John Wiley & Sons, Ltd.

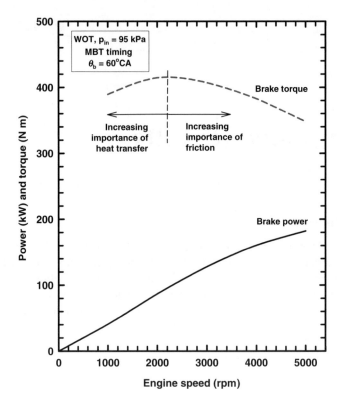

Figure 8.1   Total brake engine power and torque as functions of engine speed at wide-open throttle

of engine speed for a wide-open throttle (WOT) condition. For purposes of the computations, the inlet pressure was assumed constant with a value of 95 kPa. For each engine speed, the combustion timing was adjusted to provide the highest brake torque (MBT timing). For these conditions, the torque is between about 350 and 420 N-m, and has a maximum value at about 2200 rpm. A maximum torque at some intermediate engine speed is a result of the increasing importance of friction at the higher speeds, and the increasing importance of the heat transfer at the lower speeds. The characteristics of heat transfer and friction are explored later in this section. Due to the fairly constant torque, the power increases almost linearly with engine speed.

The results in Figure 8.1 capture the major thermodynamic aspects of engine operation for WOT. As mentioned in several places in this book, the simulation does not explicitly consider knock or other combustion irregularities. In practice, an engine may be designed with various strategies to avoid these combustion difficulties including operating at other than MBT timing, and operating with equivalence ratios above stoichiometric ($\phi > 1.0$). In addition, any unsteady fluid flow features in the intake or exhaust system would alter the character of these results. These unsteady flow features are often called "wave" effects, and designing intake and exhaust systems to take advantage of these features is known as "tuning" the flow systems [2]. These features, although not used in these calculations, can be incorporated into the simulation if the data are known.

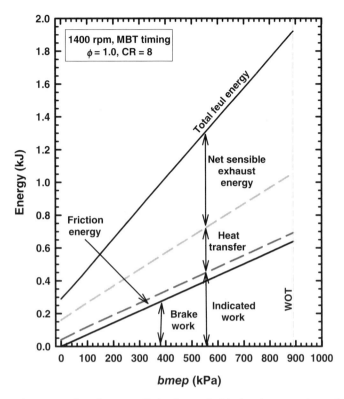

Figure 8.2 Actual energy values for one cylinder for work, friction, heat transfer, and exhaust energy as functions of *bmep* for 1400 rpm

Figures 8.2–8.4 explore the distribution of the fuel energy among the work, friction, heat transfer, and exhaust as functions of engine load for one cylinder. Engine load is represented by *bmep*. These three figures are different ways to examine the same data. Figure 8.2 shows the actual energy quantities (kJ) for these conditions. Starting from the bottom of the figure, the energy assigned to the brake work increases linearly from zero (idle) to its maximum value at WOT. Friction is assigned the next increment of energy, and (on this scale) appears fairly constant with load. Heat transfer is the next quantity, and it increases (in terms of "kJ") as load increases. The net sensible energy contained in the exhaust gas is the final increment, and this energy increases (in terms of "kJ") with load. The net sensible exhaust energy is the sensible energy in the exhaust gases minus the sensible energy in the intake gases. The small unburned energy (about 0.7%) is not shown due to the scale of the figure. The final line on top represents the total fuel energy.

Figure 8.3 shows the percentage of the fuel energy for the brake work, heat transfer, and exhaust energy as functions of load. On this basis, the values for the brake work represent the brake thermal efficiency. The brake work starts at zero (no shaft work out) and increases until WOT. This increase is largely the result of the decreasing importance of the heat transfer and pumping work as load increases. The percentage of the fuel energy devoted to heat transfer decreases as load increases. Although the actual (kJ) heat transfer increases with load (see Figure 8.2), the percentage decreases since the fuel energy increases more rapidly. The percentage of the fuel energy that eventually leaves with the exhaust gas remains high (~44%)

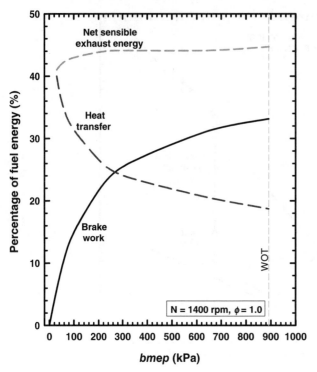

Figure 8.3 Percentages of brake work, heat transfer, and exhaust energy as functions of *bmep* for 1400 rpm

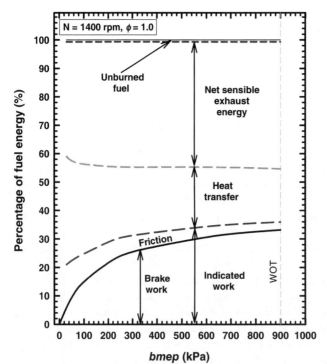

Figure 8.4 Distribution of the fuel energy as functions of *bmep* for 1400 rpm

and is fairly constant as a function of load. For this figure, the total of the three quantities are typically in the mid-90%. The totals will be exactly 100% with the addition of the percentages for friction energy and the energy associated with the unburned fuel. This is shown more clearly in the next figure.

Figure 8.4 shows the distribution of the fuel energy on an "accumulated" basis as functions of load for this condition. From the bottom of the figure, the items are brake work, friction, heat transfer, net sensible exhaust energy, and the unburned fuel. Again, on a percentage basis, the net sensible exhaust energy is fairly constant, the heat transfer and friction decrease in importance, and the brake work increases as engine load increases.

The results in Figures 8.2–8.4 illustrate a few general features related to engine performance as functions of load. First, for all loads, a large portion (~40%) of the fuel energy remains in the exhaust and is eventually expelled from the engine. This is a consequence of the difficulty of converting thermal energy into work. This aspect will be discussed in other parts of this book in more detail. Second, for low loads, the percentage of the fuel energy that is converted to work is low. For engine loads greater than about 50% of the WOT load, the conversion percentage is higher.

Figures 8.5–8.8 illustrate the distribution of the fuel energy among the work, friction, heat transfer, and exhaust as functions of engine speed for one cylinder. As with Figures 8.2–8.4,

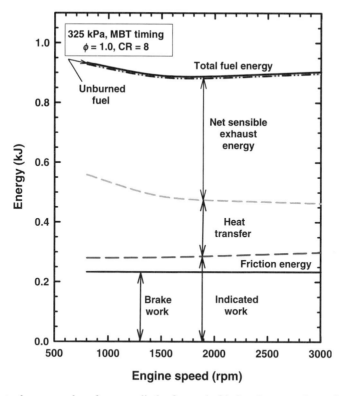

Figure 8.5 Actual energy values for one cylinder for work, friction, heat transfer, and exhaust energy as functions of engine speed for a *bmep* of 325 kPa

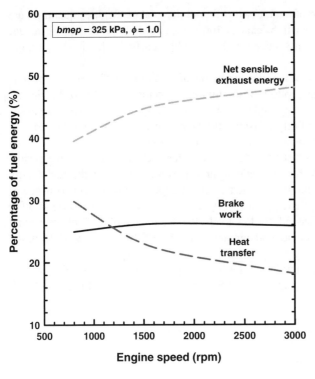

Figure 8.6   Percentages of brake work, heat transfer, and exhaust energy as functions of engine speed for a *bmep* of 325 kPa

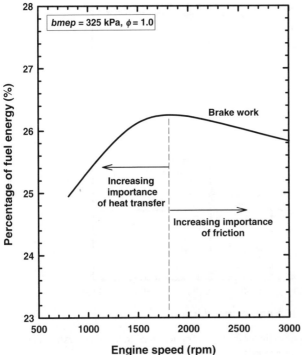

Figure 8.7   Brake work as a percentage of the fuel energy as functions of engine speed for a *bmep* of 325 kPa

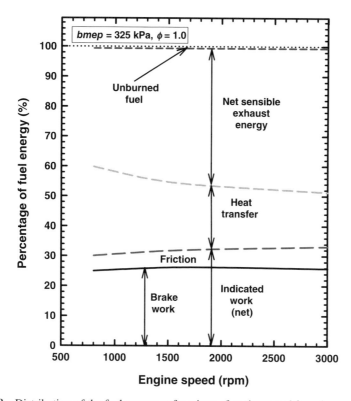

Figure 8.8 Distribution of the fuel energy as functions of engine speed for a *bmep* of 325 kPa

these figures are based on the same data, but provide different ways to examine the results. Figure 8.5 shows the actual energy quantities (kJ) for these conditions. Starting from the bottom of the figure, the energy assigned to the brake work remains fairly constant on this scale from 800 to 3000 rpm. Friction is assigned the next increment of energy, and increases slightly with increases of speed. Heat transfer is the next quantity, and it decreases (in terms of "kJ") as speed increases. Although the actual rate (kJ/s) of heat transfer increases with speed, the heat loss (kJ) decreases with increasing speed because of the decreasing real time available for the heat transfer. The net sensible energy contained in the exhaust gas is the next increment, and this energy increases (in terms of "kJ") with load. The small unburned energy (about 0.7%) is shown next to the top line. The final line on top represents the total fuel energy.

Figure 8.6 shows that the percentage of the fuel energy for the brake work, heat transfer and exhaust energy as functions of engine speed. On this basis, the values for the brake work represent the brake thermal efficiency. The brake work percentage is fairly constant (on this scale), but actually has a slight maximum at an intermediate speed. Figure 8.7 shows the brake work as a function of engine speed with a reduced scale. For this scale, a maximum of the brake work percentage is observed for an engine speed of about 1800 rpm for these conditions. This maximum is the result of the decreasing importance of the relative heat transfer and increasing importance of friction as engine speed increases.

Figure 8.6 also shows that the percentage of the fuel energy devoted to heat transfer decreases as engine speed increases. This is largely due to the smaller amount of real time for heat transfer as engine speed increases (as mentioned above). The percentage of the fuel energy that leaves with the exhaust gas increases from about 40% to 48% as engine speed increases from 800 rpm to 3000 rpm. For this figure, the totals of the three quantities are typically in the mid-90%. The totals will be exactly 100% with the addition of the percentages for friction energy and the energy associated with the unburned fuel. This is shown more clearly in Figure 8.8.

Figure 8.8 shows the distribution of the fuel energy on an "accumulated" basis as functions of engine speed for this condition. From the bottom of the figure, the items are brake work, friction, heat transfer, net sensible exhaust energy, and the unburned fuel. Again, on a percentage basis, the net sensible exhaust energy increases, the heat transfer decreases, friction increases slightly, and the brake work remains roughly constant as engine speed increases.

The results in Figures 8.5–8.8 illustrate a few general features related to engine performance as functions of engine speed. The major conclusion is that, for these conditions, the brake work as a percentage of the fuel energy will typically be a maximum for an intermediate speed. This, as mentioned above, is a result of the balance between increasing friction and decreasing relative heat transfer as engine speed increases. The case study on cylinder heat transfer has more information on these items.

The next set of Figures 8.9–8.13 will illustrate various quantities as functions of engine speed for several loads. Figure 8.9 shows the brake thermal efficiency as functions of engine speed for three values of *bmep* and for WOT. The brake efficiency is a strong function of load

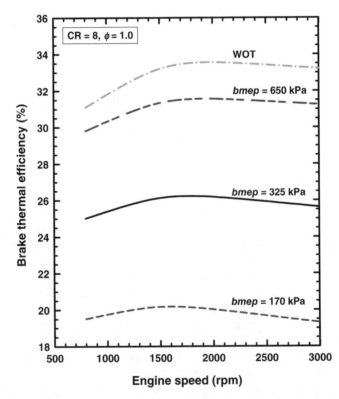

Figure 8.9   Brake thermal efficiency as functions of engine speed for three *bmep* levels and WOT

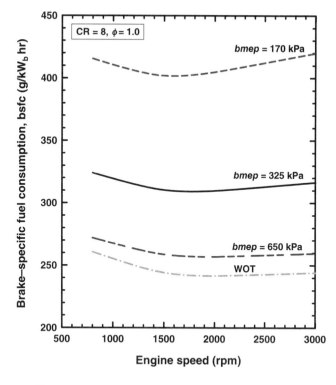

Figure 8.10   Brake specific fuel consumption as functions of engine speed for three *bmep* levels and WOT

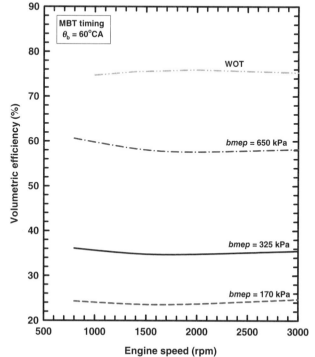

Figure 8.11   Volumetric efficiency as functions of engine speed for WOT and three *bmep* levels

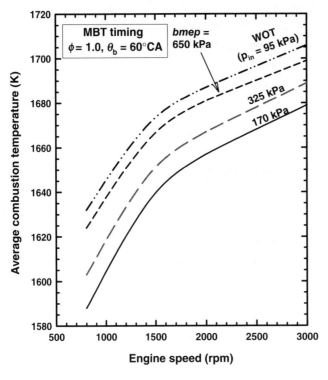

Figure 8.12 Average (one-zone) combustion temperature as functions of engine speed for WOT and three *bmep* levels

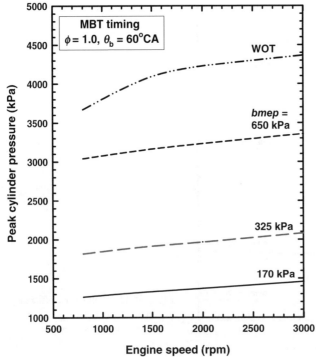

Figure 8.13 Peak cylinder pressure as functions of engine speed for WOT and three *bmep* levels

and a more modest function of speed. The strong function of load is partly due to the need for throttling for the lower loads and the resulting pumping work, and due to the increasing importance of heat transfer for the low loads (see Figure 8.3). For each load, the thermal efficiency is maximum for one engine speed as mentioned above.

For completeness, Figure 8.10 shows the brake-specific fuel consumption (bsfc) as functions of engine speed for three values of *bmep* and for WOT. As defined in Chapter 2, the bsfc is essentially the inverse of the thermal efficiency. The comments above concerning the thermal efficiencies (Figure 8.9) apply to the results of Figure 8.10.

Figure 8.11 shows the volumetric efficiency as functions of engine speed for three values of *bmep* and for WOT. The volumetric efficiency is based on ambient air density at 300 K and 100 kPa. As described in Chapter 2, other definitions could be used for the volumetric efficiency. A common alternative definition might use the manifold conditions. Since the manifold air temperature will probably be higher than the ambient, this definition (based on the manifold temperature) will result in slightly higher volumetric efficiencies than the definition used here. In any case, the trends of the results will be similar. As shown in Figure 8.11, the volumetric efficiency is only slightly a function of engine speed and a more significant function of engine load. Again, the significance of load is largely due to the throttling for the low loads. Also, again, the computations do not include any "wave" effects which could increase or decrease the actual volumetric efficiency.

Figure 8.12 shows the average, one-zone, gas temperatures during the combustion period as functions of engine speed for three values of *bmep* and for WOT. For the conditions examined, the average combustion temperatures range from about 1590 K to about 1710 K. The temperatures increase the most up to about 1600 rpm and then increase at a somewhat slower rate with speed. The combustion temperatures increase with increases of load with the highest temperature for WOT at the highest speed. For any load (*bmep*), the gas temperatures increase with increases of speed due to the decreasing importance of heat transfer.

Figure 8.13 shows the peak cylinder pressure as functions of engine speed for three values of *bmep* and for WOT. The peak cylinder pressure ranges from about 1260 kPa to about 4370 kPa for these conditions. The peak pressure is a strong function of engine load and a modest function of engine speed. The highest cylinder peak pressure is for WOT at the highest engine speed. The crank angle of the peak cylinder pressure is a function of the engine operating and design parameters. These crank angle values are provided in the case study on combustion.

8.4   Performance Results (Part II)—Functions of Operating/Design Parameters

The rest of this chapter will examine engine performance as functions of engine operating and design parameters.

8.4.1   Combustion Timing

The timing of the combustion event is critical to obtaining maximum performance. For a spark ignition engine, the spark timing can be adjusted to advance (earlier) or retard (later) the start of combustion. After the spark, chemical reactions begin, but the energy release does not immediately increase the cylinder pressure. After a period known as the "ignition delay,"

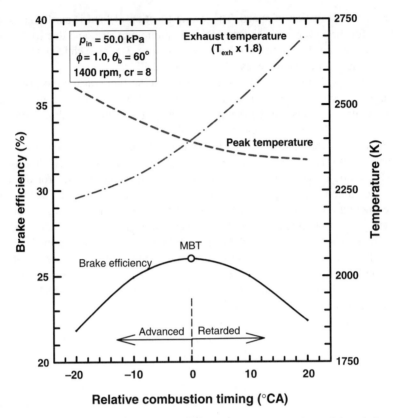

Figure 8.14 Brake thermal efficiency and peak gas temperature as functions of the relative combustion timing for an inlet pressure of 50 kPa

reactions proceed which are characterized as the start of combustion. This ignition delay is a complex function of the local (near the spark plug) conditions including turbulence, temperature, species compositions and pressure. For purposes of this work, the timing will be related to the start of combustion and not the spark timing.

Figure 8.14 shows the brake thermal efficiency, peak one-zone gas temperature, and average energy exhaust temperature (times a factor of 1.8) as functions of the relative start of combustion for an inlet pressure of 50.0 kPa and 1400 rpm. The relative combustion timing is related to the actual MBT timing. So if the MBT timing is −22° aTDC, then a relative timing of −10° is an actual timing of −32°aTDC. The relative combustion timing for MBT is denoted on the figure as 0.0°CA. The brake thermal efficiency (and brake torque, *bmep*, …) decreases as the timing is advanced or retarded from the MBT timing.

At the top of Figure 8.14 is the peak gas temperature which is highest for the most advanced timing and decreases as the timing is retarded. This is because for advanced (early) start of combustion, the gases are compressed for a longer period which increases the peak pressure and peak temperature. Also, Figure 8.14 shows the average energy exhaust temperature (which is described in more detail below) which is the lowest for the most advanced timing and increases as timing is retarded. This is because during expansion, the temperatures

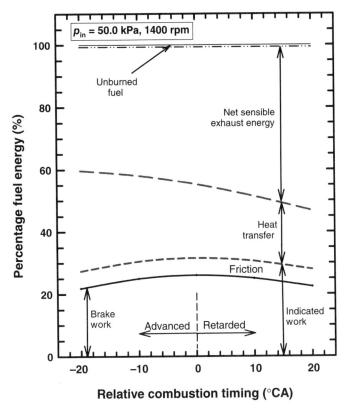

Figure 8.15 Distribution of the fuel energy as functions of the relative combustion timing for an inlet pressure of 50 kPa

associated with the most advanced timings are expanded to temperatures which are lower than the temperatures for the more retarded timings. In addition, for the advanced timings, the gases have more time for heat transfer.

Figure 8.15 shows the distribution of the fuel energy among the work, friction, heat transfer, and net exhaust energy as functions of the relative combustion timing. The brake work was described for Figure 8.14 and has the same character in Figure 8.15 (but the scale is larger). Friction remains about the same as a function of combustion timing. The relative cylinder heat decreases as the timing is retarded from the most advanced timings. This is consistent with the higher gas temperatures for the more advanced timings. On the other hand, the net sensible energy in the exhaust increases as the timing is retarded from the most advanced timings. The exhaust gas temperatures (not shown) are consistent with the exhaust energy: the exhaust temperatures are highest for the most retarded timings (see Figure 8.14).

8.4.2   Compression Ratio

Figure 8.16 shows the net indicated and brake thermal efficiencies as functions of compression ratio for conditions which included a *bmep* of 325 kPa, 1400 rpm, stoichiometric mixture,

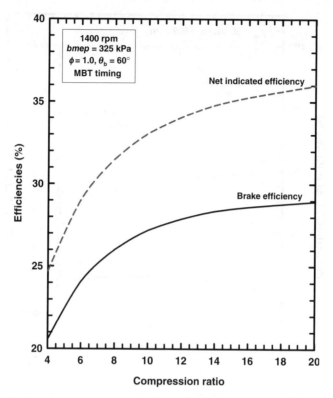

Figure 8.16 Net indicated and brake thermal efficiency as functions of compression ratio for the part load condition. *Source:* Caton 2007. Reproduced with permission from ASME

and MBT combustion timing. The thermal efficiencies increase with increasing compression ratio. This is a well-known and appreciated feature of reciprocating engines. The simple "Otto" ideal, air standard cycle, shows that the thermal efficiency increases with compression ratio (r),

$$\eta = 1 - r^{1-k}$$

Although this expression is not quantitatively accurate [2], the trend of increasing efficiencies with increasing compression ratio is correct. This feature is a result of the mechanical advantage for higher compression ratios. As is well known, however, maximum compression ratios are limited by spark knock for spark ignition engines, and limited by maximum pressures for all engines. In spite of these constraints, designs for engines continue to strive to use the highest compression ratio that is consistent with providing the highest performance.

Figure 8.16 shows that although the thermal efficiencies increase with compression ratio, the rate of this increase becomes lower at the higher compression ratios. Further, the increase of the brake thermal efficiency as compression ratio increases is somewhat mitigated at the higher compression ratios due to increasing friction. For these conditions, the brake thermal efficiency does not increase significantly for compression ratios greater than about 16.

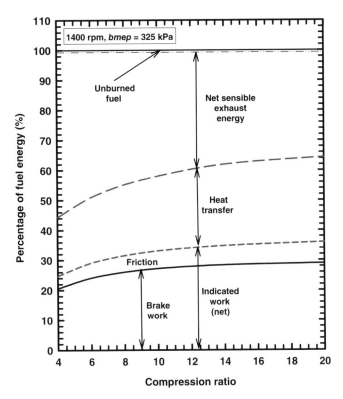

Figure 8.17   Distribution of the fuel energy as functions of compression ratio for a *bmep* of 325 kPa and 1400 rpm

Figure 8.17 shows the distribution of the fuel energy as a function of compression ratio for these conditions. The brake work increases, the friction increases, the heat transfer increases, and the exhaust energy decreases as compression ratio increases. The major conclusion from these results is that, for higher compression ratios, the engine is more successful in converting the thermal energy into work and less of the fuel energy remains in the exhaust gases. This is true even though the friction and heat transfer increase as compression ratio increases.

8.4.3   Equivalence Ratio

Figure 8.18 shows the net indicated and brake thermal efficiencies as functions of equivalence ratio for a *bmep* of 325 kPa and 1400 rpm. Thermal efficiencies increase as the equivalence ratio decreases. The increase is more rapid for the rich mixtures since less fuel energy is unburned. As discussed next, increasing efficiencies are largely a result of decreasing heat transfer and decreasing use of the throttle (for constant load) as the equivalence ratio decreases.

Figure 8.19 shows the distribution of the fuel energy as a function of equivalence ratio for these conditions. As the equivalence ratio decreases from stoichiometric, the brake work increases, the friction (percentage) increases, and the heat transfer decreases. As the

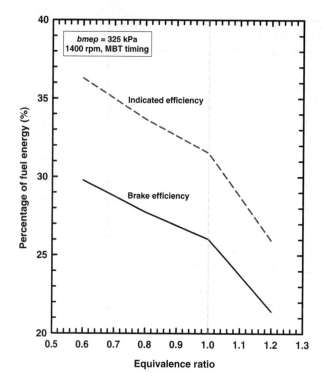

Figure 8.18 Net indicated and brake thermal efficiency as functions of equivalence ratio for a *bmep* of 325 kPa and 1400 rpm

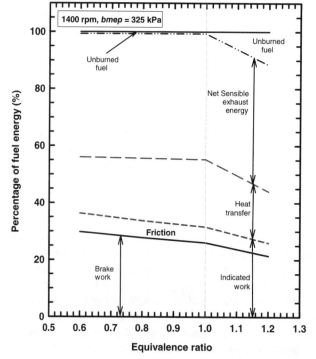

Figure 8.19 Distribution of the fuel energy as functions of equivalence ratio for a *bmep* of 325 kPa and 1400 rpm

equivalence ratio changes, the percentage of the fuel energy in the sensible energy component of the exhaust gases is nearly constant. For the rich mixtures, the heat transfer decreases as the equivalence ratio increases from stoichiometric. The heat transfer decreasing from stoichiometric for both lean and rich mixtures is largely a consequence of the decreasing temperatures on both sides of stoichiometric.

Also, for the rich mixtures, as the equivalence ratio increases from stoichiometric, the unburned amount of fuel increases. This, of course, is due to the lack of sufficient oxygen to completely oxidize the fuel.

8.4.4   Burn Duration

Figure 8.20 shows the net indicated and brake thermal efficiencies as functions of the burn duration for a *bmep* of 325 kPa, 1400 rpm, and MBT combustion timing. The thermal efficiencies increase as the burn duration decreases. For burn durations less than about 30°CA, however, the gains are small. In practice, very short burn durations may be problematic due to roughness, noise and engine durability issues.

8.4.5   Inlet Temperature

Figure 8.21 shows the net indicated and brake thermal efficiencies and the *bmep* as functions of the inlet temperature for an inlet pressure of 95 kPa (WOT) and 1400 rpm. As the inlet

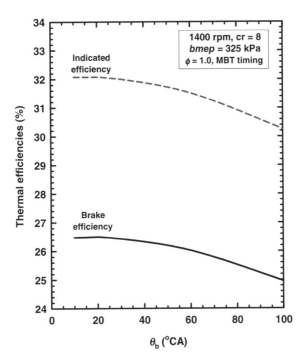

Figure 8.20   Net indicated and brake thermal efficiency as functions of burn duration for a *bmep* of 325 kPa and 1400 rpm

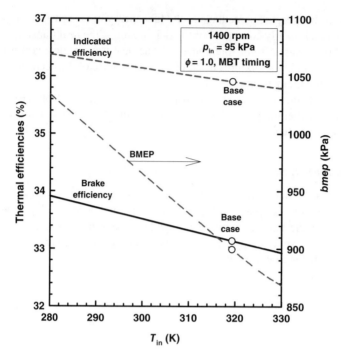

Figure 8.21 Net indicated and brake thermal efficiency, and *bmep* as functions of inlet temperature for an inlet pressure of 95.0 kPa (WOT) and 1400 rpm

temperature increases, the efficiencies decrease due to increasing heat losses and decreasing volumetric efficiencies. Also, as the inlet temperature increases, the *bmep* decreases due to the decreasing inlet charge density. Note that, for a constant part load condition, this effect may be somewhat less. For a part load case, as the inlet temperature increases, the inlet pressure would need to increase to provide the same output and this may result in modest or no reductions of the efficiencies.

8.4.6 Residual Mass Fraction

Figure 8.22 shows the residual mass fraction as functions of engine speed for three values of *bmep* and for WOT. As engine speed and load increase, the residual fraction decreases. For the conditions examined, the residual fraction ranges from about 5% to 17%. These are typical values, but for lower loads and lower speeds, the residual fraction could be even higher [2].

8.4.7 Exhaust Pressure

The next two figures show the exhaust pressure as functions of speed and load. Although the exhaust pressure was determined from an algorithm [3] (and not from any fundamental thermodynamics), the values do affect the final results. Figure 8.23 shows the exhaust pressure

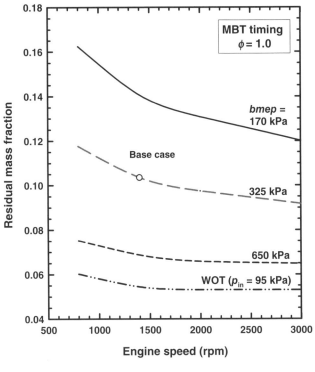

Figure 8.22 Residual mass fraction as functions of engine speed for three *bmep* values and for WOT

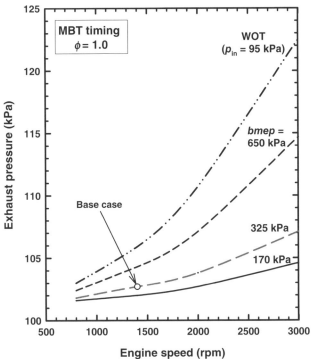

Figure 8.23 Exhaust pressure as functions of engine speed for three *bmep* values and for WOT

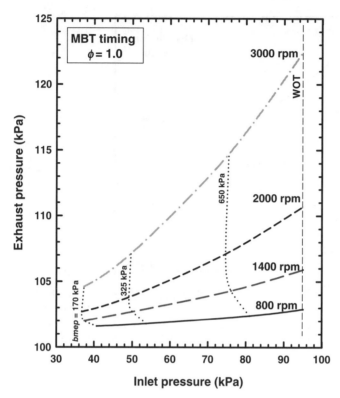

Figure 8.24   Exhaust pressure as functions of inlet pressure for four engine speeds for WOT and three *bmep* levels

as a function of engine speed for three values of *bmep* and for WOT. The exhaust pressure increases with increases of engine speed and load. Overall, for these conditions, the exhaust pressure ranged from about 102 kPa to 122 kPa. As listed in Chapter 7, the exhaust pressure for the case with a *bmep* of 325 kPa at 1400 rpm is 102.6 kPa.

Figure 8.24 shows the exhaust pressure as a function of the inlet pressure for four engine speeds. Dotted lines are included in the figure to designate three engine loads. As the inlet pressure increases, the exhaust pressure increases for each engine speed. The highest exhaust pressures are for WOT operation at the highest speed.

Figure 8.25 shows the brake thermal efficiencies as functions of the load (*bmep*) for 1400 rpm for two constant exhaust pressures: 95 kPa and 120 kPa. The exhaust pressures used in the previous calculations (see Figures 8.23 and 8.24) were between these two values. The effect of the exhaust pressure is somewhat modest on the brake thermal efficiency and is most significant for the conditions near or at WOT.

Figure 8.26 shows the brake and net indicated thermal efficiencies as functions of the exhaust pressure for WOT (p_{in} = 95 kPa) for 1400 rpm. The efficiencies decrease nearly linearly as exhaust pressure increases, but the effect is modest. For these conditions, as the exhaust pressure increases from 96 kPa to 120 kPa (about a 25% increase), the efficiencies decrease about 3% (relative).

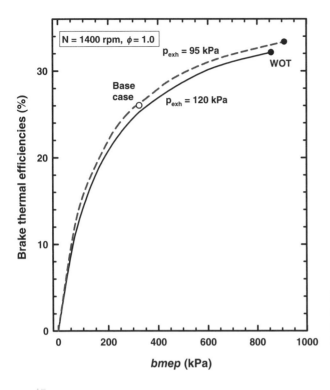

Figure 8.25   Brake thermal efficiencies as functions of load (*bmep*) for two exhaust pressures for 1400 rpm

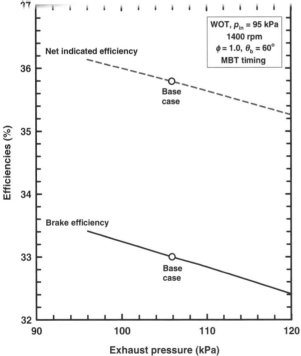

Figure 8.26   Net indicated and brake thermal efficiencies as functions of the exhaust pressure for 1400 rpm and WOT

8.4.8   Exhaust Gas Temperature

Knowledge of the temperature of the exhaust gases is important to understand the energy distribution, for turbocharging applications, and for applications of exhaust emission reduction systems. From an energy utilization perspective, the exhaust gas temperature that corresponds to the exhaust gas energy is the most useful. The energy average exhaust gas temperature is,

$$T_{\text{exh}} = \frac{\int \dot{m}_{\text{exh}} C_{p,\,\text{exh}} T_{i,\text{exh}} d\theta}{\int \dot{m}_{\text{exh}} C_{p,\,\text{exh}} d\theta}$$

where $T_{i,\text{exh}}$ is the instantaneous exhaust gas temperature. In other words, the average energy exhaust gas temperature is proportional to the exhaust gas energy. This is different than a time-average exhaust gas temperature which may not capture the full implications of the energy flow [4]. A comparison of these two exhaust temperatures is provided below. In some cases, this temperature is known as a "mass average" exhaust temperature. This would be exact where the specific heats in the numerator and denominator are the same.

Figure 8.27 shows the energy average exhaust gas temperature as functions of engine speed for WOT and three *bmep* levels. This figure shows that the average energy exhaust gas temperature increases with speed. This is largely due to the higher temperatures and the

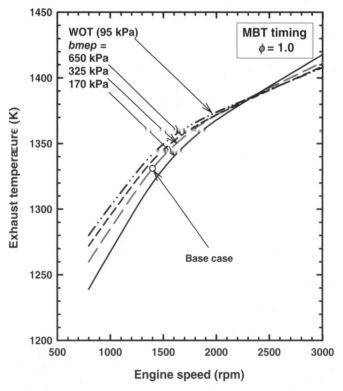

Figure 8.27   Energy average exhaust temperatures as functions of engine speed for three *bmep* values and for WOT

reduction of time for heat transfer as speed increases. The effects of engine load are different for lower temperatures and higher temperatures. For these conditions, the change occurs at about 2300 rpm. At lower speeds, the temperature difference dominates and the higher loads possess the higher exhaust temperatures for the same engine speed. For the higher speeds, the greater mass flow for the higher loads results in low exhaust temperatures. The change in the cylinder gas temperature as functions of engine speed may be reviewed in Figure 8.12.

Often exhaust gas temperatures are measured with one or more small thermocouples inserted into the exhaust gas stream. The thermocouples will respond with an average temperature which may be related to a "time average." A time average exhaust temperature for the *exhaust valve open period* may be defined as,

$$T_{\text{exh,time,avg}} = \frac{\int T_{\text{i,exh}} d\theta}{\int d\theta} = \frac{\int T_{\text{i,exh}} d\theta}{\Delta\theta_{\text{EVOP}}}$$

where $T_{\text{i,exh}}$ is the instantaneous exhaust gas temperature and $\Delta\theta_{\text{EVOP}}$ is the exhaust valve open period. For the base case conditions (325 kPa *bmep*), Figure 8.28 shows the energy average and time average exhaust gas temperatures as functions of engine speed. In this case, the time average exhaust temperature is about 50–90 K lower than the corresponding energy average exhaust temperature.

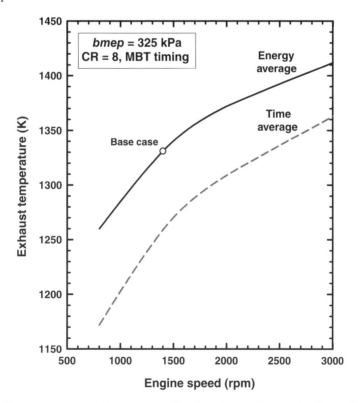

Figure 8.28   Energy average and time average exhaust gas temperatures as functions of engine speed for a *bmep* of 325 kPa

Note that the thermocouple will respond to the exhaust gases during the exhaust valve open period as well as during the exhaust valve close period. The exhaust valve close period is typically on the order of 65% of the total time, and so, the thermocouple is not measuring only the flowing gases but also the "stagnant" gases from the previous cycle. The thermocouple temperatures are closer to the time average exhaust gas temperature for the complete 720°CA than to the energy average exhaust temperatures. Some experimental work has shown that the time average exhaust gas temperatures may be about 100 K lower than the energy average exhaust gas temperature [4]. In addition, thermocouple temperatures in the exhaust port or manifold will be lower than the temperatures of the gas at the exhaust valve. This latter location is important for precise energy accounting (unless sufficient detail of the heat loss of the port/manifold is included).

The calculation above for the time average exhaust temperature was for the exhaust valve open (EVO) period. To complete the calculation for the complete 720°CA requires information or assumptions about the valve closed portion. In particular, an exhaust port heat transfer submodel is needed to accurately complete this calculation. In addition, any residual motion in the port should be included as well. Although not necessarily difficult, these features were not implemented and so the above was not completed for the full 720°CA.

8.4.9   Exhaust Gas Recirculation

Exhaust gas recirculation has been used since the late 1960s to reduce combustion temperatures and thereby reduce the formation of nitric oxides (NO_x). The use of EGR is quite effective for nitric oxide reductions since it is readily available and somewhat inert with little oxygen. Over the years, applications of EGR have become more and more successful. Thermodynamic engine cycle simulations can be used to evaluate the characteristics of EGR for a range of operating and design variables. Several studies [5–7] have been published with many more details than can be presented here. Since the case study on nitric oxide emissions includes results for the effects of EGR, this aspect of the use of EGR is omitted here. In this subsection, the results will focus on the thermal efficiencies and overall engine performance.

The use of EGR typically involves some degree of cooling of the recirculated exhaust gases before entering the inlet system. Since this amount of cooling is engine-specific and is generally unknown, the current work will examine two configurations which bracket the levels of cooling. For the highest level of cooling, the exhaust gas will be assumed to be cooled to the inlet temperature (319.3 K). This configuration will be called the "cooled" configuration. For zero cooling, the exhaust gas will be assumed to retain all its energy and enter the inlet system at the exhaust temperature. This latter configuration will be called the "adiabatic" configuration. These two EGR configurations result in inlet mixture temperatures that bracket the possible temperatures in practice.

The use of EGR alters several aspects of the thermodynamics compared to not using EGR. With EGR, the inlet mixture will contain species such as carbon dioxide and water vapor. These species and differences in gas temperatures throughout the cycle will change the thermodynamic properties. All of these effects are automatically included in the following computations.

Figure 8.29 shows the inlet mixture temperature as functions of the EGR level for both EGR configurations. For the cooled EGR configuration, the inlet temperature is always at 319.3 K.

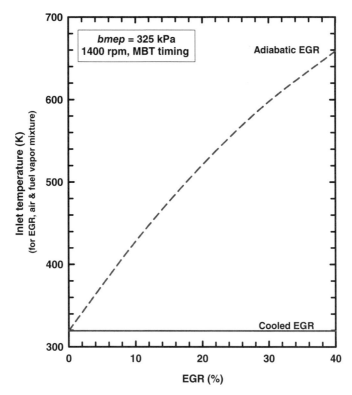

Figure 8.29   Inlet temperatures as functions of EGR level for both the adiabatic and cooled EGR configurations. *Source:* Caton 2006. Reproduced with permission from ASME

For the adiabatic EGR configuration, the inlet temperature increases in temperature as the EGR level increases. For the use of EGR, the exhaust gas temperature decreases as the EGR level increases since combustion temperatures decrease due to the dilution. For the adiabatic EGR configuration, this means that as the EGR level increases, the exhaust gas temperature decreases—therefore, the rate of increase of the inlet mixture temperature decreases for the higher EGR levels.

Figure 8.30 shows the net indicated and brake thermal efficiencies as functions of the EGR level for both EGR configurations for a *bmep* of 325 kPa (part load case) and 1400 rpm. Both efficiencies increase with increasing EGR level, and the results are similar for the two EGR configurations. Such similar results for the two configurations are limited to cases where the inlet pressure is less than atmospheric (throttled). In addition, these results are based on successful combustion for all EGR levels. For conventional engines, combustion deteriorates at some level of EGR (say, ~20%). These two aspects are discussed below.

The increases of the thermal efficiencies with increasing EGR levels for part load are the result of lower gas temperatures, lower heat transfer, increased ratio of specific heats, reduced dissociation near top dead-center, and lower pumping work. This latter advantage is due to the fact that to attain the same *bmep*, the inlet pressure increases (less throttle) for increasing EGR level. These items are quantified in the related publications [5–7].

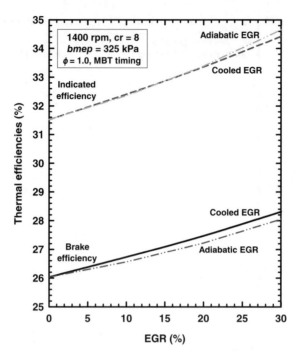

Figure 8.30   Computed net indicated and brake thermal efficiencies as functions of EGR for both EGR configurations for a *bmep* of 325 kPa and 1400 rpm. *Source:* Caton 2006. Reproduced with permission from ASME

The results in Figure 8.30 illustrate an interesting feature of the trade-offs for throttled (inlet pressure less than atmospheric) cases which the engine cycle simulation captures. The cooled EGR configuration has the higher gross indicated efficiency (not shown here) largely due to the lower heat losses relative to the adiabatic configuration. The adiabatic configuration, however, has lower pumping work compared to the cooled EGR configuration since less throttle is needed. These two effects are about equal, and the overall result is that the net indicated efficiencies are about the same for the two EGR configurations.

The results in Figure 8.30 also show that the cooled EGR configuration results in slightly higher brake thermal efficiencies compared to the adiabatic EGR configuration. This is largely a result of the higher mechanical friction for the adiabatic configuration due to the higher cylinder pressures. So although the adiabatic configuration has an advantage of lower pumping work, it has higher mechanical friction. One conclusion from these results is that cooling the exhaust gas as much as practical leads to the best thermodynamic performance (and as shown in a Chapter 16, will lead to the highest level of nitric oxide reductions).

As mentioned above, the two EGR configurations result in similar efficiencies because a part load (*bmep* = 325 kPa) case was examined. The case study on high efficiency engines, on the other hand, is based on conditions which include inlet pressures above atmospheric. For these cases, the two EGR configurations result in much different efficiencies. The adiabatic EGR configuration results in decreasing efficiencies as EGR increases. Further discussion of this may be found in the case study on high efficiency engines.

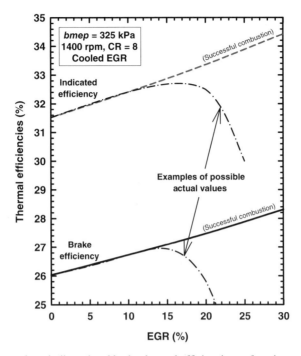

Figure 8.31 Computed net indicated and brake thermal efficiencies as functions of EGR for the cooled EGR configuration for a *bmep* of 325 kPa and 1400 rpm. Examples of possible actual values for unsuccessful combustion also included

One aspect of using EGR that needs to be discussed is that most conventional combustion systems will exhibit poor performance as the EGR level increases [8,9]. As the EGR level increases, the burn duration and cycle-to-cycle fluctuations will increase. Eventually a level of EGR is reached where engine operation becomes rough and unstable and hydrocarbon emissions increase rapidly. Obviously, these aspects have not been included in the previous computations. Figure 8.31 shows the thermal efficiencies as functions of EGR level for the cooled EGR configuration. Also included in this figure are examples of possible actual values of efficiencies for cases where the combustion process is not completely successful. The actual curves will depend on the specific engine, but most engines will result in similar trends to that represented in Figure 8.31. Much more on this topic is available in the open literature [5–9].

8.4.10 Pumping Work

Pumping work is often not explicit in energy balance considerations, but deserves to be mentioned for completeness. Recall that in Chapter 2, the pumping work was identified as the work required to move gases into and out of the control system (cylinder contents). For the cases examined in this book where the exhaust pressure is greater than the inlet pressure, this work is always *into* the system. For the four-stroke cycle, the pumping work is, of course, for only two strokes or one revolution. Note that in all the above results all the fuel energy was

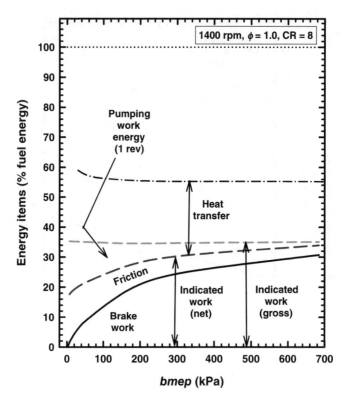

Figure 8.32   Gross indicated, net indicated, and brake thermal efficiencies as functions of load (*bmep*) for 1400 rpm

accounted for without introducing the pumping work. As explained below, pumping work is an indirect consideration of the energy balance.

The following two figures will illustrate values for the pumping work as functions of load and speed. Figure 8.32 shows energy items such as the heat transfer, the gross indicated work, the net indicated work, and the brake work as percentages of the fuel energy. This figure shows these items as functions of load (*bmep*) for an engine speed of 1400 rpm. The gross indicated work represents the potential work from the compression and expansion strokes, and decreases very slightly as the load increases. The curve below the one for the gross work represents the net indicated work, and the difference between these two curves is due to the pumping work. As the load increases, the pumping work decreases due to less throttling. The bottom curve represents the brake efficiency. The heat transfer energy is indicated by the top curve and the curve for the indicated work. Note that as more throttle is used (lower *bmep*), the heat transfer percentage increases slightly. This is discussed in more detail below.

Figure 8.32 is essentially the same as Figure 8.4 with the curve for the gross indicated work added. This allowed the pumping work to be clearly denoted. Of particular importance, however, is that Figure 8.4 shows the complete distribution of the fuel energy without the consideration of the pumping work.

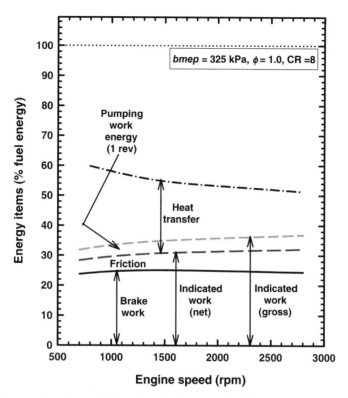

Figure 8.33   Gross indicated, net indicated, and brake thermal efficiencies as functions of engine speed for 325 kPa *bmep*

Figure 8.33 shows similar items as shown in Figure 8.32, but as functions of engine speed. The gross indicated work increases slightly as engine speed increases. The next curve is the indicated efficiency, and the difference between the two is the pumping work. The bottom curve is the brake efficiency, and this is the same curve as in Figure 8.8. The slight increase of the pumping work as engine speed increases is largely due to the increasing mass flows as speed increases. Note that for this case the inlet pressure (throttle) was nearly constant. As before, the heat transfer energy is indicated by the top curve and the curve for the indicated work. As discussed above, the heat transfer percentage is greatest for the lower speeds due to the longer available times for the heat transfer (relative to higher speeds).

The next discussion illustrates the accumulated work for each of the four strokes. These results are related to Figure 7.18 in Chapter 7 which showed the instantaneous work as a function of crank angle for the case with a *bmep* of 325 kPa and at 1400 rpm. Figure 8.34 shows the various accumulated work values as percentages of the fuel energy for each stroke. In this figure, the positive values are work *out* of the system. The expansion and compression strokes combine to provide the gross indicated work. The intake stroke provides a positive work out, but the exhaust stroke provides a greater amount of work into the system. The combination of the intake and exhaust strokes (the next bar) represents the pumping work which is into

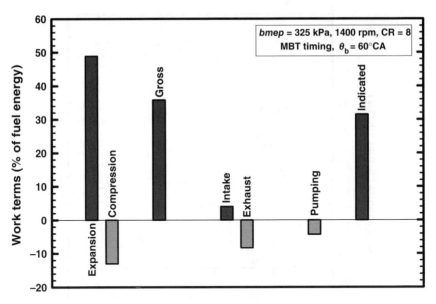

Figure 8.34   Individual work terms as a percentage of the fuel energy for 325 kPa *bmep* and 1400 rpm

the system. Finally, the last bar represents the net indicated work (the gross work minus the absolute value of the pumping work). The net indicated work as illustrated in Figure 8.34 has a value of 31.5% which is the net indicated thermal efficiency for this case. An important feature of these results is that the pumping work is already integrated into the value of the indicated work. This is another reason why the pumping work is not a direct part of the fuel energy distribution.

Although, as mentioned above, the pumping work is not a direct part of the fuel energy distribution, the pumping work does affect the overall performance. The pumping work affects the final performance in two major ways. First, a part of the gross work must be used to move the gases into and out of the cylinder—this is the pumping work. Although this is not a direct use of fuel energy (the pumping work is into the system), it does reduce the potential of the gross work. Second, the throttling (reduced inlet pressure) associated with the pumping work results in a lower inlet charge density, and therefore, less fuel energy is available.

The next figure is designed to quantify the importance of the pumping work. Figure 8.35 shows a comparison of the fuel energy distribution for a part load case (*bmep* = 325 kPa) and a WOT case (*bmep* = 891 kPa) for 1400 rpm. The pumping work is shown above the 100% line for each case since this work is an input to the cylinder (and not part of the fuel energy). As expected, the pumping work is much higher for the part load case compared to the WOT case—about nine times higher. Also note that the average ratio of specific heats for the compression and expansion strokes for the two cases are 1.228 and 1.230. Since these values are almost the same, the role of the ratio of specific heats for this comparison is negligible.

For this comparison, note that the relative heat transfer is significantly lower for the WOT case compared to the part load case. As explained elsewhere, even though the actual heat

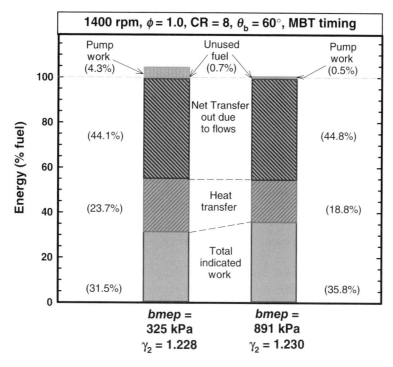

Figure 8.35   Energy items as a percentage of the fuel energy for a part load and WOT case at 1400 rpm

transfer energy is higher for the WOT case, the relative heat transfer is lower since the fuel energy is much higher for the WOT case. This is an important aspect of this comparison. In other words, the throttling has reduced the incoming charge density and therefore the incoming fuel mass. This means that even for the same absolute heat transfer, the relative heat transfer would decrease. This is quantified next.

By examining the energy distribution for these two cases, a subtle but important aspect of the pumping work can be ascertained. As shown, the net indicated thermal efficiency increased by about 4.3% (absolute) as the part load case transition to the WOT case. For these conditions, the relative heat transfer decreased by about 4.9% (absolute). This suggests that the WOT case is more efficient than the part load case, at least in part, because of a favorable heat transfer situation. Other engine conditions have been examined, and this comparison consistently shows that the favorable heat transfer is a major reason engine conditions with higher inlet pressures are more efficient than part load cases.

8.5   Summary and Conclusions

This chapter has presented a fairly complete set of overall engine performance results as functions of speed, load, timing, equivalence ratio, compression ratio, and inlet temperature. In addition, the effects of the use of EGR on performance were provided. These results illustrate the value of the thermodynamic engine cycle simulations. From these results, some general findings and conclusions can be stated regarding the thermodynamics of engines.

- With respect to engine speed, heat transfer is more important at low speeds and friction is more important at higher speeds. This most often leads to a maximum of performance at some intermediate speed.
- Although heat transfer rates increase with engine speed, the time available for the heat transfer decreases with increasing speed. The net result is that the fraction of the fuel energy associated with the heat transfer decreases with increasing speed.
- For a throttled engine, the pumping work is dominant for low loads (less than 50% of the WOT condition).
- In general, the energy remaining in the exhaust gases is significant (for the conditions examined, about 40% of the fuel energy). This is indicative of the difficulty of converting thermal energy into work.
- The highest cylinder pressures and temperatures exist for WOT at the highest engine speeds.
- For each operating condition, maximum performance is obtained for start of combustion (MBT timing) that minimizes the compression work and maximizes the expansion work.
- As the start of combustion retards from the most advanced timings, peak cylinder gas temperatures decrease and exhaust gas temperatures increase.
- As is well known, increasing compression ratio increases power, torque, and efficiencies. These gains are greatest for the increasing low compression ratios, and diminish for increasing higher compression ratios. Further, brake parameters increase less than net indicated parameters because of the increasing friction for higher compression ratios.
- Operation with lean mixtures provides higher thermal efficiencies due to lower gas temperatures, lower heat losses, lower pumping work (for the same engine load), and higher ratios of specific heats.
- Shorter burn durations provide higher performance since the energy release can occur closer to TDC. For the conditions examined, burn durations shorter than about 30°CA did not provide significant improvements.
- For WOT, increasing inlet air temperatures decreases performance.
- The residual mass fraction is lowest for WOT and high engine speeds.
- The exhaust pressure increases with engine speed and load.
- The average exhaust temperature increases with engine speed. For lower speeds, the exhaust temperature increased with increases of load; while for the higher speeds, the exhaust temperature decreased slightly with increases of load.
- The average energy exhaust temperature is typically higher than a time average exhaust gas temperature.
- The use of moderate EGR levels improves performance due to lower gas temperatures, lower heat losses, lower pumping work (for the same engine load), and higher ratios of specific heats. In actual applications, higher levels of EGR will result in combustion difficulties and deteriorating performance.
- Pumping work has an indirect influence on engine performance, and is not part of the fuel energy inventory. The pumping work reduces the potential of the gross work. Also, these results showed that for part load operation, a large part of the efficiency decrease is due to increases of the relative heat transfer.

- The lower inlet pressures associated with part load reduce the inlet charge density (reducing the inlet fuel). The reduced inlet fuel is, of course, responsible for the reduced work output at part load.

References

1. Woschni, G. (1968). A universally applicable equation for the instantaneous heat transfer coefficient in the internal combustion engine, SAE Transactions, SAE paper no. 670931, vol. 76, pp. 3065–3083.
2. Heywood, J. B. (1988). *Internal Combustion Engine Fundamentals*, McGraw-Hill book Company, New York.
3. Sandoval, D. and Heywood, J. B. (2003). An improved friction model for spark-ignition engines, Society of Automotive Engineers, SAE paper no. 2003–01–0725.
4. Caton, J. A. (1982). Comparisons of thermocouple, time-averaged and mass-averaged exhaust gas temperatures for a spark-ignited engine, 1982 Society of Automotive Engineers International Congress and Exposition, Cobo Hall, Detroit, MI, paper No. 820050, February.
5. Caton, J. A. (2006). Utilizing a cycle simulation to examine the use of EGR for a spark-ignition engine including the second law of thermodynamics, in Proceedings of the 2006 Fall Conference of the ASME Internal Combustion Engine Division, Sacramento, CA, 5–8 November.
6. Shyani, R. G., and Caton, J. A. (2009). Results from a thermodynamic cycle simulation for a range of inlet oxygen concentrations using either EGR or Oxygen enriched air for a spark-ignition engine, 2009 SAE International Congress and Exposition, Society of Automotive Engineers, SAE paper no. 2009–01–1108, Cobo Hall, Detroit, MI, 20–23 April.
7. Shyani, R. G., and Caton, J. A. (2009). A thermodynamic analysis of the use of EGR in SI engines including the second law of thermodynamics, *Proceedings of the Institution of Mechanical Engineers, Part D, Journal of Automobile Engineering*, **223** (1), 131–149.
8. Kuroda, H., Nakajima, Y. I., Sugihara, K., Takagi, Y., and Ashby, H. A. (1978). The fast burn with heavy EGR – a new approach for low NO_x and improved fuel economy, Society of Automotive Engineers, SAE paper no. 780006.
9. Cha, J. Y., Kwon, J. H., Cho, Y. J., and Park, S. S. (2001). The effect of exhaust gas recirculation (EGR) on combustion stability, engine performance and exhaust emissions in a gasoline engine, *KSME International Journal*, **15** (10), 1442–50.

9

Second Law Results

9.1 Introduction

The majority of the previous results were based solely on the use of the first law of thermodynamics (i.e., on the use of energy conservation concepts). This chapter is focused on results mainly from the second law of thermodynamics. An important thermodynamic property related to these analyses is exergy. This chapter will introduce, define, and use the property exergy. A unique aspect of these analyses is that exergy (unlike energy) can be destroyed. Processes that destroy exergy include combustion, mixing and heat transfer across a temperature difference. Comprehensive results from the second law will be provided. Most of these results will illustrate the distribution of the fuel exergy among work, heat transfer exergy, exhaust exergy, and destroyed exergy. Basic background information on the second law of thermodynamics and exergy may be found in numerous references (e.g., [1–3]).

While the use of a second law analysis is not necessary for general engine performance computations, the insight provided by a second law analysis is invaluable in understanding the details of the overall thermodynamics of engine operation. The second law of thermodynamics is a rich and powerful statement of related physical observations that has a wide range of implications with respect to engineering design and operation of thermal systems. For example, the second law can be used to determine the direction of processes, to establish the conditions of equilibrium, to specify the maximum possible performance of thermal systems, and to identify those aspects of processes that are detrimental to overall performance.

The engine cycle simulation was used to explore the performance of a spark-ignition engine from the perspective of the second law. The next subsection will review the concept of exergy. This will be followed by results as functions of engine operating and design parameters.

9.2 Exergy

Related to the analysis based on the second law of thermodynamics is the concept of exergy (which is also known as essergy (essence of energy), availability, and available energy

An Introduction to Thermodynamic Cycle Simulations for Internal Combustion Engines, First Edition.
Jerald A. Caton © 2016 John Wiley & Sons, Ltd. Published 2016 by John Wiley & Sons, Ltd.

(see e.g., 1). Exergy, a thermodynamic property of a system and its surroundings, is a measure of the maximum useful work that a given system may attain as the system is allowed to reversibly transition to a thermodynamic state which is in equilibrium with its environment. In other words, only a portion of a given amount of energy is "available" to produce useful work (the exergy), while the remaining portion of the original energy is "unavailable" for producing useful work [1].

In general, the processes of interest are the thermal, mechanical, and chemical processes. An example of the thermal aspect of exergy is a case where the system temperature is above the environmental temperature. By utilizing an ideal heat engine (such as a Carnot engine), a portion of the energy (the exergy) from the system could be converted to work until the system temperature equaled the environmental temperature (the remaining energy is, therefore, the unavailable portion of the energy). An example of the mechanical aspect of exergy is a system which is at a pressure above the environment. By utilizing an ideal expansion device (such as an ideal turbine), the energy of the system could be converted to work until the system pressure equaled the environmental pressure.

A final consideration is the chemical aspect of exergy.[1] This aspect considers the potential to complete work by exploiting the concentration differences of the various species relative to the related concentrations in the environment. The consideration of the species concentration component of exergy is often neglected (particularly when considering mobile engine applications) due to the practical difficulties of implementing such a system and the relatively small amounts of work produced. The current study will also neglect this contribution. Other authors have also recommended neglecting the species concentration aspect of exergy for engine applications [4,5].

9.3   Previous Literature

A large number of previous investigations employing the second law of thermodynamics or exergy analyses with respect to internal combustion engines have been completed. Several literature reviews are available that summarize these previous publications (e.g., References 6,7). These investigations have used thermodynamic cycle simulations, experimental data, or a combination of simulations, and experiments. Investigations that have used the second law of thermodynamics to study internal combustion engines in a detailed manner date back to the late 1950s, but the majority of these investigations have been completed since the 1980s. Descriptions of some of these previous investigations may be found in References 6,7. Also, the results presented below are related to a number of publications (e.g., References 8–10).

9.4   Formulation of Second Law Analyses

The second law analysis depends on a determination of the instantaneous values of entropy (see Chapter 4). In addition, to determine the entropy production due to irreversibilities,

---

[1] The chemical aspect of exergy by convention refers to the concentration differences between the species in the system and in the environment [3]. In contrast, the (chemical) fuel energy is included in the exergy terms since the total (chemical and sensible) energy is used for the internal energy and enthalpy.

entropy balances must be completed. For any portion of the cycle, an entropy balance may be constructed. In general, the balance is

$$\Delta S = S_{\text{end}} - S_{\text{start}} = S_{\text{in}} - S_{\text{out}} + S_Q + \sigma \tag{9.1}$$

where S_{end} is the total entropy at the end of the period, S_{start} is the total entropy at the start of the period, S_{in} is the total entropy transferred into the system due to flows, S_{out} is the entropy transferred out of the system due to flows, S_Q is the entropy transferred with the heat transfer process, and σ is the total entropy generated by any internal irreversibilities. The entropy transferred into the system with any heat transfer (defined positive into the system) is given by

$$S_Q = \int \frac{\dot{Q}}{T} dt \tag{9.2}$$

where T is the temperature on the boundary of the system where the heat transfer occurs. For the single-zone portion of the simulation, this may be replaced with the system temperature, T. For the multiple-zone portion of the simulation, this temperature would be either the unburned gas temperature (T_{u}), the burned gas temperature (T_{b}), or the boundary layer temperature (T_{bl}).

The terms S_{in} and S_{out} are given by

$$S_{\text{in}} = \int \dot{m}_{\text{in}} s_{\text{in}} dt$$
$$S_{\text{out}} = \int \dot{m}_{\text{out}} s_{\text{out}} dt \tag{9.3}$$

For completeness, the entropy generated by any internal irreversibilities is often described by

$$\sigma = \int \dot{\sigma}_{\text{cv}} dt \tag{9.4}$$

where $\dot{\sigma}_{\text{cv}}$ is the rate of entropy generation for the control volume. The entropy generated by irreversibilities will be obtained from the entropy balance given by eq. 9.1

$$\sigma = S_{\text{end}} - S_{\text{start}} - S_{\text{in}} + S_{\text{out}} - S_Q \tag{9.5}$$

In certain cases, the rate of entropy generation may be needed for a given portion of the cycle. This may be determined by differentiating eq. 9.5

$$\dot{\sigma} = \frac{d\sigma}{dt} = \frac{dS_{\text{sys}}}{dt} - \dot{S}_{\text{in}} + \dot{S}_{\text{out}} - \dot{S}_Q \tag{9.6}$$

where $d\sigma/dt$ is the rate of entropy generation due to irreversibilities, dS_{sys}/dt is the rate of change of the system entropy, $\dot{S}_{\text{in}}$ is the rate of entropy transferred into the system via mass flow in, $\dot{S}_{\text{out}}$ is the rate of entropy transferred out of the system via mass flow out, work out or destruction, and $\dot{S}_Q$ is the rate of entropy that is transferred with the heat transfer according to

$$\dot{S}_Q = \frac{\dot{Q}}{T} \tag{9.7}$$

Once the thermodynamic properties (including entropy) are known for a given cycle simulation, the exergy may be computed. In this development, the kinetic and potential energies are neglected (and can be shown to be negligible). Since the overall engine operation includes both closed system and open system portions, two forms of exergy (system and flow exergy) are needed. At all times, for the complete system, the system specific exergy is

$$b = (u - u_o) - (-p_o(v - v_o)) - T_o(s - s_o) \tag{9.8}$$

where b is the specific exergy (or exergy for closed systems), u, v, and s are the specific internal energy, the specific volume, and the specific entropy, respectively, u_o, v_o and s_o are the specific internal energy, specific volume and specific entropy for the restricted dead state (described below), respectively, and p_o and T_o are the pressure and temperature of the dead state, where the subscript "o" is for the dead state.

The term, $p_o(v-v_o)$, represents the work completed against the atmosphere at p_o, and hence, is not useful. The two negative signs for this term have been used to emphasize that for cases where the system-specific volume is less than the specific volume of the restricted dead state then this work is subtracted since it cannot be used. The second negative sign accounts for the fact that as the volume expands, the work is done by the system.

The dead state is defined as the conditions of the environment at a temperature of T_o and a pressure of p_o. This is chosen since when a system is in complete equilibrium with its local environment, there is no opportunity for the system to produce any further useful work and the system is at its "dead" state. Since the contents of the system are not allowed to mix or react with the environment, this dead state is not complete and is often referred to as a "restricted" dead state [3]. The recommended [1] temperature and pressure for the dead state conditions are 298.15 K and 101.325 kPa (1 atm). In addition to the dead state temperature and pressure, the composition of the dead state must be specified.

The selection of the composition of the dead state is important when considering the chemical exergy component. Since the current work and most of the previous work have neglected this component, the selection of the composition is not important for the relative values of exergy. On the other hand, the composition of the dead state does affect the absolute values of the exergy. This has an implication on the magnitude of these values and whether they are positive or negative. This work has used standard air for the dead state composition.

The second form of exergy is for the flow periods (open system), and the flow exergy (or exergy for flows), b_f, is given by

$$b_f = (h - h_o) - T_o(s - s_o) \tag{9.9}$$

where h is the specific enthalpy, h_o and s_o are the specific enthalpy and specific entropy of the restricted dead state, respectively, and s is the specific entropy of the flowing matter. For flows out of the system, the flowing matter is the cylinder contents, and for flows into the system, the flowing matter must be specified.

The total system exergy (B) is determined from the specific exergy (b) and the system mass (m),

$$B = m\, b \tag{9.10}$$

Exergy is not a conserved property, and hence, may be destroyed by irreversibilities such as heat transfer through a finite temperature difference, combustion, friction, or mixing processes. To determine the destroyed exergy, an exergy balance is used. Between any end states, the change in the exergy may be related to the relevant processes:

$$\Delta B = B_{end} - B_{start}$$
$$\Delta B = B_{in} - B_{out} + B_Q - B_W - B_{dest} \tag{9.11}$$

where ΔB is the change of the total system exergy for a process, B_{end} is the total exergy at the end of the period, B_{start} is the total exergy at the start of the period, B_{in} is the total exergy transferred into the system accompanying flow into the system, B_{out} is the exergy transferred out of the system accompanying flow out of the system, B_Q is the exergy transferred accompanying the heat transfer, B_W is the exergy transfer due to work, and B_{dest} is the exergy which is destroyed by irreversible processes. This relation may be used to ascertain the destruction of exergy by solving eq. 9.11 to find B_{dest}. That is

$$B_{\text{dest}} = B_{\text{start}} - B_{\text{end}} + B_{\text{in}} - B_{\text{out}} + B_Q - B_W \qquad (9.12)$$

For work interactions, the exergy is equal to the useful work (the work minus the work done against the surroundings):

$$B_W = W - W_{\text{surr}} \qquad (9.13)$$

where the work done against the surroundings is given by

$$W_{\text{surr}} = p_o(V_{\text{end}} - V_{\text{start}}) \qquad (9.14)$$

For heat transfer, the exergy which is transferred out of the system is equal to the "available" portion of the heat transfer

$$B_Q = \int\left(1 - \frac{T_o}{T}\right)\delta Q \qquad (9.15)$$

where B_Q is the exergy of the heat transfer, δQ is the differential heat transfer which is transferred at a system (boundary) temperature of T. The exergy that transfers into the system (B_{in}) and out of the system (B_{out}) due to flows is given as

$$B_i = \int\left(\dot{m}_i\, b_{f,i}\right)dt \qquad (9.16)$$

where $b_{f,i}$ is the specific flow exergy, the subscript "i" refers to each individual flow (for this study, intake or exhaust).

In addition to the above exergy balance (eq. 9.12), the exergy destroyed may also be determined from the change in entropy due to irreversibilities as follows:

$$B_{\text{dest}} = T_o\sigma = I \qquad (9.17)$$

where σ is the change in entropy due to irreversibilities as determined for the entropy balance described above (eq. 9.5), and I is the exergy destruction due to irreversibilities (this is the common nomenclature). This approach (eq. 9.17) was used in the current study to serve as an internal consistency check.

Similar to the discussion in the above section on entropy, the rate of exergy destroyed may be needed for certain portions of the cycle. This may be obtained by differentiating eq. 9.12 to obtain

$$\frac{dB_{\text{dest}}}{dt} = \dot{I} = -\frac{dB_{\text{sys}}}{dt} + \dot{B}_{\text{in}} - \dot{B}_{\text{out}} + \dot{B}_Q - \dot{B}_W \qquad (9.18)$$

where dB_{dest}/dt is the rate of destroyed exergy, dB_{sys}/dt is the rate of change of exergy of the system, $\dot{B}_{\text{in}}$ is the rate of exergy transferred into the system accompanying any mass flow into the system, and $\dot{B}_{\text{out}}$ is the rate of exergy transferred out of the system accompanying any mass flow out of the system, $\dot{B}_Q$ is the rate of exergy transferred accompanying the heat transfer, and $\dot{B}_W$ is the rate of exergy transferred out of the system due to work done by the system.

For completing the exergy balances, values are needed for the exergy of the fuel. Recall that for the energy of the fuel, the lower heating value (*LHV*) evaluated for a constant pressure process is the accepted standard for these purposes [5].

$$LHV = -(\Delta H)_{p_o,T_o} = \sum_P n_i h_{f,i}^{\circ} - \sum_R n_j h_{f,j}^{\circ} \tag{9.19}$$

where $\Delta H_{p_o,T_o}$ is the change in enthalpy for reactants (R) to products (P) at the reference pressure (p_o) and temperature (T_o), and h_f° is the enthalpy of formation for the species i or j.

By definition, the fuel exergy $\left(B_{\text{fuel}} \right)$ is given by the Gibbs free energy[2]

$$B_{\text{fuel}} = -(\Delta G)_{T_o,p_o} \tag{9.20}$$

and

$$(\Delta G)_{T_o,p_o} = \left\{ (g_P)_{T_o,p_o} - (g_R)_{T_o,p_o} \right\} \tag{9.21}$$

where $\Delta G_{T_o,p_o}$ is the change in the Gibbs free energy for reactants (R) to products (P) at the reference conditions. In a manner consistent with the definition of the LHV, the Gibbs free energy is defined for a composition change from reactants (liquid fuel and air) to gaseous combustion products.

For a given fuel, the ratio of these numbers is a constant,

$$\frac{(\Delta G)_{T_o,p_o}}{(\Delta H)_{T_o,p_o}} \tag{9.22}$$

For isooctane, this constant is 1.0286 [5]. This constant is used to determine the value of the fuel exergy by multiplying this number with the fuel's heating value

$$B_{\text{fuel}} = \frac{(\Delta G)_{T_o,p_o}}{(\Delta H)_{T_o,p_o}} (LHV) m_f \tag{9.23}$$

The exergy destroyed during combustion (*CD*) will be expressed as a percentage of the fuel exergy

$$CD = \frac{B_{\text{dest}}}{B_{\text{fuel}}} \times 100\% \tag{9.24}$$

9.5   Results from the Second Law Analyses

9.5.1   Basic Results

Once the first law analysis has been completed, the thermodynamic conditions of all state points during the cycle are known. Quantities such as entropy and exergy then can be obtained, and a second law analysis can be completed. The following results are based on the same

---

[2] Recall that the specific Gibbs free energy is given by $g = h - T_o s$.

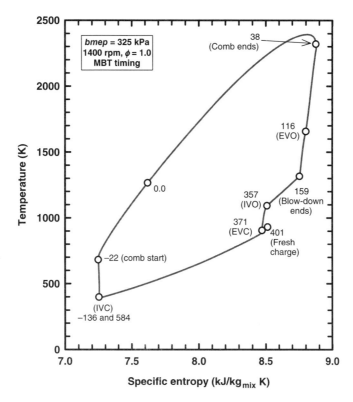

Figure 9.1   The temperature-specific entropy property diagram for the base case conditions.
Source: Caton 1999. Reproduced with permission from ASME

base case condition examined in Chapter 7 and Chapter 8. This was a part load condition and includes a *bmep* of 325 kPa at 1400 rpm. Figure 9.1 shows the cylinder one-zone gas temperature as a function of the specific entropy for the base case conditions. In this figure, specific crank angles are indicated. Starting with the intake valve close (IVC) crank angle (136°bTDC), the gas temperature increases while the specific entropy remains nearly constant (a nearly isentropic process). The entropy actually increases slightly during the beginning of the compression process due to slight heating of the gases from the hot cylinder walls. During the last stages of the compression process, the entropy decreases slightly as the hot gases lose energy to the cylinder walls. The scales necessary for this figure, however, do not permit these modest effects to be clearly shown.

At the start of combustion (22°bTDC), the specific entropy increases proportionately with the increase in temperature (due to the irreversibilities associated with the combustion process). The entropy reaches a maximum value near the end of combustion (38°aTDC), and then the entropy decreases slightly as temperatures decrease more rapidly during the expansion stroke. At a crank angle of 116°aTDC, the exhaust valve opens (EVO), and a blow-down process begins (and the gases in the cylinder continue the expansion process). Blow-down ends at about 159°aTDC (see Figure 7.9), and the decrease in entropy becomes more significant (more like a constant pressure process).

At 357°aTDC, the intake valve opens (IVO), and due to the intake manifold pressure being lower than the cylinder pressure, cylinder mass flows back into the inlet manifold. This is a near constant volume process where the entropy change is modest. At 371°aTDC, the exhaust valve closes (EVC), and the flow into the intake manifold ends. At 401°aTDC, fresh intake charge begins entering the cylinder, and the specific entropy steadily decreases. Finally, at 584°aTDC (IVC), the thermodynamic state returns to the exact same conditions as at the start (136°bTDC) of the cycle.

Figure 9.2 shows the total system entropy as a function of crank angle. Although the data is similar to that in Figure 9.1, Figure 9.2 includes both the effects of changes of specific entropy as well as any effects of mass changes. As shown, during compression the total entropy remained almost constant reflecting the nearly reversible compression process (with negligible heat loss for this scale). Once combustion starts, the total entropy increases rapidly. At the end of combustion, the total entropy decreases slightly due to heat loss. Once the exhaust valve opens, however, the total entropy decreases rapidly due mainly to the mass flow out. Total entropy only begins to increase after the exhaust valve closes (EVC). Once fresh charge enters the cylinder, the total entropy increases rapidly as cylinder mass increases. At about 538°aTDC, the flow reverses back into the inlet manifold, and the total entropy decreases a small amount. As required, the final value of the total entropy equals the initial value.

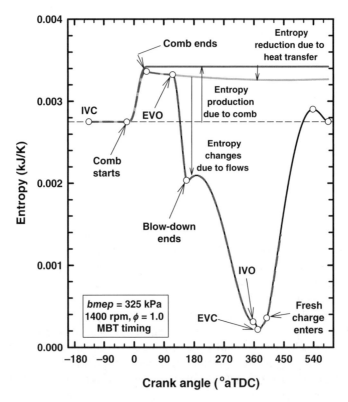

Figure 9.2   Total entropy as a function of crank angle for the base case conditions. *Source:* Caton 1999. Reproduced with permission from ASME

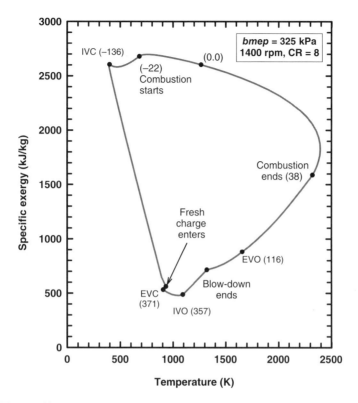

Figure 9.3   The specific exergy as a function of the one-zone, gas temperature for the base case conditions. *Source:* Caton 1999. Reproduced with permission from ASME

With the properties of entropy, internal energy, and specific volume, eq. (9.8) can be used to determine the specific exergy for these conditions. Figure 9.3 shows the specific exergy as a function of gas temperature. As shown, starting with IVC, the specific exergy increases slightly due to the addition of compression work. As combustion begins (22°bTDC), the specific exergy decreases due to combustion irreversibilities, heat transfer, and work extraction. The specific exergy continues to decrease even after the maximum temperature. The exhaust valve opens and exhaust blow-down occurs. The specific exergy continues to decrease while the temperature decreases. Finally, the new charge enters and the specific exergy begins to increase almost linearly back to its starting value.

With knowledge of the specific exergy and the cylinder mass, the total system exergy may be determined from eq. (9.10). Figure 9.4 shows the total system exergy as a function of crank angle for this case. As shown, the total exergy increases during the compression stroke due to the addition of exergy through the work done during compression. Then as combustion starts, the total exergy begins to decrease due to the combustion irreversibilities, the work being extracted, and the heat transfer. When the exhaust valve opens, the total exergy continues to decrease due to the extracted work, heat loss, and mass flow out. Finally, fresh charge enters the cylinder and the exergy increases back to the original value at IVC. Note that during the

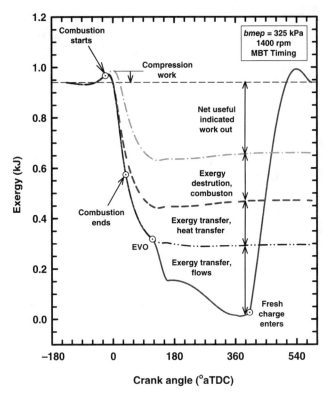

Figure 9.4 Total exergy as a function of crank angle for the base case conditions. *Source:* Caton 1999. Reproduced with permission from ASME

last portion of the intake stroke, a bit of mass returns to the inlet manifold thus reducing the total exergy back to the initial value.

Figure 9.4 also shows the time evolution of the work, exergy destruction, exergy moved with the heat transfer, and the exergy transfer due to the flows. The final values of these quantities are the values at the right of the figure for IVC. These final values are presented next.

Figure 9.5 shows the percentage of the fuel energy and fuel exergy that is assigned to each of the energy and exergy items for the base case. The left bar is for the energy quantities and the right bar is for the exergy quantities. From the bottom, the first item is the indicated work. This is the same value on a *"kJ"* basis for both the energy and exergy bars, but the percentages are slightly different since the lower heating value (energy) and the fuel exergy values are slightly different (see Table 7.2, Chapter 7). The next item is the heat transfer. The energy percentage is 23.7%, and the exergy percentage is 19.2%. This difference is, of course, due to the fact that only a portion of the energy is exergy. This is even more dramatic for the net sensible exhaust gas — the values for the energy and exergy are 44.1% and 27.2%, respectively. In other words, even though a significant amount of energy is in the exhaust, not all of this energy is available for doing useful work.

On the right bar (exergy), additional quantities are shown. These are for the destruction of exergy. The largest is due to the combustion process and equals 20.7% for these conditions.

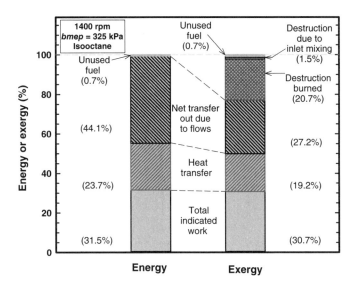

Figure 9.5   The percentage of fuel energy and exergy for work, heat transfer, exhaust and destroyed for the base case conditions. *Source:* Caton 1999. Reproduced with permission from ASME

As explained above, the combustion process is highly irreversible and this is reflected in the high level of exergy destruction. The next item is the exergy destruction due to the inlet charge mixing with the cylinder contents. For this case, this was 1.5%. This mixing process contributes much less to the degradation of the exergy than the combustion process. The final item for both the energy and exergy bars is the unburned fuel (0.7%).

9.5.2   Parametric Results

The remaining results in this chapter will focus on the effects of engine design and operating parameters from a second law perspective. Most of the results will focus on the distribution of the exergy among the energy items (such as work, heat transfer, and exhaust), and on the contributions to the destruction of exergy.

Figure 9.6 shows the percentage of the fuel exergy for work, heat transfer, exhaust, and destroyed exergy as functions of engine load for 1400 rpm. This figure can be compared to Figure 8.4 in Chapter 8 which was for the energy values. Starting at the bottom of the figure, the brake and indicated work increase with increasing *bmep*. Since "work" is 100% exergy, these curves are the same as shown earlier for the energy values. The percentage of the fuel exergy associated with the heat transfer decreases as *bmep* increases. The net sensible exhaust exergy increases very slightly as *bmep* increases. The exergy destroyed during the combustion process is nearly constant as *bmep* increases. The final item at the top of the figure is the percentage of the fuel exergy associated with the unburned fuel and with the exergy destruction during the inlet process due to the mixing of the fresh charge with the existing cylinder contents.

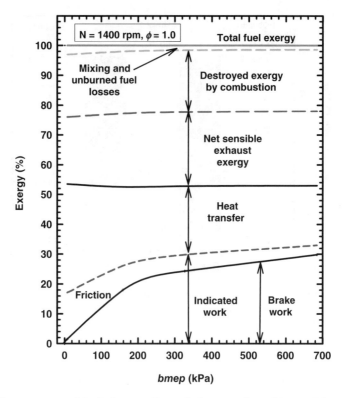

Figure 9.6   The percentage of the fuel exergy for work, heat transfer, exhaust and destroyed as functions of engine load for 1400 rpm

Figure 9.7 is similar to Figure 9.6, but shows all the items as a function of engine speed for a *bmep* of 325 kPa. Figure 9.7 (for exergy) can be compared to Figure 8.8 (for energy) in Chapter 8. At the bottom of Figure 9.7, the brake work is nearly constant as engine speed increases. This is mainly due to the constraint of constant *bmep*. The friction percentage increases slightly with increases of engine speed. The net indicated work increases slightly as speed increases. The percentage of the fuel exergy associated with the heat transfer decreases slightly as speed increases. As mentioned in Chapter 8, this is due to shorter "real" time available for the heat transfer as engine speed increases. The percentage of the fuel exergy associated with the exhaust increases slightly as speed increases. The percentage of the fuel exergy destroyed during the combustion process is nearly constant with increases of speed. The percentage of the fuel exergy associated with the unburnt fuel and with the destruction during the inlet process increase very slightly as speed increases.

Some of the above items are now examined in more detail. Figure 9.8 shows the percentage of the fuel exergy moved to the cylinder walls due to heat transfer as functions of engine speed for three values of *bmep*. This percentage ranges from about 15% to 34%. The percentage of exergy moved decreases with increases of speed and with increases of load. For the defined system, as long as the temperature of the energy does not change, no exergy is destroyed during the cylinder heat transfer process — the exergy is only transferred out of the system.

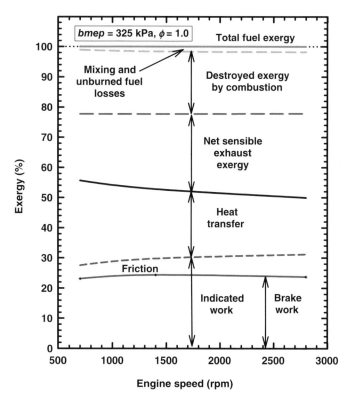

Figure 9.7 The percentage of the fuel exergy for work, heat transfer, exhaust and destroyed as functions of engine load for a *bmep* of 325 kPa

Comments at the end of this chapter will discuss the destruction of exergy as the heat transfer energy reaches the cylinder walls.

Figure 9.9 shows the exergy transferred out of the system due to the exhaust flow as functions of engine speed for three values of *bmep*. This percentage ranges between about 23% and 31%, and increases with engine speed. This increase with engine speed is largely due to the decrease of the heat transfer with engine speed. The effect of engine load is modest. The exergy that is transferred out of the system due to the exhaust flow is largely the net result of the exergy balance. This means that the exergy not assigned to work, heat transfer or destroyed may be transferred out of the system.

Figure 9.10 shows the percentage of the fuel exergy that is destroyed during the combustion process as functions of engine speed for two values of *bmep*. This percentage decreases slightly with increases of speed and even much less with increases of load. The percentage for these conditions ranges from about 20.4% to 21.0%. For an engine speed change between about 1000 and 2500 rpm, the change of the exergy destruction is less than about 2% (relative percentage). A major conclusion of these second law analyses is that the exergy destruction during combustion is fairly constant (at about 20.5%) for a range of engine parameters.

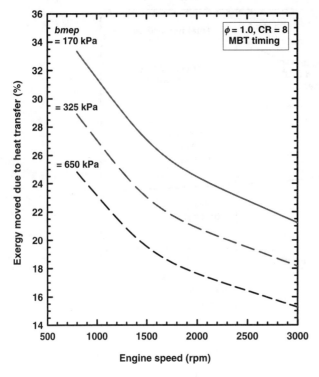

Figure 9.8 The percentage of the fuel exergy moved to the cylinder walls due to heat transfer for three values of *bmep*

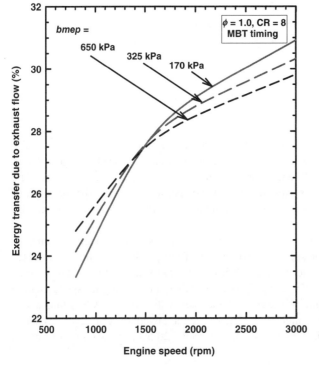

Figure 9.9 The percentage of the fuel exergy transferred with the exhaust flow for three values of *bmep*

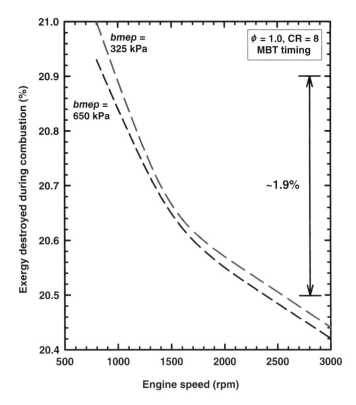

Figure 9.10   The percentage of the fuel exergy destroyed during the combustion process as functions of engine speed for two values of *bmep*

Figure 9.11 shows the percentage of the fuel exergy that is destroyed during the inlet flow process due to the mixing of the fresh charge with the existing cylinder contents. The percentage is small and ranges from about 1.0% to 2.5% for these conditions. This percentage decreases with increases of load, and has a modest and mixed response to increases of engine speed.

Figure 9.12 shows the percentage of the fuel exergy destroyed during the combustion process as functions of equivalence ratio for a *bmep* of 325 kPa and 1400 rpm. Stoichiometric (an equivalence ratio of 1.0) is identified in the figure. For the conditions examined, the percentage ranges from about 19% to 25%. Unlike most of the other engine parameters, equivalence ratio has a relatively significant impact on the exergy destruction during combustion. This is primarily due to the significant effect that equivalence ratio has on combustion temperatures. For mixtures leaner than stoichiometric, the percentage increases as equivalence ratio decreases. Of course, the combustion temperatures decrease as the mixture becomes leaner due to the dilution.

For mixtures richer than stoichiometric, the results can be presented in two fashions. Since for rich mixtures, some of the fuel is either unburned or only partially burned, the basis of the "percentage of fuel exergy" may have two values. One approach is to consider only fuel that participates in the reaction, and the other approach is to consider all the fuel. The figure

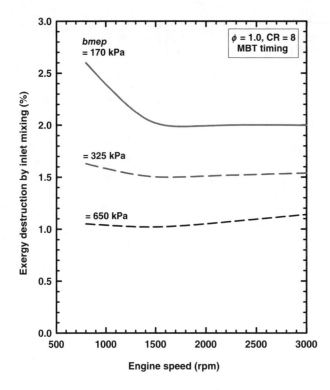

Figure 9.11   The percentage of the fuel exergy destroyed due to inlet mixing

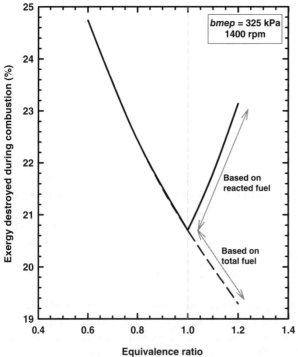

Figure 9.12   The percentage of the fuel exergy destroyed during the combustion process as functions of equivalence ratio

shows both results. For mixtures richer than stoichiometric, the solid line represents the percentage based on the reacted fuel. For this case, the percentage increases as equivalence ratio increases. Based on reacted fuel, the minimum destruction of exergy during the combustion process is for stoichiometric mixtures. This is largely a result of the high combustion temperatures related to stoichiometric operation.

For the second approach, the dashed line in Figure 9.12 represents the results for the exergy destroyed percentage based on the total fuel. Based on the total fuel, the percentage decreases as equivalence ratio increases. This is largely due to the fuel amount used in the denominator of this percentage continually increasing as equivalence ratio increases. From the perspective of reducing exergy destruction, using the total fuel amount (dashed line) is somewhat misleading. These two approaches are further illustrated in the next two figures.

Figures 9.13 and 9.14 show the percentage of the fuel exergy for each of the exergy items as functions of equivalence ratio for a *bmep* of 325 kPa and 1400 rpm. Stoichiometric conditions (an equivalence ratio of 1.0) are identified on the figures. The two figures show the same data except for the manner in which the destroyed exergy is described for the rich mixtures (see above discussion with Figure 9.12). Indicated work increases as equivalence ratio decreases — this is largely a result of decreasing relative heat transfer and increases of the ratio of specific

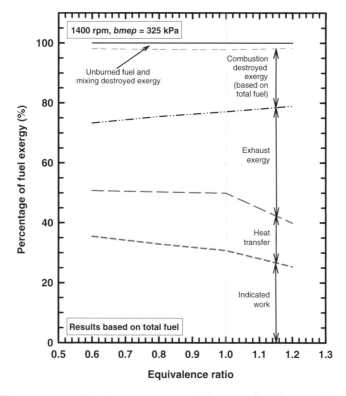

Figure 9.13   The percentage of the fuel exergy for work, heat transfer, exhaust and destroyed as functions of equivalence ratio for a *bmep* of 325 kPa. This version bases the destroyed exergy on the total fuel exergy

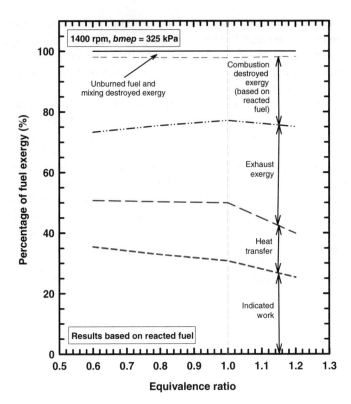

Figure 9.14 The percentage of the fuel exergy for work, heat transfer, exhaust and destroyed as functions of equivalence ratio for a *bmep* of 325 kPa. This version bases the destroyed exergy on the reacted fuel exergy

heats. From stoichiometric and leaner, the percentage of the fuel exergy associated with the heat transfer decreases. Also, from stoichiometric and leaner, the percentage of the fuel exergy associated with the exhaust decreases. The percentage of the fuel exergy destroyed during the combustion process increases as the equivalence ratio decreases (as shown in Figure 9.12). The only difference between Figures 9.13 and 9.14 is for the exergy destroyed during combustion for the rich mixtures. Again, this difference is described above with reference to Figure 9.12.

Figure 9.15 shows the percentage of the fuel exergy destroyed during the combustion process as a function of compression ratio. For these conditions, this percentage ranged between about 20% and 22%, and decreased as compression ratio increased. For compression ratios between about 15 and 20, the percentage was nearly constant. A subtle aspect to these results is that as the compression ratio increases, the conversion of thermal energy to work is more effective. These results illustrate the importance of the conversion of thermal energy to work. An important observation is that the results from the simulation for the exergy destruction during combustion represent the net result during the combustion process. That is, the destruction of exergy due only to combustion is not possible to isolate. What is reported is "during" combustion and reflects all that occurs during this portion of the cycle. So the results in Figure 9.15 reflect the greater conversion of thermal energy to work for increasing compression ratios.

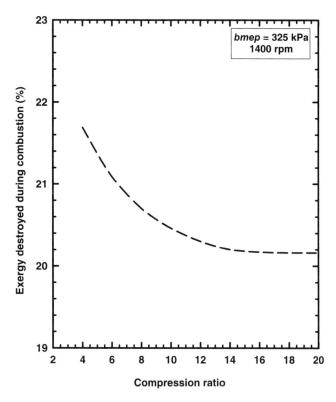

Figure 9.15   The percentage of the fuel exergy destroyed during the combustion process as functions of compression ratio

This, in turn, results in a net decrease of the exergy destruction during combustion for the higher compression ratios.

Figure 9.16 shows the percentage of the fuel exergy for each of the exergy items as functions of compression ratio for a *bmep* of 325 kPa and 1400 rpm. The indicated work increases with increases in compression ratio, and this follows the reasons provided in Chapter 8. The percentage associated with the cylinder heat transfer and with the exhaust gases increases as compression ratio increases. As shown in Figure 9.15, the percentage due to the destroyed exergy during the combustion process decreases slightly as the compression ratio increases. The percentage associated with the unburned fuel and with the inlet process decreases slightly as compression ratio increases.

Figure 9.17 shows the percentage of the fuel exergy destroyed during the combustion process and the one-zone peak gas temperature as functions of the relative combustion timing. The relative combustion timing is the CA° relative to MBT timing. For these conditions, this percentage ranged between about 20% and 23%. This percentage increased slightly as the timing was advanced or retarded from MBT timing. For the retarded timings, this was at least partly due to the decreasing combustion temperatures. For the advanced timings, this may be partly due to the non-optimal phasing — increasing compression work.

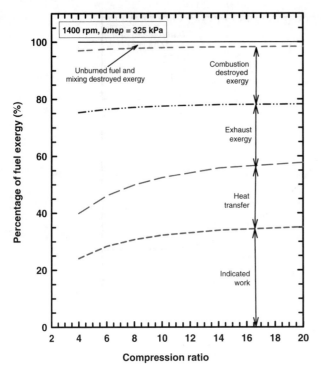

Figure 9.16 The percentage of the fuel exergy for work, heat transfer, exhaust, and destroyed as functions of compression ratio for a *bmep* of 325 kPa

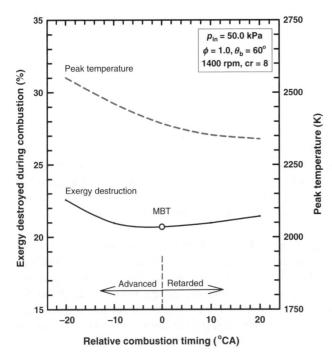

Figure 9.17 The percentage of the fuel exergy destroyed during the combustion process and the peak, one-zone gas temperature as functions of the relative combustion timing

Figure 9.18 shows the percentage of the fuel exergy for each of the energy items as functions of the relative combustion timing for a *bmep* of 325 kPa and 1400 rpm. The indicated work is maximum for MBT timing (relative timing of zero). The percentage associated with the cylinder heat transfer decreases as the combustion timing is retarded from the most advanced timings. This is due to the decreasing temperatures as the combustion timing is retarded. The percentage associated with the exhaust flow increases as the combustion timing is retarded from the most advanced timings. This is due to the increasing exhaust temperatures as the combustion timing is retarded. The percentage due to the destroyed exergy during the combustion process is shown more clearly in Figure 9.17. The percentage associated with the unburnt fuel and with the inlet process is fairly constant for these conditions for all combustion timings.

A slightly more complete tabulation of exergy destruction is available for another engine condition [11]. Although the trends are the same as reported above, additional parameters including seven additional fuels are examined. For the study reported in [11], exergy destruction during combustion ranged between about 20% and 23% for most of the engine operating and design parameters. The three parameters that resulted in exergy destruction outside this range were equivalence ratio, EGR, and oxygen-enriched inlet air.

Finally, a separate study has provided correlations of the exergy destruction during combustion with engine parameters [12]. This study found a high correlation ($r^2 = 0.98$) between

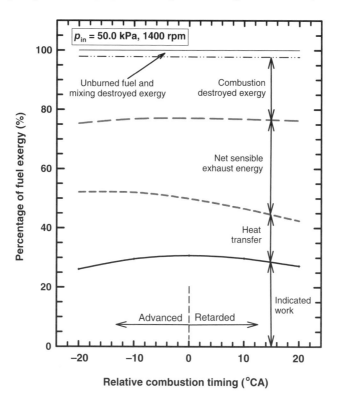

Figure 9.18   The percentage of the fuel exergy for work, heat transfer, exhaust, and destroyed as functions of the relative combustion timing for an inlet pressure of 50.0 kPa

the exergy destruction and the average burned gas temperature for five engine parameters (EGR, *htm*, *bmep*, rpm, and θ_b). A slightly higher correlation ($r^2 = 0.99$) was found between the exergy destruction and the ratio of specific heats for six engine parameters (the above plus the equivalence ratio). This latter correlation also contains the dependence on the temperatures since the ratio of specific heats vary monotonically with temperature.

9.5.3 Auxiliary Comments

In this subsection, a number of comments are presented regarding the above second law results with emphasis on the combustion, flow, mixing, and heat transfer processes. As shown, the combustion irreversibilities are a major source of exergy destruction. The irreversibilities associated with the combustion process are (at least somewhat) necessary for successful engine operation under today's engine operating and design constraints. No other process is known which allows the utilization of the fuel's exergy in an equally rapid and successful fashion. This said, however, any modifications to the combustion process which would minimize the exergy destruction while retaining the good implementation characteristics would be desirable.

Another possible use of the fuel is by a fuel cell where the fuel energy is used directly to produce electrical energy, and in principle, would not be subject to the same level of irreversibilities associated with the combustion process. The fuel cell is, however, still subject to second law constraints, but in different ways compared to combustion processes. Unfortunately, at this time, the current fuel cell technology is not competitive with the internal combustion engine due to mechanical and chemical difficulties. Also, a fuel cell may not be able to use standard liquid fuels as readily as an engine.

9.5.3.1 Processes Outside the Control System

Another aspect that deserves a comment is regarding a number of irreversible processes which have not been considered for the defined system, because these processes occur outside of the system boundaries. Recall that the defined system is the cylinder contents. Some of the important irreversible processes which are outside the defined system include flow losses in the manifolds and past the valves, other mixing losses, and the exergy destroyed by heat transfer across a finite temperature difference. Each of these will be discussed next for completeness.

Flows past restrictions are a source of irreversibility and the associated destruction of exergy. These irreversibilities were not considered here since the chosen thermodynamic system excluded all flow restrictions. Flow restrictions include the inlet and exhaust passageways, and the inlet and exhaust valves. The losses associated with these items may be determined using procedures similar to the ones used in the current study.

Finally, mixing of dissimilar matter is yet another source of irreversibilities. Some mixing has been considered here: the mixing of the unburned and burned gases, and the mixing of fresh inlet charge and residual exhaust gases. Some mixing processes, however, occurred outside of the defined thermodynamic system. For example, the mixing of the fuel vapor and air occurs upstream of the defined system. Also, if EGR was used, the mixing of these gases with the fresh charge would happen upstream of the given system.

The heat transfer is an irreversible process if the temperature difference is finite, but the exergy loss is realized only at the point where the temperature of the energy decreases.

For example, for the hot cylinder gases, the energy is transferred to the cylinder walls which are at a lower temperature. The exergy of the gases at the cylinder walls, therefore, is less than the exergy of the gases in the cylinder. This is described in more detail next.

To continue this discussion, the concept of unavailable energy is described. The heat transfer (Q) is composed of both an available (exergy) and an unavailable component:

$$Q = Q_A + Q_U = B_Q + Q_U \tag{9.25}$$

where Q_A is the exergy (available portion) of the energy associated with the heat transfer (also denoted by the symbol, B_Q), and Q_U is the unavailable portion of the energy associated with the heat transfer. The unavailable energy is given by

$$Q_U = \int \frac{T_o}{T} \delta Q \tag{9.26}$$

where T_o is the dead state temperature, T is the gas (boundary) temperature, and δQ is the differential amount of heat transfer.

The base case will be examined to illustrate the above concepts on the irreversibilities associated with heat transfer with a finite temperature difference.[3] The data for the base case provides the following distribution of energy and exergy (for one cylinder) for the heat transfer as it crosses the control system boundaries (but not at the wall):

$$
\begin{aligned}
Q_{total} &= 0.212 \text{ kJ} \\
B_Q &= 0.176 \text{ kJ} \\
(Q_{total} - B_Q) &= 0.036 \text{ kJ}
\end{aligned}
$$

Now, once the energy has transferred to the cylinder walls, the quantity of the energy is retained, but the exergy decreases. This may be determined by computing the destroyed exergy due to this heat transfer process (which is outside the thermodynamic system as defined above),

$$\left(B_{dest}\right)_{HT} = \int \left(\frac{T_o}{T_{wall}} - \frac{T_o}{T_{gas}} \right) \delta Q \tag{9.27}$$

where $(B_{dest})_{HT}$ is the destroyed exergy due to the heat transfer process, T_o is the dead state temperature, T_{wall} is the cylinder wall temperature, T_{gas} is the instantaneous cylinder gas temperature, and δQ is the differential heat transfer at this time. The result of the heat transfer process from the gas temperature to the wall temperature for the base case leads to the following:

$$
\begin{aligned}
\text{Energy: heat transfer to the wall} \quad & Q_{total} &&= 0.212 \text{ kJ} \\
\text{Exergy: of the heat transfer} \quad & B_Q &&= 0.176 \text{ kJ} \\
\text{Exergy: at the wall temperature} \quad & B_{Q,wall} &&= 0.072 \text{ kJ} \\
\text{Exergy: destroyed at the wall} \quad & (B_{dest})_{HT} &&= 0.104 \text{ kJ}
\end{aligned}
$$

In other words, the 0.176 kJ of exergy is degraded to 0.072 kJ of exergy at the wall temperature — resulting in 0.104 kJ of destroyed exergy at the wall. Clearly, for this case, the heat transfer process is responsible for a significant destruction of exergy. As shown, the heat

---

[3] Again, the irreversibilities of the heat transfer process occur outside the defined thermodynamic system for the current work. The discussion here is to assist in quantifying the destruction of exergy in a related process.

transfer process has destroyed 59% of the original exergy associated with the heat transfer. Again, this process occurred outside the defined thermodynamic system of the current work, and hence, this irreversibility is not included in the tabulations above.

This subsection will end with a brief discussion of the overall engine heat transfer process from the second law perspective. The energy that is lost as cylinder heat transfer starts at the high gas temperatures, is moved to the cylinder wall at their intermediate temperature (for these calculations, that temperature was 450 K), and finally is moved to the cooling system. The original energy value is the same throughout this process, but obviously the quality of the energy has degraded. The exergy values quantify this degradation. The energy associated with the heat transfer was 0.212 kJ. Of this, 0.176 kJ was exergy. Once this exergy was moved to the cylinder walls at 450 K, the exergy had degraded to 0.072 kJ. And of course, once the energy reaches the cooling system, the exergy is near zero (near ambient conditions). The cylinder heat transfer process is a good example of a highly irreversible process that eventually degrades all the value (exergy) of the energy.

9.6   Summary and Conclusions

This chapter described the use of a comprehensive thermodynamic cycle simulation to explore the implications of the second law of thermodynamics on engine operation and design. Exergy was introduced as the thermodynamic property that quantifies the potential of energy to provide work. Results were presented as functions of engine parameters. Most of the results illustrated the distribution of the fuel exergy among work, heat transfer exergy, exhaust exergy, and destroyed exergy.

The findings of this work with respect to the exergy terms include the following:

- The exergy represented by the net indicated work (as a percentage of the fuel exergy) ranged between 24.5% and 34.6%, and was highest for the highest speeds and highest loads. These numbers differ slightly from the indicated efficiency numbers since the fuel exergy is slightly higher than the lower heating value.
- The exergy displaced to the cylinder wall via heat transfer (as a percentage of the fuel exergy) ranged between 15.9% and 31.5%. This fraction was lowest for the highest speeds and highest loads for the same reasons described above.
- The net exergy expelled with the exhaust gases (as a percentage of the fuel exergy) ranged between 21.0% and 28.1%. This fraction was lowest for the lowest speeds and loads.
- The exergy destroyed by the combustion process (as a percentage of the fuel exergy) ranged between 20.3% and 21.4%. This fraction did not vary much for the conditions of this study. This fraction was lowest for the highest speeds and loads, since these conditions had the highest gas temperatures (which preserved the exergy).
- The exergy destroyed by the mixing process of the fresh charge with the existing cylinder gases (as a percentage of the fuel exergy) ranged between 0.9% and 2.3%. This fraction was much smaller than the other terms examined, and did not vary much for the conditions of this study. This fraction was lowest for the highest loads, and fairly insensitive to engine speed.
- Although outside of the defined thermodynamic system, additional destruction of exergy was described as the energy came in equilibrium with the cylinder walls. For the base case, 59% of the exergy associated with the cylinder heat transfer was destroyed due to this temperature decrease.

References

1. Wark, K. and Richards, D. E. (1999). *Thermodynamics*, sixth edition, McGraw-Hill Company, New York.
2. Atkins, P. W. (1984). *The Second Law*, Scientific American Books, New York.
3. Moran, M. J. (1989). *Availability Analysis – A Guide to Efficient Energy Use* (Corrected Edition), ASME Press, New York.
4. Flynn, P. F., Hoag, K. L., Kamel, M. M., and Primus, R. J. (1984). A new perspective on diesel engine evaluation based on second law analysis, Society of Automotive Engineers, SAE Paper no. 840032.
5. Heywood, J. B. (1988). *Internal Combustion Engine Fundamentals*, McGraw-Hill Book Company, New York.
6. Caton, J. A. (2001). A review of investigations using the second-law of thermodynamics to study internal combustion engines, *Transactions of the Society of Automotive Engineers – Journal of Engines*, **109**, 1242–1266, Paper No. 2000–01–1081.
7. Rakopoulos, C. D. and Giakaoumis, E. G. (2006) Second law analyses applied to internal combustion engines operation, *Progress in Energy and Combustion Sciences*, **32**, 2–47.
8. Caton, J. A. (1999). Results from the second-law of thermodynamics for a spark-ignition engine using a thermodynamic engine cycle simulation, in In-Cylinder Flows and Combustion Processes, Volume 3, Proceedings of the 1999 Fall Technical Conference, ed. S. R. Bell, the ASME Internal Combustion Engine Division, American Society of Mechanical Engineers, ICE–Vol. 33–3, Paper No. 99–ICE–239, 35–49, University of Michigan, Ann Arbor, MI, 16–20 October.
9. Caton, J. A. (2000). On the destruction of availability (exergy) due to combustion processes – with specific application to internal-combustion engines, *Energy, the International Journal*, **25** (11), 1097–1017.
10. Caton, J. A. (2003). A cycle simulation including the second law of thermodynamics for a spark-ignition engine: implications of the use of multiple-zones for combustion, *Transactions of the Society of Automotive Engineers – Journal of Engines*, Paper No. 2002-01-0007, Vol. **111-3**, 281–299, September.
11. Caton, J. A. (2010). The destruction of exergy during the combustion process for a spark-ignition engine, in proceedings of the ASME Internal Combustion Engine Division 2010 Fall Technical Conference, paper no. ICES2010–35036, San Antonio, TX, 12–14 September.
12. Caton, J. A. (2015). Correlations of exergy destruction during combustion for internal combustion engines, *International Journal of Exergy*, **16** (2), 183–213.

10

Other Engine Combustion Processes

10.1 Introduction

For completeness, this brief chapter will comment on other engine combustion processes and describe the connection with engine cycle simulations. This book has been largely based on engines using spark-ignition combustion.

The most common alternative to spark-ignition combustion is compression ignition (diesel) combustion. In addition to spark ignition and diesel engine combustion processes, many alternative combustion processes have been and are being examined for engine applications. The primary motivations to examine these alternative combustion processes are the need to meet stringent exhaust emission regulations and to increase the fuel efficiency of engines. Most of these alternative combustion processes attempt to use the best aspects of the spark ignition and diesel combustion processes while minimizing the weaknesses of each of these processes.

First, this chapter will provide a brief discussion of the diesel engine combustion process, and then highlight the best features from spark ignition (SI) and compression ignition (CI) engines. Next, this chapter will include descriptions of a few of the various alternative engine combustion processes that are possible. Finally, this chapter will end with a brief discussion of the application of thermodynamic cycle simulations to study engines with these alternative combustion processes.

10.2 Diesel Engine Combustion

The combustion process of a diesel engine is significantly different than the combustion process of a spark ignition engine. The classic diesel engine combustion process is based on the injection of liquid fuel into the combustion chamber slightly before the top dead center (TDC), the vaporization of the fuel, the mixing of the fuel vapor with the cylinder air, and the autoignition (no forced ignition) of the fuel vapor–air mixture. For the above to happen, the fuel needs to have appropriate qualities of vaporization and autoignition. In fact, much of

An Introduction to Thermodynamic Cycle Simulations for Internal Combustion Engines, First Edition.
Jerald A. Caton © 2016 John Wiley & Sons, Ltd. Published 2016 by John Wiley & Sons, Ltd.

the optimization of the diesel engine is dependent on the diesel fuel quality and particularly the cetane number[1] specification.

Because of the highly stratified diesel combustion process, the diesel engine must operate with excess air—typically, with overall equivalence ratios less than about 0.8 (or, equivalently, with an overall lambda (λ) of more than about 1.25). In addition, due to this stratification, the conventional diesel engine may be inclined to have incomplete combustion. This incomplete combustion often results in soot or particulate matter in the exhaust gases. In spite of the over-all lean operation, combustion often occurs in specific locations at or near stoichiometric for the conventional engine. This results in high levels of nitric oxides.

On the other hand, diesel engines are not subject to "end gas knock" because the "end gases" consist mostly of air. This lack of a knock constraint allows the diesel engine to operate at higher compression ratios than typical spark ignition engines. In fact, often diesel engines must be designed with a compression ratio greater than the optimum for efficiency to insure reliable ignition.

Another consideration with diesel engines is that, since the engine operates with a lean mixture, these engines possess a lower power density than a stoichiometric spark ignition engine. To compensate, diesel engines are often designed with turbochargers or superchargers. In addition, load control for the diesel engine is by fuel metering rather than throttling.

The advantages of the diesel engine relative to the spark ignition engine include the following: the diesel engine operates without a throttle (low to zero pumping losses), operates lean (lower temperatures and heat losses), and is designed with higher compression ratios (potential for higher thermal efficiencies). The disadvantages of the conventional diesel engine include higher initial cost, lower power density (for naturally aspirated engines), and higher emissions.

As the above indicates, the diesel engine is much different than the spark ignition engine. Even so, many of the thermodynamic considerations described for the spark-ignition engine are applicable to the diesel engine. These will be summarized at the end of this chapter. As may be clear, the two engines both have positive and negative features, and the next subsection summarizes this observation.

10.3   Best Features from SI and CI Engines

From the descriptions of the two main engine combustion processes, a number of features can be identified which have the potential for higher thermal efficiencies, lower emissions, and other favorable attributes. Some of these features are conflicting and the optimum set of such features is not unique. The following is a list of the engine features that are suggested from the two combustion processes which have the potential to improve the internal combustion (IC) engine:

- Operate with no intake throttle for load control. (This suggests the use of fuel metering for load control).
- Operate as much as possible with low combustion temperatures.

---

[1] The cetane number represents the ignition quality of the fuel and is determined from a comparison of ignition time delays. The time delay used is defined as the time between the start of the injection process and the time when the fuel ignites using a specially designed test procedure. The results for actual fuels are compared to reference fuels, and a cetane number is assigned to the actual fuels.

- Operate with a well-mixed charge to avoid rich combustion regions and subsequent soot.
- Operate with a high enough compression ratio to achieve high efficiencies. (This implies that the opportunity for spark knock is eliminated or at least minimized).
- For full load, operate near stoichiometric for highest specific power.
- For part load, operate lean for highest efficiency.
- Design the combustion process such that the required fuel specifications are modest or eliminated.
- Design with lower pressure fuel injection systems.

Throughout the history of the reciprocating, IC engine, engineers have sought to blend the best qualities of the SI and CI engines (or to minimize the worst qualities of each engine). This has led to the development of a wide variety of proposed schemes to accomplish combustion in these engines. This variety will be illustrated in the following discussion of the various approaches for different combustion processes.

10.4   Other Combustion Processes

10.4.1   Stratified Charge Combustion

One of the first approaches (from at least the 1920s) for merging the best features of the two combustion processes was to use a stratified charge with spark ignition. For example, Ricardo [1] was issued a patent published in 1926 which incorporated some of these ideas. One concept related to this approach is to divide the charge into (i) a relatively rich mixture for easy ignition, and (ii) an overall lean mixture (which is ignited from the rich portion). The overall lean condition provides for high thermal efficiency and minimizes nitric oxide formation. The rich mixture is a small part of the overall mixture to minimize soot formation or incomplete combustion. These engine designs have used divided and single combustion chambers. Various versions of stratified charge engines with spark ignition have been developed, documented and even commercialized [2–5].

10.4.2   Low Temperature Combustion

Since at least the mid-1970s, a number of engine manufacturers and research institutions have been working on developing new engine combustion processes which appear to have possibilities for lower nitric oxide and particulate emissions while maintaining or increasing thermal efficiencies. These new combustion processes are often based on some degree of fuel and air premixing, and on the use of high levels of dilution. Due to the high levels of dilution, a common aspect of these combustion modes is the relatively low temperatures of combustion—hence, many of these processes may be grouped into the category of *low temperature combustion*. One example of obtaining these new combustion processes is by the use of high levels of exhaust gas recirculation (either internal or external). In the past, such high levels of exhaust gas recirculation (EGR) were not investigated or were considered impractical.

Some of the first works on these approaches were reported for a two-stroke cycle engine by Onishi et al. [6] in 1979. They used retained exhaust gases (internal EGR) to support a special combustion mode that they termed "active thermo-atmospheric combustion" (ATAC).

In a related fashion, Noguchi et al. [7] described their work on using retained exhaust gases to promote gasoline combustion for a two-stroke cycle engine by the use of reactive intermediate species. Najt and Foster [8] in 1983 reported on similar work applied to a four-stroke cycle, single cylinder engine. These works [6–8] were motivated by the idea that retaining high temperature exhaust gases would promote ignition of lean, dilute mixtures. Beginning in the 1980s, many of these combustion processes were called homogeneous charge compression ignition (HCCI) combustion.

The common aspects of HCCI combustion engines include having a (more or less) homogeneous air-fuel mixture (typically lean) that would autoignite throughout the mixture due to the high temperatures at the end of compression. One of the major advantages of these approaches is that the high combustion temperatures of conventional SI and CI engine combustion are avoided (and the associated nitric oxides are minimized). Some of the difficulties with this concept include control of the timing of ignition, avoiding rapid combustion, and avoiding misfires due to the dilution. To remedy these disadvantages, dozens of related combustion processes have been proposed and developed over the last several decades. As examples, a couple of these combustion processes are described next.

Spark Assisted Compression Ignition (SACI) has been and continues to be investigated as a form of HCCI which has theoretically more control about the start of combustion [9,10]. Results have demonstrated that the SACI combustion mode is difficult to implement and may be subject to cyclic variations. Partially Premixed Combustion (PPC) is a blend of a conventional compression ignition combustion process and an HCCI combustion process [11–14]. For a PPC implementation, the fuel is injected during the compression stroke and some partial mixing of fuel and air occurs prior to autoignition. With many of these combustion processes, various fuels have been used with a spectrum of results.

A final example of an alternative combustion mode involves the use of multiple fuels. One version of this concept is based on a main mixture of gasoline and uses a small amount of diesel fuel as a pilot for ignition. The main mixture is a highly diluted charge, usually with a high level of EGR, which will not ignite on its own. The pilot injection timing and the ratio of the two fuels provides a method to control the combustion phasing. This has been called Reactivity-Controlled Compression Ignition (RCCI) combustion [15]. More on the impact of fuels on low temperature combustion engine concepts is presented by Lu et al. [16].

The above brief descriptions of low temperature combustion processes for internal combustion engines possess some common elements which are shared among many of the engine concepts in this chapter. Most of these concepts utilize high levels of EGR and lean mixtures. In addition, these engines are often designed with high compression ratios since spark knock is not a constraint. Finally, the combustion processes may occur relatively fast. These features have demonstrated low emissions and high efficiencies. A case study is presented in this book on high efficiency engines which is based on these features that are related to low temperature combustion.

10.5   Challenges of Alternative Combustion Processes

For most of the alternative combustion processes, no completely successful implementation has been documented. Most of these approaches provide some significant advantages, but most still involve some weaknesses. The following are some brief comments on these weaknesses of the low temperature combustion processes.

The major challenge by far is the control and timing of the autoignition. By its very definition ("auto"), it is difficult to precisely control and time the autoignition for a range of operating conditions. In addition to the timing of the autoignition event, some control of the rate of heat release is needed for a range of operating conditions.

Current successful technology for these low temperature combustion processes is mostly limited to moderate to low loads. At high loads (with near stoichiometric operation), the event is much like spark knock with cylinder pressure oscillations, and the emissions increase. To counter the difficulties at high load operation, some designs are based on switching to more conventional (either SI or CI) engine combustion for the high load conditions. To accomplish this transition in a smooth and transparent manner is another challenge.

Preparing the homogeneous mixture at low ambient temperatures with low volatility fuels (such as diesel fuel) is a challenge. Some degree of mixture non-uniformity may be tolerated by this approach, and indeed, may be advantageous. Finally, depending on the technology adopted, the low temperature combustion engine could be more expensive and more complicated.

10.6   Applications of the Simulations for Other Combustion Processes

As described above, the types of combustion processes for IC engines can span a wide range. In general, all IC engines (regardless of the combustion process) may be studied with the thermodynamic cycle simulations presented in this book if the combustion process is known (e.g., mass fraction burned as a function of crank angle) or can be modeled with a combustion sub-model. Often the difficulty is that the alternative combustion processes are not understood fully. In some cases, only limited data are available and extrapolation to other operating conditions is problematic.

Another limitation is that some of the basic simulations will not be able to provide the level of detail necessary to satisfy all requirements. For example, for diesel and related combustion processes, the one-zone average gas temperature is not representative of the local high temperatures due to the stratification of the combustion process. This deficiency would mean that nitric oxide emissions could not be correctly estimated. The case study on nitric oxide emissions provides examples of the sensitivity of nitric oxide calculations on the gas temperature.

The literature has many examples of thermodynamic cycle simulations being adapted for cases with alternative combustion processes. For example, Eichmeier et al. [17] describe the development and use of a zero-dimensional phenomenological cycle simulation to study an RCCI combustion engine. Yao et al. [18] provides a good summary of many examples of thermodynamic cycle simulations (including CFD simulations) used to investigate a range of engines using many of these alternative combustion processes.

As a further example of the applicability of the cycle simulations, an engine using a low temperature combustion process similar to that described above has been approximated with the current engine cycle simulation with surprisingly good results [19–21]. Much of the case study on high efficiency engines is based on this success.

Finally, many of the thermodynamic results obtained and presented in this book for SI engines are generally valid for engines using these other combustion processes. For example, in general, the trend related to cylinder heat transfer applies to all IC engines. This means that the heat transfer is most significant at low speeds and high loads. Another example is that the conversion of thermal energy to work is difficult and is a significant function of the specific heats. For all engines, for lower specific heats, a greater fraction of the thermal energy can be converted to work. Lastly, most of the results from the second law of thermodynamics are valid for all IC engines.

10.7 Summary

This chapter described alternative combustion processes relative to spark ignition combustion. Some detail was provided on the compression ignition (diesel) combustion process. Certain features were identified from the SI and CI engines that were advantageous for low emissions and high thermal efficiency. Combining these advantageous features into one engine design has been a goal since at least the 1920s.

Dozens of engine combustion concepts have evolved over the last several decades which are aimed at providing low emissions and high efficiency with highly dilute mixtures with novel combustion processes for fast burns. Most of these approaches may be characterized as employing low temperature combustion. Although these concepts are promising, as of this date (2015) no wide-spread commercial application is known [22].

The main purpose of this chapter was to demonstrate that the thermodynamic cycle simulations are appropriate for studying these types of engines regardless of the combustion mode. As described, once a burn rate is known (or a sub-model exists for the combustion process), the engine cycle simulation may be used for these novel combustion processes. For those zero-dimensional thermodynamic cycle simulations with no spatial information, however, local temperatures may not be represented correctly and items such as nitric oxide emissions will be incorrect. On the other hand, cylinder pressures and overall engine performance may often be valid estimates. A case study presented in this book on high efficiency engines demonstrates this application.

References

1. Ricardo, H. R. (1926). Internal-combustion engine, patent no. US 1594755 A, 03 August.
2. Alperstein, M., Schafer, G. H., and Villforth, F. J. (1974). Texaco's stratified charge engine – multifuel, efficient, clean, and practical, Society of Automotive Engineers, SAE paper no. 740561.
3. Urlaub, A. G. and Chmela, F. G. (1974). High-speed, multifuel engine: L9204 FMV, Society of Automotive Engineers, SAE paper no. 740122.
4. Date, T. and Yagi, S. (1974). Research and development of the Honda CVCC engine, Society of Automotive Engineers, SAE paper no. 740605.
5. Scussel, A. J., Simco, A., and Wade, W. (1978). The Ford PROCO update, Society of Automotive Engineers, SAE paper no. 780699.
6. Onishi S., Jo, S. H., Shoda, K., Jo, P. D., and Kato, S. (1979). Active thermo-atmosphere combustion (ATAC) – a new combustion process for IC engines, Society of Automotive Engineers, SAE paper no. 790501.
7. Noguchi, M., Tanaka, Y., and Takeuchi, Y. (1979). A study on gasoline engine combustion by observation of intermediate reactive products during combustion, Society of Automotive Engineers, SAE paper no. 790840.
8. Najt, P. M. and Foster, D. E. (1983). Compression-ignited homogenous charge combustion, Society of Automotive Engineers, SAE paper no. 830264.
9. Willand, J., Nieberding, R., Vent, G., and Enderle, C. (1998). The knocking syndrome – its cure and its potential, Society of Automotive Engineers, SAE paper no. 982483.
10. Lavy, J., Dabadie, J., Angelberger, C., Duret, P., Willand, J., Juretzka, A., Schäflein, J., Ma, T., Lendresse, Y., Satre, A., Schulz, C., Kramer, H., Zhao, H., and Damiano, L. (2000). Innovative ultra-low NO$_x$ controlled

auto-ignition combustion process for gasoline engines: the 4–SPACE project, Society of Automotive Engineers, SAE paper no. 2000–01–1837.

11. Kimura, S., Aoki, O., Ogawa, H., Muranaka, S., and Enomoto, Y.(1999). New combustion concept for ultra-clean and high-efficiency small di diesel engines, Society of Automotive Engineers, SAE paper no. 1999–01–3681.

12. Akihama, K., Takatori, Y., Inagaki, K., Sasaki, S., and Dean, A. M. (2001). Mechanism of the smokeless rich diesel combustion by reducing temperature, Society of Automotive Engineers, SAE paper no. 2001–01–0655.

13. Manente, V., Johansson, B., Tunestal, P., and Cannella, W. (2010). Effects of different type of gasoline fuels on heavy duty partially premixed combustion, Society of Automotive Engineers, *SAE International Journal of Engines*, **2**(2), 71–88.

14. Manente, V., Zander, C., Johansson, B., Tunestal, P., and Cannella, W. (2010). An advanced internal combustion engine concept for low emissions and high efficiency from idle to max load using gasoline partially premixed combustion, Society of Automotive Engineers, SAE paper no. 2010–01–2198.

15. Kokjohn, S., Hanson, R., Splitter, D., and Reitz, R. (2010). Experiments and modeling of dual-fuel HCCI and PCCI combustion using in-cylinder blending, Society of Automotive Engineers, *SAE International Journal of Engines*, **2**(2), 24–39.

16. Lu, X., Han, D., and Huang, Z. (2011). Fuel design and management for the control of advanced compression-ignition combustion processes, *Progress in Energy and Combustion Science*, **37**, 741–83.

17. Eichmeier, J. U., Reitz, R., and Rutland, C. (2014). A zero-dimensional phenomenological model for RCCI combustion using reaction kinetics, Society of Automotive Engineers, SAE paper no. 2014–01–1074.

18. Yao, M., Zheng, Z., and Liu, H. (2009). Progress and recent trends in homogenous charge compression ignition (HCCI) engines, *Progress in Energy and Combustion Science*, **35**, 398–437.

19. Caton, J. A. (2010). An assessment of the thermodynamics associated with high-efficiency engines, proceedings of the ASME 2010 Internal Combustion Engine Division Fall Technical Conference, paper no. ICEF2010–35037, San Antonio, TX, 12–15 September.

20. Caton, J. A. (2012). The thermodynamic characteristics of high efficiency, internal-combustion engines, *Energy Conversion and Management*, **58**, 84–93.

21. Caton, J. A. (2013). Thermodynamic considerations for advanced, high efficiency IC engines, *Journal of Engineering for Gas Turbines and Power*, **136** (10), 101512–101512–6, 2014; also in proceedings of the 2013 Fall Technical Conference of the ASME Internal Combustion Engine Division, paper no. ICEF2013–19040, Dearborn, MI, 13–16 October 2013.

22. Zhao, H. (ed.) (2007). *HCCI and CAI Engines for the Automotive Industry*, Woodhead Publishing Limited, CRC Press, Cambridge, England.

11

Case Studies: Introduction

The second part of this book is devoted to a number of case studies using the engine cycle simulation described in the first part of the book. This brief introduction will describe the various case studies, the common elements associated with the studies, and the general methodology of these studies.

11.1   Case Studies

To illustrate the use of engine cycle simulations, a number of case studies are described in the following pages. One of the goals of these studies is to demonstrate the utility of the cycle simulations by examining a variety of engine topics from a thermodynamic perspective. In addition, the results presented form a basic foundation regarding engine operation and performance as functions of a wide range of independent design and operating variables.

The following, then, is a list and brief description of the case studies:

1. *Combustion (Chapter 12):* An important aspect of IC engines is the heat release schedule and the phasing of the heat release associated with the fuel reaction. The thermodynamic engine simulation provides fundamental understanding about the sensitivity of the engine performance relative to the heat release schedule and timing.
2. *Cylinder heat transfer (Chapter 13):* The cylinder heat transfer is examined for both conventional and low heat rejection engine configurations. The sensitivity of engine performance to the cylinder heat transfer is described.
3. *Fuels (Chapter 14):* Most of the work reported in this book is based on isooctane as the fuel. This case study examines seven additional pure fuels—methane, propane, hexane, ethanol, methanol, hydrogen, and carbon monoxide.
4. *Oxygen-enriched inlet air (Chapter 15):* The simulation was used to examine the use of oxygen-enriched air. Detailed cylinder conditions and overall engine performance parameters are determined for inlet oxygen concentrations of 21–40%.
5. *Over-expanded engines (Chapter 16):* This case study examines a different engine architecture. For this study, the engine possesses a longer expansion stroke than compression

An Introduction to Thermodynamic Cycle Simulations for Internal Combustion Engines, First Edition.
Jerald A. Caton © 2016 John Wiley & Sons, Ltd. Published 2016 by John Wiley & Sons, Ltd.

stroke (Atkinson cycle). Engine performance is determined for both part load and wide-open throttle as functions of a range of expansion and compression ratios.

6. *Nitric oxides (Chapter 17):* The emissions of nitric oxides are examined for a range of engine operating and design parameters. The sensitivity of the results to the chemical kinetics of nitric oxide formation and destruction is obtained. Detailed (instantaneous) and overall results are provided.

7. *High efficiency engines (Chapter 18):* Engines with high compression ratios, lean mixtures, high levels of EGR, short burn durations and other such features are explored. Both the mechanical and thermodynamic reasons for the improvements of thermal efficiency are obtained.

11.2   Common Elements of the Case Studies

The engine and some of the engine operating conditions are common among the case studies. The engine is the same as used in the earlier portions of this book: a 5.7 liter, V-8 configuration. For convenience, Table 11.1 lists the engine specifications which are unchanged in most of the following studies. In addition, Table 11.2 lists some of the operating and fuel input parameters, and their typical values. For example, the values for the Wiebe constants are as recommended by Heywood [1]: $m = 2.0$ and $a = 5.0$. Obviously, in the case study concerning the heat release schedule, the values of these parameters are varied. Similar comments could be stated for all of the engine, fuel and operating parameters.

A majority of these case studies are based on a combustion (burn) duration of 60°CA, a compression ratio of 8:1, an equivalence ratio of 1.0, 0% EGR, and using isooctane. For the different case studies, one item that is often different is the cylinder heat transfer correlation

Table 11.1   Engine specifications

Item	Value
Number of cylinders	8
Bore (mm)	101.6
Stroke (mm)	88.4
Crank rad/Con rod ratio	0.305
Inlet valves:	
Diameter (mm)	50.8
Max lift (mm)	10.0
Opens (°CA aTDC)	357
Closes (°CA aTDC)	−136
Exhaust valves:	
Diameter (mm)	39.6
Max lift (mm)	10.0
Opens (°CA aTDC)	116
Closes (°CA aTDC)	371
Valve overlap (degrees)	14°

Table 11.2 Typical engine and fuel input parameters

Item	Value used	How obtained
Displaced volume (dm^3)	5.733	Computed
Compression ratio	8.0	Input
AF$_{stoich}$	15.07	Computed
Equivalence ratio	1.0	Input
Inlet (air-fuel) temperature	319.3 K	Input
Start of combustion (°bTDC)	Typical at MBT	Determined for MBT
Combustion duration (°CA)	60	Input
Cylinder wall temp (K)	450	Input
Fuel LHV (kJ/kg)	44,400	For isooctane [1]
Fuel exergy (kJ/kg)	45,670	For isooctane [1]
Heat transfer multiplier	1.00	Input
Heat transfer correlations (typical)	Woschni [2] Hohenberg [3]	Input
Wiebe constant, m	2.0	Input
Wiebe constant, a	5.0	Input

and the value of the *htm*. The two heat transfer correlations used in most of these case studies are from Woschni [2] and from Hohenberg [3]. Other differences between the various case studies include the base *bmep* and engine speed.

11.3   General Methodology of the Case Studies

In general, the following case studies are based on a similar methodology and subject to a set of similar constraints. As described throughout this book, these engine cycle simulations are based on successful combustion.[1] This is advantageous for clearly determining the role of the thermodynamics. The disadvantage is that items such as combustion stability, cycle-to-cycle variations, knock, preignition, or other combustion issues are not considered. That is because the simulations do not contain any fundamental information on the combustion processes such as flammability limits, flame propagation rates, or chemical kinetics. In general, combustion is assumed successful and proceeds according to the Wiebe function. This approach, therefore, provides results reflective of the thermodynamics, but (due to unsuccessful combustion or knock) these results may not always be obtained in practice.

One main aspect of many of the following evaluations is that all parameters remain the same except for the one that is varied. This provides a consistent and unambiguous exami-nation of the various effects. As an example, for sufficiently lean operation the burn duration

---

[1] One exception to this statement that the combustion process is always successful relates to the use of EGR. For high levels of EGR, an "*ad hoc*" model is used to capture the deteriorating combustion process as the charge becomes highly dilute. This was illustrated in Chapter 8.

might be expected to be longer than for stoichiometric operation. To ascertain the influence of only the equivalence ratio, however, the burn duration may be kept constant as equivalence ratio is varied.

In most of the following cases, the combustion timing has been adjusted to provide the maximum *bmep* (and brake thermal efficiency). This combustion timing that provides the maximum *bmep* timing is referred to as MBT timing. Consistent with the above, combustion timing was adjusted for MBT performance with no concern for knock.

This general methodology is specific to the work described in this book. Other investigators using other simulations have elected to proceed in different ways. For example (as discussed in Chapter 3), quasi-dimensional thermodynamic engine cycle simulations have empirical relations to describe the combustion process. In theory, these types of simulations have the potential to describe some of the combustion issues described above. For these types of simulations to have this capability, they need to have a significant amount of empirical input. In some cases, the fundamental thermodynamics may be obscured.

In summary, the following case studies are examples of how the engine cycle simulation can be used to explore various topics regarding internal combustion engines. These examples span a range of topics including investigating novel engine designs (over-expanded engines), alternative fuels and oxidizers, combustion, heat transfer, exhaust gas emissions, and high efficiency engines. Although these topics span a wide range, a vast number of other topics that could be investigated in similar fashion can be envisioned.

References

1. Heywood, J. B. (1988) *Internal Combustion Engine Fundamentals*, McGraw-Hill Book Company, New York.
2. Woschni, G. (1968) A universally applicable equation for the instantaneous heat transfer coefficient in the internal combustion engine, *SAE Transactions*, SAE paper no. 670931, **76**, 3065–3083.
3. Hohenberg, G. F. (1979) Advanced approaches for heat transfer calculations, Society of Automotive Engineers, SAE Paper No. 790825.

12

Combustion: Heat Release and Phasing

12.1 Introduction

This case study deals with the combustion event. The combustion process is complex, involves many interrelated sub-processes (e.g., chemical kinetics, turbulence, and heat transfer), and is probably the most important process for the engine. For purposes of this case study, only two aspects will be evaluated: the importance of the heat release schedule and the phasing (timing) of the combustion process.

The material in this chapter (as with most of this book) is based mainly on the thermodynamics of the engine processes. As illustrated below, a wealth of information is obtained by this approach. On the other hand, aspects of the combustion process that involve details such as that associated with the combustion chemistry and turbulence are not explicitly included. These details are important, but are beyond the scope of the current work.

12.2 Engine and Operating Conditions

The engine selected for this study is the automotive, 5.7 liter, V–8 configuration outlined above with a compression ratio of 8:1 using isooctane. Each of the two parts of this chapter is based on slightly different operating conditions, and these are described below in the respective sub-section. In particular, items that are different include the heat transfer correlations, and the inlet and exhaust pressures.

This chapter is divided into two parts: (i) heat release schedule and (ii) combustion phasing.

12.3 Part I: Heat Release Schedule

To describe the combustion process in internal combustion engines, the time-dependent release of thermal energy from the chemical reaction has often been used [1]. This has been referred to as "heat release" or "apparent heat release." The "apparent" is often used since the determination of the heat release is approximate due to a number of simplifications. The heat release is a function of the fuel burning rate or the fuel mass fraction burned, and therefore, the heat release (as will be shown below) is proportional to the time derivative of the mass fraction burned.

A related topic is the conversion of measured engine cylinder pressure data to the heat release schedule or to the fuel mass fraction burned as functions of time or crank angle. These experimentally determined heat release schedules are useful for detailed engine diagnostics concerning the combustion process. A number of techniques have been proposed for this purpose (e.g., [2–4]). This topic is beyond the scope of this chapter.

For the first part of this chapter (heat release schedule), Table 12.1 lists some of the operating conditions. These operating conditions include an inlet pressure of 50.0 kPa, and an exhaust pressure determined from an algorithm presented by Sandoval and Heywood [5]. The heat transfer correlation used for this set of computations was from Woschni [6] with an *htm* of 1.0.

The fuel mass fraction burned is often given as a simple analytical expression which may be only a function of the fraction of the duration of the combustion process. The mass fraction burned, x_b, is defined as the ratio of the fuel mass burned and the total fuel mass.

$$x_b = \frac{m_{f, \text{burned}}}{m_{f, \text{total}}} \tag{12.1}$$

As is obvious, the mass fraction burned begins at zero and should reach 1.0 for complete consumption of the fuel. For this work, the fuel mass fraction burned is determined by the Wiebe function [7]

$$x_b = 1 - \exp\{-ay^{m+1}\} \tag{12.2}$$

Table 12.1 Summary of results Part I: Heat release schedules (1400 rpm, p_{in} = 50 kPa, ϕ = 1.0, CR = 8, p_{exh} from Reference 5) (based on Woschni [6] heat transfer correlation)

"a"	"m"	$(x_b)_{final}$	θ_o MBT	η_b (%)	*bmep* (kPa)
3	2	0.9502	−26.5	24.63	307.0
4	2	0.9817	−24.0	25.60	319.3
5	2	0.9933	−22.0	26.02	324.6
6	2	0.9975	−20.5	26.32	327.1
7	2	0.9991	−19.0	26.34	328.5
5	0.1	0.9933	−3.5	26.06	324.9
5	0.5	0.9933	−8.5	25.96	323.8
5	1	0.9933	−14.0	25.92	323.3
5	2	0.9933	−22.0	26.02	324.6
5	3	0.9933	−27.5	26.17	326.5

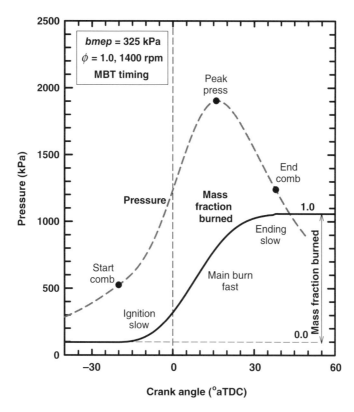

Figure 12.1 Cylinder pressure and mass fraction burned as functions of crank angle

where "*a*" and "*m*" are parameters that are selected to provide agreement with experimental information. The values that provide good agreement with the data depend on complex functions of the turbulence, chemistry, fuel, equivalence ratio, chamber geometry, and a number of other features. For typical automotive spark-ignition engines, Heywood [3] has recommended values of *m* = 2.0 and *a* = 5.0. The sensitivity of the results to these parameters and other information are provided in this section.

Figure 12.1 shows the cylinder pressure and the mass fraction burned as functions of crank angle for the base case conditions. The start and end of combustion are noted on the figure. As this demonstrates, the Wiebe function for the mass fraction burned provides a good representation of the combustion process. In the beginning, the mass fraction burned increases slowly which is representative of the ignition and start of combustion processes. During the main part of the combustion event, the mass fraction burned increases more rapidly. Finally, at the end, the mass fraction burned increases more slowly which represents a slower burning and the extinguishment of the combustion process.

Figure 12.2 shows the mass fraction burned as a function of the fraction of duration of the combustion period. Two common definitions of the burn duration are indicated on this figure. The 0–100% period is from 0 to 1.0 mass fraction burned. This is the burn duration used in this book. A second duration is defined from 0.1 to 0.9 mass fraction burned. This second burn

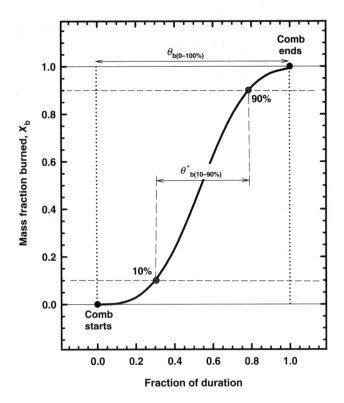

Figure 12.2 Mass fraction burned as functions of fraction of the burn duration for $a = 5$ and $m = 2$ – with burn durations of 0–100% and 10–90% denoted

duration is often useful in experiments since the exact start and end of combustion are difficult to detect, and the 10% and 90% times are easier to identify. These two burn durations are related, and this relation depends on the mass fraction burned curve. For the Wiebe function and constants listed above, the two burn durations are related by,

$$\theta^*_{b(10-90\%)} = 0.48\,\theta_{b(0-100\%)} \tag{12.3}$$

That is, the burn duration for 10% to 90% is a factor of 0.48 smaller than the burn duration for 0% to 100%. For example, for the 60°CA burn duration used in this work,

$$\theta^*_{b(10-90\%)} = 28.8°\,\text{CA}$$
$$\theta_{b(0-100\%)} = 60.0°\,\text{CA} \tag{12.4}$$

The mass fraction burned schedules for a range of values of "a" and "m" for the Wiebe function will be examined next. Figures 12.3 and 12.4 show the results for the Wiebe function for a number of values of "a" and "m". For the range of values for "a" and "m" selected, the shapes of the curves are quite different. This variation of the mass fraction curves provides the opportunity to show agreement with a range of combustion processes.

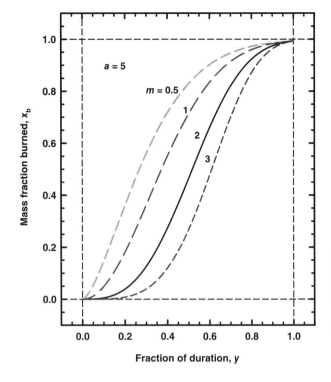

Figure 12.3 Mass fraction burned as functions of fraction of burn duration for several sets of Wiebe parameters "a" and "m". *Source:* Caton 2000. Reproduced with permission from ASME

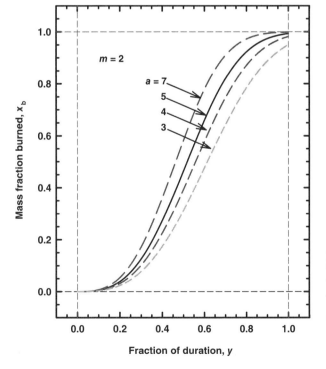

Figure 12.4 Mass fraction burned as functions of fraction of burn duration for several sets of Wiebe parameters "a" and "m". *Source:* Caton 2000. Reproduced with permission from ASME

Figure 12.3 shows the mass fraction burned as a function of the fraction of the combustion duration for different values of the constant "m" for $a = 5.0$. As "m" increases, the initial rate of increase of x_b decreases, and the final rate increases. For "m" values of about 1.0 or less, the shape of the curve is less representative of the expected SI engine combustion process.

Figure 12.4 shows the mass fraction burned as a function of the fraction of the combustion duration for different values of the constant "a" for $m = 2.0$. As "a" increases, the rate of increase of x_b increases and the final rate decreases. In general, these expressions provide a reasonable representation of the desired results (see Figure 12.1). In particular, the base case constants for the Wiebe expression provide a good description of a successful combustion process.

Depending on the constant "a," the Wiebe function does not necessarily provide a value of 1.0 for a "y" value of 1.0 (end condition). In other words, for some values of "a," the final mass fraction does not represent 100% fuel consumption. Figure 12.5 shows the final mass fraction computed from eq. (12.2) as a function of the constant "a." For values of constant "a" greater than about 7, the final burned value is nearly 1.0; but for values less than 7, the final value will be less than 1.0. For example, for the case of "a" equal to 5.0, this results in a final burned mass fraction of 0.9933 or about 0.67% unused fuel.

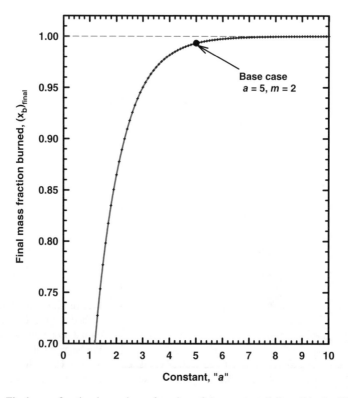

Figure 12.5 Final mass fraction burned as a function of the constant "a" used in the Wiebe function

12.3.1 Results for the Heat Release Rate

As formulated here, the cycle simulation does not need an explicit heat release rate expression. Rather, the simulation uses the prescribed fuel mass fraction burned as a function of crank angle as input. The "heat release" from the simulation is a result of the energy properties (internal energy and enthalpy) which include both the chemical and sensible parts of the energy. Knowledge of the apparent heat release, however, is informative. Once the mass fraction burned is specified as a function of time (crank angle), a "heat release rate" may be obtained from the appropriate time derivative. The apparent heat release rate is related to this derivative as follows:

$$\left(\frac{\delta Q}{dt}\right)_{HR} = \dot{x}_b (LHV) m_f \tag{12.5}$$

where $\left(\dfrac{\delta Q}{dt}\right)_{HR}$ is the apparent heat release rate, LHV is the lower heating value of the fuel on an energy basis, and m_f is the total mass of fuel burned. For cases where the LHV and m_f are constant, examining the derivative of the mass fraction burned is proportional to examining the heat release rate. For the Wiebe function, the derivative is

$$\dot{x}_b = \frac{a(m+1)}{\theta_b} y^m \exp\left\{-ay^{m+1}\right\} \tag{12.6}$$

For this expression, if θ_b has units of radians then the units of $\dot{x}_b$ are

$$\dot{x}_b \equiv \left\{\frac{\text{mass burned / total mass}}{\text{radians}}\right\} \text{ or } \left\{\frac{kg_b / kg_t}{\text{radians}}\right\} \tag{12.7}$$

Results for this expression (eq. 12.6) are provided below in later parts of this section.

Next, the results for the instantaneous heat release rate are presented for one engine operating condition. The heat release rate from the Wiebe function was given above (eqs. 12.5 and 12.6):

$$\left(\frac{\delta Q}{d\theta}\right)_{HR}^{analytic} = (LHV) m_f \frac{a(m+1)}{\theta_b} y^m \exp\left\{-ay^{m+1}\right\} \tag{12.8}$$

Most often, the heat release rate is expressed in terms of crank angle or radians. If the combustion duration has units of "crank angle" then the heat release has units of kJ/CA; if the combustion duration has units of radians then the heat release has units of kJ/rad. This version of the heat release is labeled "analytic" since it is based on the mass fraction burned relation, and does not include the consideration of any chemical dissociation or other such items which are a part of the simulation.

Figure 12.6 shows the gross and net instantaneous heat release rates from the analytical expression (eq. 12.8) as a function of crank angle. The difference between the gross and net heat release is the cylinder heat transfer. Since the heat transfer is defined as positive into the system and since during combustion the heat transfer is actually out of the system, the energy

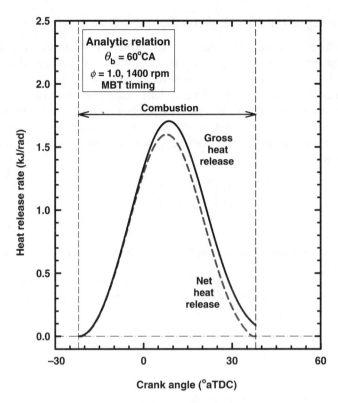

Figure 12.6 Instantaneous gross and net heat release rates from an analytical expression (see text) as functions of crank angle for the base case conditions

equivalent of the heat transfer increases the computed value of the heat release rate. Hence, the heat release rate discussed here is the "gross" heat release (including heat transfer). In other words, the gross heat release supplies the energy for the work, heat transfer, and the change of the sensible internal energy of the cylinder contents. Figure 12.7 shows the accumulated heat release from the analytical expression for both the gross and net versions as a percentage of the total fuel energy.

For comparison, the simulation may provide an approximate quantity that relates to the analytic heat release by considering the first law of thermodynamics. This derivation is for the closed valve portion of the cycle,

$$\left(\frac{dU_{sen}}{d\theta}\right) = \left(\frac{\delta Q}{d\theta}\right)^{sim}_{HR} + \left(\frac{\delta Q}{d\theta}\right)_{ht} - \left(\frac{\delta W}{d\theta}\right) \tag{12.9}$$

where $\left(\dfrac{dU_{sen}}{d\theta}\right)$ is the change of the sensible component of the internal energy. Note that the signs are correct: heat transfer is defined as positive into the system and work is defined as

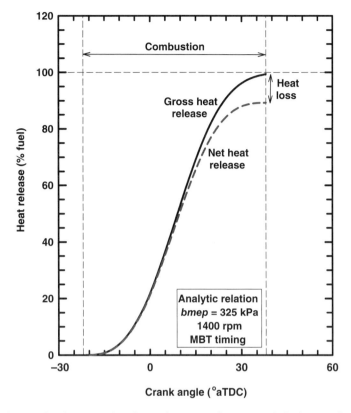

Figure 12.7 Accumulated gross and net heat release rates from an analytical expression as functions of crank angle for the base case conditions

positive out of the system. Solving for the heat release rate and substituting for the boundary work yields

$$\left(\frac{\delta Q}{d\theta}\right)_{HR}^{sim} = \left(\frac{dU_{sen}}{d\theta}\right) + \left(p\frac{dV}{d\theta}\right) - \left(\frac{\delta Q}{d\theta}\right)_{ht} \tag{12.10}$$

Recall that the change in the sensible component of the internal energy may be given by removing the chemical energy component

$$\left(\frac{dU_{sen}}{d\theta}\right) = \left(\frac{dU_{tot}}{d\theta}\right) - \left(\frac{dU_{chem}}{d\theta}\right) \tag{12.11}$$

This sensible internal energy may be estimated by using the average mixture specific heat and the rate of change of temperature (per radian). Therefore, eq. (12.10) is equivalent to

$$\left(\frac{\delta Q}{d\theta}\right)_{HR}^{sim} = m_{mix}C_{v,mix}\frac{dT}{d\theta} + p\frac{dV}{d\theta} - \frac{\delta Q_{ht}}{d\theta} \tag{12.12}$$

The heat release from the simulation, $\left(\dfrac{\delta Q}{d\theta}\right)^{sim}_{HR}$, involves the chemical energy bonded with any dissociated product species (such as carbon monoxide and hydrogen). Further, the mixture specific heats are based on the complete species description including equilibrium species. As shown in Chapter 4, the specific heats will be higher for the complete species description compared to the frozen species approximation. For these reasons, a comparison to the analytic heat release (eq. 12.8) with the simulation is best done with the frozen species assumption.

Figure 12.8 shows the instantaneous gross and net heat release from the simulation using the frozen species assumption. Figure 12.9 shows the accumulated gross and net heat release as a percentage of the fuel energy as functions of crank angle using the frozen species properties. The results from the simulation using the frozen species assumption (Figures 12.8 and 12.9) are essentially the same as from the analytical expression (Figures 12.6 and 12.7).

Again, the simulation calculations do not use a "heat release," but rather the energy balance uses energy properties that include both the sensible and chemical parts. The simulation using the complete products is, of course, more representative of the actual processes. Finally, the energy for the complete cycle is fully accounted for (as described elsewhere in this book).

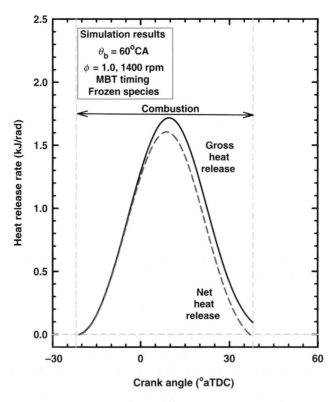

Figure 12.8 Instantaneous gross and net heat release rates from the simulation using frozen species as functions of crank angle for the base case conditions

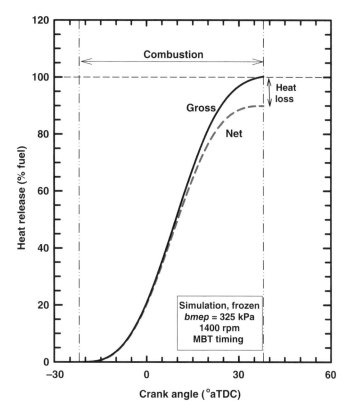

Figure 12.9 Accumulated gross and net heat release rates from the simulation for frozen species as functions of crank angle

Next, results as functions of the mass fraction burn parameters ("*a*" and "*m*") were obtained for a part load condition: 1400 rpm and 50.0 kPa (inlet pressure). Table 12.1 is a list of a few of these results as a function of the values of the parameters "*a*" and "*m*" for MBT timing. These results include the final value of the mass fraction burned, the MBT timing for the start of combustion, the brake thermal efficiency, and the *bmep*.

Figures 12.10 and 12.11 show the derivative (multiplied by θ_b) of the mass fraction burn as a function of the fraction of the combustion duration for various combinations of "*a*" and "*m*." For the range of "*a*" and "*m*" selected, the resulting curves are quite different. The derivative increases from zero, reaches a maximum roughly midway through the combustion process, and then decreases to a value near zero at the end. Figure 12.10 shows this derivative (multiplied by θ_b) as a function of the fraction of the combustion duration for $m = 2$ for four values of "*a*." As "*a*" increases, the maximum heat release increases and this maximum occurs earlier.

Figure 12.11 shows this derivative (multiplied by θ_b) as a function of the fraction of the combustion duration for $a = 5$ for four values of "*m*." The lowest value for the maximum value of this derivative is obtained for $m = 1.0$. For "*m*" values greater or less than 1.0, the maximum value of this derivative increases. For "*m*" values less than 1.0, the maximum occurs earlier, and for "*m*" values greater than 1.0, the maximum occurs later in the combustion process.

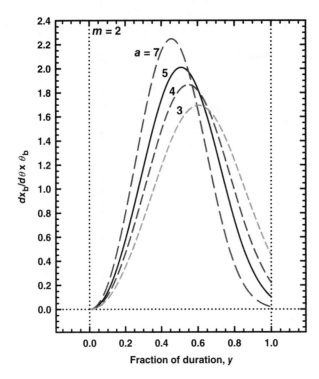

Figure 12.10 Rate of change of the mass fraction multiplied by the combustion duration for the Wiebe law for $m = 2$ for four values of "a". *Source:* Caton 2000. Reproduced with permission from ASME

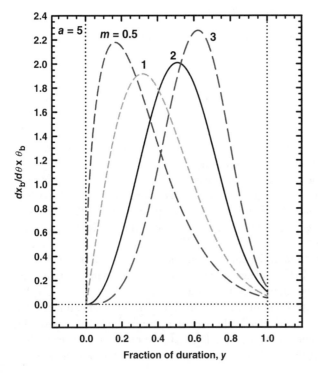

Figure 12.11 Rate of change of the mass fraction multiplied by the combustion duration for the Wiebe law for $a = 5$ for four values of "m". *Source:* Caton 2000. Reproduced with permission from ASME

Figures 12.12 and 12.13 show the brake mean effective pressure (*bmep*) as a function of the start of combustion for different sets of the constants using the Wiebe function for the mass fraction burn for an inlet pressure of 50 kPa. As shown, the *bmep* is maximum for one particular combustion start and is lower for timings advanced or retarded from this timing. As the value of "*a*" is reduced, the maximum *bmep* decreases, and the start of combustion must be advanced to achieve the maximum brake torque (which is the same as maximum *bmep*). For these values (3 < *a* < 7), the maximum *bmep* varies by less than about 7%.

As shown in Figure 12.13, the lowest value of the maximum *bmep* occurs for an "*m*" of 1.0. As the value of "*m*" is reduced below or above 1.0, the maximum *bmep* increases. As "*m*" increases, the start of combustion must be advanced (earlier) to achieve maximum *bmep*. In general, the maximum *bmep* is not too sensitive to variations of "*m*" for these conditions.

Figures 12.14 and 12.15 show the brake mean effective pressure and the brake thermal efficiency, respectively, as functions of the values of "*a*" for three values of "m" for the MBT timing cases for the base case inlet pressure (50.0 kPa). As "*a*" increases, the *bmep* and brake efficiency increase, but at a decreasing rate. The overall increase of the *bmep* and brake thermal efficiency as "*a*" increases from 3.0 to 7.0 is about 6.8%.

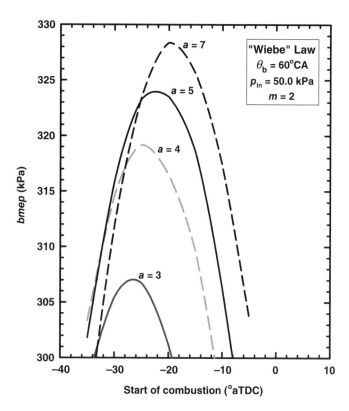

Figure 12.12   *bmep* as functions of the start of combustion for Wiebe parameters *m* = 2.0 and four values of "*a*" for the base case conditions

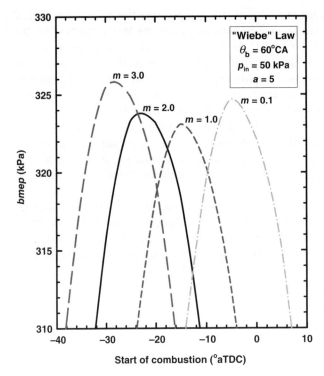

Figure 12.13 *bmep* as functions of the start of combustion for Wiebe parameters $a = 5.0$ and five values of "*m*" for the base case conditions

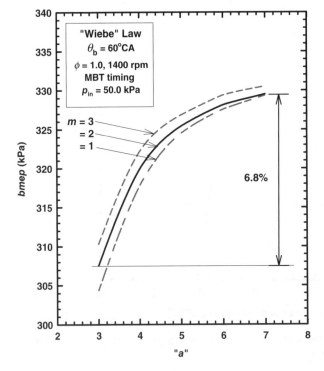

Figure 12.14 *bmep* as functions of Wiebe parameters "*a*" and "*m*" for the base case conditions

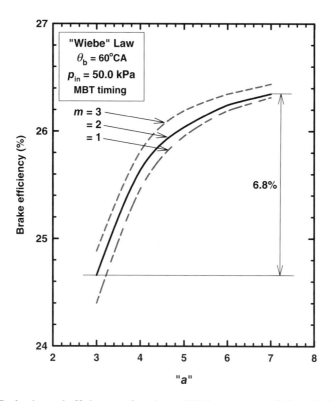

Figure 12.15 Brake thermal efficiency as functions of Wiebe parameters "*a*" and "*m*" for the base case conditions

The above then is a brief example of the sensitivity of engine performance on the heat release schedule (which is proportional to the fuel mass fraction burned schedule). The combustion process may be assumed to follow a Wiebe function for the appropriate constants. The resulting engine performance is not strongly dependent on the schedules, but the selected constants should be within reasonable agreement with accepted values. Related information on this topic may be found in Reference 8.

12.4 Part II: Combustion Phasing[1]

The importance of the phasing of the combustion event for internal-combustion engines is well appreciated, and the use of engine cycle simulations can provide detail information. One of the advantages of the use of a simulation as opposed to an experiment is that the effects of individual variables may be determined. The objective of the current work was to examine the optimum phasing (based on maximum *bmep*) as functions of engine design and operating

---

[1] A version of the material in this part of the chapter is available in Reference 16.

variables. Both the crank angle for 50% fuel mass burned (CA_{50}) and the crank angle for peak pressure (CA_{pp}) are reported as functions of the engine variables. In contrast to common statements in the literature, CA_{50} and CA_{pp} vary depending on the design and operating variables. Examples of previous research on combustion phasing are available [9–15], and more details on the work reported in this part may be found in [16].

For the second part of this chapter (combustion phasing), the operating condition examined is a moderate load (*bmep* = 325 kPa), moderate speed (1400 rpm) condition. To achieve the constant load condition, the inlet pressure was adjusted as needed. Results were obtained for two versions of this engine: (i) conventional and (ii) high efficiency.[2] The operating parameters for the conventional engine includes many of the values described above—including a combustion (burn) duration of 60°CA, a compression ratio of 8:1, an equivalence ratio of 1.0, and 0% EGR. The high efficiency engine is operated with a combustion duration of 30°CA, a compression ratio of 16:1, an equivalence ratio of 0.7, and 45% EGR. To maintain the 325 kPa *bmep*, the inlet pressure for the high efficiency engine was significantly increased. The fuel used in both cases is isooctane.

The phasing of the combustion event (and the related heat release) is known to have a direct impact on the thermal efficiency of internal combustion engines. The combustion event results in increases of cylinder pressure which is the source of the engine's work output. Advanced (early) combustion increases the compression work and heat losses, and retarded (late) combustion decreases the expansion work. By adjusting the start of combustion for the maximum brake work, an optimum (MBT) timing may be determined. This MBT timing then results in a specific CA_{50} and a specific CA_{pp}.

Some authors cite ~10°aTDC as the optimum CA_{50} for highest efficiency with the implication that this value is invariant. Although the current work will show that this approximation may be correct for some of the combinations of variables, from a thermodynamic view, it is not a universal value. Further, understanding the ways that CA_{50} and CA_{pp} are influenced by various engine variables are important insights for effective engine designs.

12.4.1 Results for Combustion Phasing

12.4.1.1 Methodology and Constraints

The approach of this study is to complete a fairly comprehensive thermodynamic evaluation of the effect of engine operating and design parameters on CA_{50} and CA_{pp}. One of the main aspects of these evaluations is that all parameters remain the same except for the one that is being varied. This provides a consistent and clear examination of the various effects. For example, for sufficiently lean operation the burn duration would be expected to be longer than for stoichiometric operation. To ascertain the influence of only the equivalence ratio, however, the burn duration was kept constant as equivalence ratio was varied. Similarly, combustion timing was adjusted for MBT performance with no concern for knock. Again, this approach provides results reflective of the thermodynamics, but (due to unsuccessful combustion or knock) these results may not be obtained in practice.

---

[2] A more complete presentation of the results for the high efficiency engine is provided in a separate case study.

12.4.1.2 Results for the Conventional Engine

Table 12.2 lists some of the input and output values for the conventional engine.[3] The two main parameters to be examined are CA_{50} and CA_{pp}. These two parameters will be shown as functions of engine design and operating parameters: spark timing, burn duration, cylinder heat transfer, compression ratio, engine speed and load, heat transfer correlation, equivalence ratio, and EGR.

As mentioned elsewhere [16,17], a variety of cylinder heat transfer correlations are available from the literature. The appropriate correlation for a given engine operating at a given set of parameters is not clear. Two popular choices for conventional engines are the correlations presented by Hohenberg [18] and Woschni [6]. The correlation by Hohenberg [18] is used for the majority of the following, but a comparison of the results using both correlations is

Table 12.2 Base case (Conventional engine) Part II: Combustion phasing (based on Hohenberg [18] heat transfer correlation)

Item	Value used	How obtained
CR = 8.0, EGR = 0%, MBT timing		
Fuel	C_8H_{18}	Input
bmep (kPa)	325.9	Output
Equivalence ratio	1.0	Input
Engine speed (rpm)	1400	Input
Mechanical frictional mep (kPa)	68.3	From algorithm [5]
Inlet pressure (kPa)	49.8	Input
Exhaust pressure (kPa)	105.0	Input
Start of combustion (°bTDC)	25.0	Determined for MBT
Combustion duration (°CA)	60	Input
Cylinder wall temperature (K)	450	Input
htm	1.0	Input
Gross indicated thermal eff (%)	36.54	Output
Net indicated thermal eff (%)	31.97	Output
Brake thermal eff (%)	26.41	Output
Exergy destruction combustion (%)	20.66	Output
Relative cylinder heat transfer (%)	22.31	Output
CA_{50} (°aTDC)	6.05	Output
CA_{pp} (°aTDC)	14.2	Output

---

[3] For the second part of the chapter, the inlet pressure is not set to 50.0 kPa, and the exhaust pressure is set at 105.0 kPa (and not determined from [1]).

provided below. (For the high efficiency engine, a heat transfer correlation developed for these types of engines is used.) Except where specifically varied, the *htm* is equal to 1.0.

Mass fraction burned—Figure 12.16 shows the mass fraction burned as functions of crank angle for three burn durations. As described in part I of this chapter, the mass fraction burned increases slowly at first, then increases more rapidly, and slowly at the end. The location of the CA_{50} is denoted in the figure. As expected, for the longer burn durations, the location of CA_{50} is slightly retarded.

Figure 12.17 shows the rate of change of the mass fraction burned (normalized to a maximum of 1.0) as functions of crank angle for three burn durations for MBT combustion timing. For the burn rate schedules examined here, the crank angle of 50% mass burn is equivalent to the highest burn rate.

Combustion timing—Figure 12.18 shows the cylinder pressure as functions of crank angle for five start of combustion timings for a burn duration of 60°CA, a compression ratio of 8, and a *bmep* of 325 kPa. The inlet pressure is adjusted for each timing to provide the same *bmep*. The combustion timings are presented relative to MBT timing—where MBT timing is the timing for maximum brake mean effective pressure or brake torque. As the timing is retarded from the maximum advance timing, the peak cylinder pressure decreases. The MBT

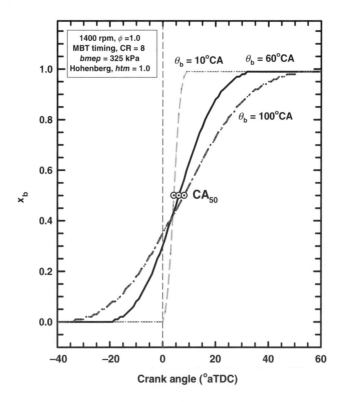

Figure 12.16   Mass fraction burned as functions of crank angle for three burn durations.
Source: Caton 2014a. Reproduced with permission from Elsevier

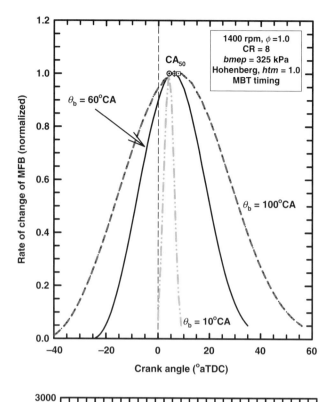

Figure 12.17   Rate of change of mass fraction burned (normalized to 1.0) as functions of crank angle for three burn durations. *Source:* Caton 2014a. Reproduced with permission from Elsevier

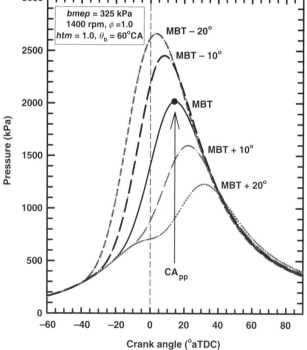

Figure 12.18   Cylinder pressures as functions of crank angle for five starts of combustion. *Source:* Caton 2014a. Reproduced with permission from Elsevier

timing results in an intermediate peak cylinder pressure. Figure 12.19 illustrates how the MBT timing was selected—the figure shows the *bmep* as a function of relative spark timing for a constant inlet pressure.

Figure 12.20 shows the CA_{50} and CA_{pp} as functions of the relative timing for this case. As timing is retarded, both parameters increase in value. The CA_{50} is a linear function of the timing. This also indicates the sensitivity of the determination of CA_{50}. The CA_{50} value will change with changes in the selection of timing. For example, if MBT timing is known ±0.25°CA, then the CA_{50} value is known within ±0.25°CA.

CA_{pp} is not as linear as CA_{50} as a function of the relative timing. For advanced timings, the CA_{pp} tends toward TDC.

Burn duration—Figure 12.21 shows the cylinder pressures as functions of crank angle for these conditions for three burn durations, and Figure 12.22 shows the corresponding CA_{50} and CA_{pp} as functions of the burn duration. As the burn duration increases, the start of combustion advances for MBT. The CA_{50} increases linearly as the burn duration increases. For burn durations between 10°CA and 100°CA, the CA_{50} increased from about 4.2°aTDC to about 7.8°aTDC.

The CA_{pp} increases, reaches a maximum, and then decreases as the burn duration increases. For burn durations approaching zero, for maximum work, the peak pressure will need to occur

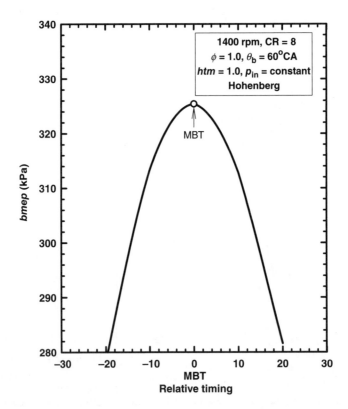

Figure 12.19 Brake mean effective pressure (*bmep*) as a function of the relative combustion timing. *Source:* Caton 2014a. Reproduced with permission from Elsevier

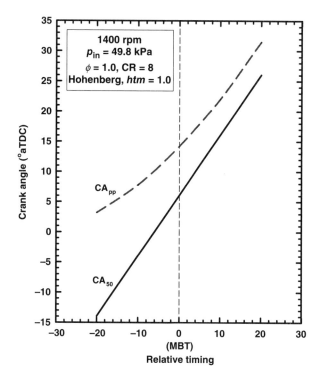

Figure 12.20 Crank angle for 50% MFB (CA$_{50}$) and for peak cylinder pressure (CA$_{pp}$) as functions of relative combustion timing. *Source:* Caton 2014a. Reproduced with permission from Elsevier

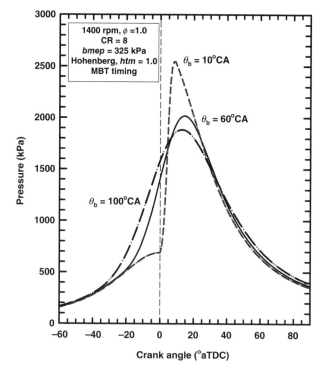

Figure 12.21 Cylinder pressure as functions of crank angle for three burn durations. *Source:* Caton 2014a. Reproduced with permission from Elsevier

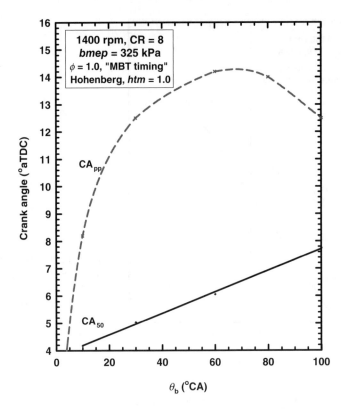

Figure 12.22 Crank angle for 50% MFB (CA$_{50}$) and for peak cylinder pressure (CA$_{pp}$) as functions of the burn duration. *Source:* Caton 2014a. Reproduced with permission from Elsevier

at TDC. As the burn duration increases from zero, the peak pressure needs to occur somewhat retarded from TDC. For most of the conditions considered here, the peak pressure is most retarded for a burn duration of about 60°CA. For the base case, the CA$_{pp}$ is 14.2°aTDC for a burn duration of 60°CA.

For burn durations greater than 60°CA, the peak pressure is located nearer TDC for maximum work. This is due to the long duration and the need to begin the combustion process earlier during the compression process. For long burn durations, heat transfer becomes more important. This results in the need to locate the peak cylinder pressure nearer TDC for long burn durations. This is discussed next.

Cylinder heat transfer —Figure 12.23 shows the CA$_{50}$ and CA$_{pp}$ as functions of a heat transfer multiplier (*htm*) for a burn duration of 60°CA. As mentioned above, the *htm* is a simple constant that increases or decreases the heat transfer coefficient. A *htm* value of zero (0.0) is an adiabatic case.

Figure 12.23 shows that the CA$_{50}$ increases essentially linearly with increases of the *htm*. For these conditions, CA$_{50}$ increases from about 1.6°aTDC to about 8.6°aTDC as *htm* increases from 0.0 to 1.67. These results indicate that without heat transfer losses the start of combustion can be advanced and the resulting cylinder pressure will produce more work than

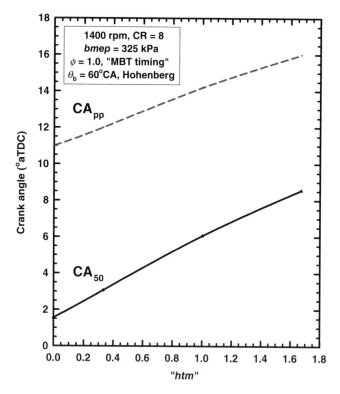

Figure 12.23 Crank angle for 50% MFB (CA$_{50}$) and for peak cylinder pressure (CA$_{pp}$) as functions of *htm*. *Source:* Caton 2014a. Reproduced with permission from Elsevier

a later combustion start. As heat losses become more important, the combustion start must be retarded for maximum work.

The CA$_{pp}$ also increases almost linearly with increases of the *htm*. For these conditions, the CA$_{pp}$ increases from about 11°aTDC to about 16°aTDC as *htm* increases from 0.0 to 1.67. Again, as heat losses become more important, the CA$_{pp}$ is retarded for maximum work. Since the lower heat losses allows the cylinder pressure to be higher, this results in a slightly advance location for the peak cylinder pressure to maximize the work.

Compression ratio—Figure 12.24 shows the CA$_{50}$ and CA$_{pp}$ as functions of compression ratio for a burn duration of 60°CA. As compression ratio increases from 6 to 20, the CA$_{50}$ increases from about 5.6 to 8.1°aTDC, and the CA$_{pp}$ decreases from about 15.8 to 8.8°aTDC. Higher compression ratios result in higher cylinder pressures and this allows the location of the peak cylinder pressure to advance. On the other hand, higher compression ratios result in a slight retarded location for the CA$_{50}$. These results are largely due to the different instantaneous cylinder pressures as compression ratio is varied.

Engine speed and load—Figures 12.25 and 12.26 show the CA$_{50}$ and CA$_{pp}$ as functions of engine speed and load (*bmep*), respectively, for a burn duration of 60°CA. The effects of speed and load on both parameters were modest. Of these effects, the only noticeable effect was a slight decrease of the CA$_{50}$ from about 6 to 4.7°aTDC as *bmep* increased from 0 to 950 kPa.

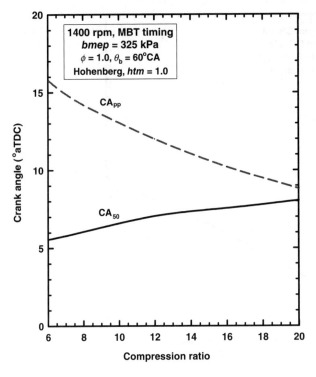

Figure 12.24 Crank angle for 50% MFB (CA_{50}) and for peak cylinder pressure (CA_{pp}) as functions of the compression ratio. *Source:* Caton 2014a. Reproduced with permission from Elsevier

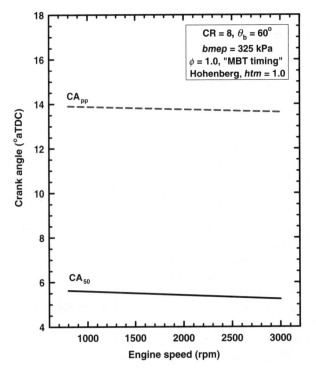

Figure 12.25 Crank angle for 50% MFB (CA_{50}) and for peak cylinder pressure (CA_{pp}) as functions of engine speed. *Source:* Caton 2014a. Reproduced with permission from Elsevier

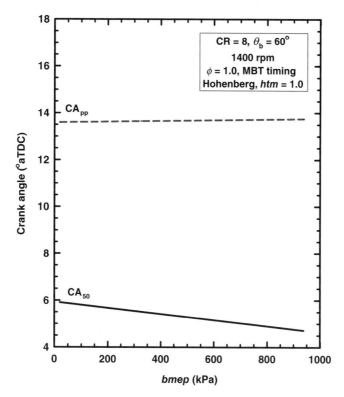

Figure 12.26   Crank angle for 50% MFB (CA_{50}) and for peak cylinder pressure (CA_{pp}) as functions of *bmep*. *Source:* Caton 2014a. Reproduced with permission from Elsevier

Heat transfer correlation—As mentioned above, to illustrate the influence of the choice of heat transfer correlation, two correlations were examined for the conventional engine: Hohenberg [18] and Woschni [6]. Figure 12.27 shows the CA_{50} and CA_{pp} as functions of burn duration for the two correlations. Although the trends are similar, the two correlations provide different values. The use of the correlation by Woschni consistently provides much more retarded values for the CA_{50} and CA_{pp}. The Woschni [6] correlation results in more heat loss compared to the use of the correlation from Hohenberg [18], and therefore, these results are consistent with the results in Figure 12.23. As with *htm*, the choice of the heat transfer correlation can significantly affect the CA_{50} and CA_{pp} values.

Equivalence ratio—Figure 12.28 shows the CA_{50} and CA_{pp} as functions of burn duration for three equivalence ratios. At best, equivalence ratio has a modest effect—but in fact, the variation is not much greater than the uncertainty of the CA_{50} and CA_{pp}.

Exhaust gas recirculation (EGR)—Figure 12.29 shows the CA_{50} and CA_{pp} as functions of burn duration for EGR levels of 0, 20 and 45%. The effect on the CA_{50} is small—a slight advance of the values as EGR increases. The effect on the CA_{pp} is slightly greater for longer burn durations. In these cases, the CA_{pp} advances for increasing levels of EGR.

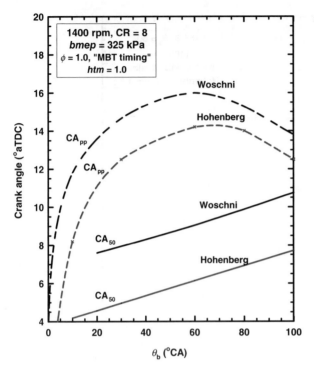

Figure 12.27 Crank angle for 50% MFB (CA_{50}) and for peak cylinder pressure (CA_{pp}) as functions of burn duration for two heat transfer correlations. *Source:* Caton 2014a. Reproduced with permission from Elsevier

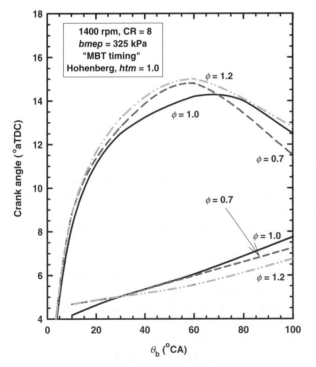

Figure 12.28 Crank angle for 50% MFB (CA_{50}) and for peak cylinder pressure (CA_{pp}) as functions of burn duration for three equivalence ratios. *Source:* Caton 2014a. Reproduced with permission from Elsevier

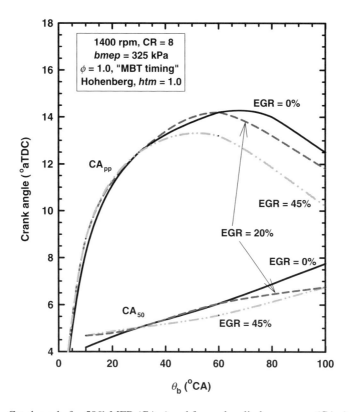

Figure 12.29 Crank angle for 50% MFB (CA_{50}) and for peak cylinder pressure (CA_{pp}) as functions of burn duration for three EGR levels. *Source:* Caton 2014a. Reproduced with permission from Elsevier

Summary for conventional engine—Figures 12.30 and 12.31 show CA_{50} and CA_{pp}, respectively, for different values of *htm*, heat transfer correlations, compression ratio, and burn duration. Changes of these parameters represent the largest changes of the CA_{50} and CA_{pp}. Changes in other parameters (such as engine speed, engine load, equivalence ratio, and EGR) resulted in much more modest changes in CA_{50} and CA_{pp}.

Figure 12.30 shows that the overall range of CA_{50} is between about 3.0°CA and 9.2°CA for these conditions. For the values examined, the variation of *htm* from 0.3 to 1.6 produced the greatest variation of CA_{50}. The other parameters that had the greatest influence on CA_{50}, in order of influence, were the burn duration (between 20°CA and 100°CA), the heat transfer correlation, and the compression ratio (between 6 and 20).

The top of Figure 12.30 shows a range of CA_{50} of 8°aTDC to 10°aTDC. As noted in the literature review, this range is commonly reported in the literature as typical values for CA_{50}. Clearly, from considerations of only the thermodynamics and for the conditions examined here, this range represents the upper bound of the values reported in this work.

Figure 12.31 shows that the overall range of CA_{pp} is between about 8.8°CA and 16.5°CA for these conditions. Unlike for CA_{50}, the engine parameter with the greatest impact on the CA_{pp} is compression ratio. The next parameters, in order of influence, were *htm*, burn duration

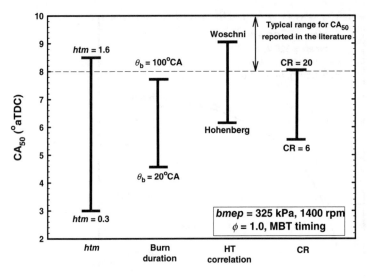

Figure 12.30   Crank angle for 50% MFB (CA_{50}) for different *htm*, burn durations, heat transfer correlations, and compression ratios. *Source:* Caton 2014a. Reproduced with permission from Elsevier

and then the selection of the heat transfer correlation. The impact of the various engine design and operating variables is different on the two parameters (CA_{50} and CA_{pp}).

12.4.1.3   Results for a High Efficiency Engine

Although high efficiency engines are discussed in another case study (Chapter 18) in this book, an examination of the combustion phasing for these engines is appropriate for this chapter. As examined in previous work [19,20] and discussed in Chapter 18, highly dilute

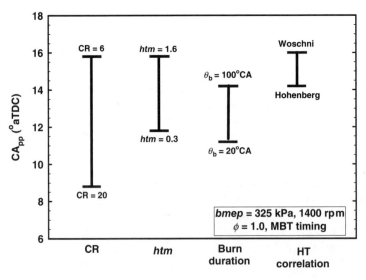

Figure 12.31   Crank angle for peak cylinder pressure (CA_{pp}) for different *htm*, burn durations, heat transfer correlations, and compression ratios. *Source:* Caton 2014a. Reproduced with permission from Elsevier

engines with relatively high compression ratios and short burn durations can achieve high thermal efficiencies. Such an engine is examined next. Table 12.3 shows the cases examined as a conventional engine is modified step-by-step until the final configuration is obtained. This approach is similar to one completed for other conditions [19,20]. Note that this evaluation is for another engine operating condition, and hence, provides additional credibility to the concept. As shown in the table, the high efficiency engine (case 6) consists of lean operation ($\phi = 0.7$), significant (cooled) EGR (45%), short burn duration (30°CA), relatively high compression ratio (16:1), and the use of a heat transfer correlation from Chang et al. [21] which has been obtained for such high efficiency engines. Table 12.4 lists some of the input and output values for case 6. All cases are based on an inlet pressure needed to obtain the same load (*bmep* = 325 kPa) and with MBT timing.

Figure 12.32 shows the net indicated and brake thermal efficiencies for each case described in Table 12.3. As each feature is added, the thermal efficiencies increase. The decrease in the burn duration from 60°CA to 30°CA provided the least incremental efficiency increase for these conditions. The other features roughly provided similar incremental efficiency increases. The brake thermal efficiency increases slightly less than the indicated thermal efficiency since mechanical losses (friction) are higher due to the higher required inlet pressures and related aspects. For the final configuration (case 6), the net indicated and brake thermal efficiencies are 50.4% and 39.9%, respectively. Table 12.3 lists some of the input and output values for the high efficiency (case 6) engine. Compared to the conventional engine (case 1), the brake thermal efficiency for the high efficiency engine has increased by a factor of 1.5—which is a major development.

Figure 12.33 shows the CA_{50} and CA_{pp} for each case. CA_{50} is between about 4.5°CA and 6.0°CA for the first five cases. The CA_{50} decreases from about 6.1°aTDC to about 2.5°aTDC from case 1 to case 6. This decrease is largely due to the reduced heat transfer associated with the Chang et al. [21] correlation. As described above, for reduced heat losses, the CA_{50} is more advanced.

Table 12.3   Description of cases

Case	Description
1 (base)	CR = 8; θ_b = 60°CA; ϕ = 1.0; EGR = 0%; Hohenberg [18]; MBT timing
2 (θ_b)	CR = 8; θ_b = 30°CA; ϕ = 1.0; EGR = 0%; Hohenberg [18]; MBT timing
3 (EGR)	CR = 8; θ_b = 30°CA; ϕ = 1.0; EGR = 45%; Hohenberg [18]; MBT timing
4 (ϕ)	CR = 8; θ_b = 30°CA; ϕ = 0.7; EGR = 45%; Hohenberg [18]; MBT timing
5 (CR)	CR = 16; θ_b = 30°CA; ϕ = 0.7; EGR = 45%; Hohenberg [18]; MBT timing
6 (HT)	CR = 16; θ_b = 30°CA; ϕ = 0.7; EGR = 45%; Chang et al. [21]; MBT timing

Table 12.4   Case 6 – high efficiency engine (based on Chang et al. [21] heat transfer correlation)

Item	Value used	How obtained
CR = 16.0, EGR (cooled) = 45%, MBT timing		
Fuel	C_8H_{18}	Input
bmep (kPa)	325.9	Output
Equivalence ratio	0.7	Input
Engine speed (rpm)	1400	Input
Mech frictional mep (kPa)	85.6	From algorithm [5]
Inlet pressure (kPa)	76.5	Input
Exhaust pressure (kPa)	105.0	Input
Start of combustion (°bTDC)	13.0	Determined for MBT
Combustion duration (°CA)	30	Input
Cylinder wall temperature (K)	450	Input
htm	1.0	Input
Gross indicated thermal eff (%)	54.13	Output
Net indicated thermal eff (%)	50.36	Output
Brake thermal eff (%)	39.89	Output
Exergy destruction combustion (%)	24.31	Output
Relative cylinder heat transfer (%)	6.89	Output
CA_{50} (°aTDC)	2.53	Output
CA_{pp} (°aTDC)	7.8	Output

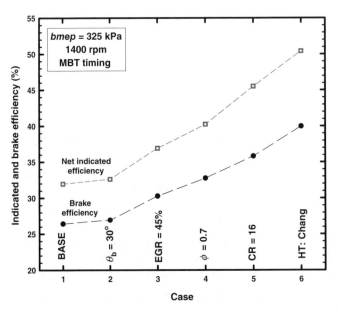

Figure 12.32   Net indicated and brake thermal efficiency for six cases leading to the high efficiency engine (case 6). *Source:* Caton 2014a. Reproduced with permission from Elsevier

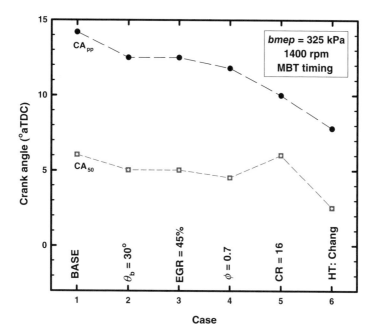

Figure 12.33 Crank angle for 50% MFB (CA_{50}) and for peak cylinder pressure (CA_{pp}) for six cases leading to the high efficiency engine (case 6). *Source:* Caton 2014a. Reproduced with permission from Elsevier

The CA_{pp} decreases steadily as each feature is added to the configuration. It decreases from 14.2°CA to 7.8°CA for these conditions. Each of these changes of CA_{50} and CA_{pp} is consistent with the changes noted earlier for the conventional engine. In general, for increasing efficiency, both CA_{50} and CA_{pp} tend to advance approaching TDC.

12.5 Summary and Conclusions

This case study considered two aspects of the combustion process: (i) the heat release schedule, and (ii) the combustion phasing. The mass fraction burn was assumed to be represented by the Wiebe function. A typical automotive spark ignition engine was selected for this study.

For the study concerning the heat release schedule, the following may be stated as a summary:

- A simple, analytical expression was derived for the apparent heat release on either a gross or net (including heat transfer) basis. A similar quantity from the simulation is difficult to obtain due to the effects of chemical dissociation which reduces the instantaneous chemical energy impact and changes the average of the specific heats.
- The simple, analytical expression for the apparent heat release agreed with the similar quantity from the simulation for the case of frozen properties (i.e., no chemical dissociation).
- As the Wiebe expression parameter "a" is increased from 3 to 7, the start of combustion for maximum performance needed to be retarded. Similarly, as the parameter "m" was increased from 0.5 to 3, the start of combustion for maximum performance needed to be retarded.

- Performance parameters such as *bmep* and brake thermal efficiency increase slightly (6.8%, relative) as the Wiebe parameter "*a*" is increased from 3 to 7. The effect of the Wiebe parameter "*m*" (from 1 to 3) is much more modest; specifically, resulting in about a 1% (relative) increase in the performance values.

For the study concerning combustion phasing, this work has examined the CA_{50} and CA_{pp} measures of combustion phasing based solely on the thermodynamics. The effects of a number of engine operating and design variables on CA_{50} and CA_{pp} were obtained. The following are some findings from this work:

- In contrast to some statements in the literature, CA_{50} and CA_{pp} are functions of engine operating and design variables. Some literature suggests that the CA_{50} is ~10°aTDC, and that this crank angle is invariant—results from this study show that this is not true.
- Both the CA_{50} and CA_{pp} are most influenced by the heat transfer level, compression ratio, and burn duration. The CA_{50} and CA_{pp} are less influenced by equivalence ratio, EGR, engine speed and engine load.
- For the conventional engine, for the conditions examined, the CA_{50} varied between about 5 and 11°aTDC, and the CA_{pp} varied between about 9 and 16°aTDC.
- A high efficiency engine was examined which was based on short burn duration, high compression ratio, lean operation and high EGR. The brake thermal efficiency of this engine was 39.9%—1.5 times higher than the conventional engine.
- For the high efficiency engine, the CA_{50} was about 2.5°aTDC and the CA_{pp} was about 7.8°aTDC. These more advanced values for the high efficiency engine were largely a result of the reduced heat losses.

References

1. Heywood, J. B. (1988). *Internal Combustion Engine Fundamentals*, McGraw-Hill Book Company, New York.
2. Gatowski, J. A., Balles, E. N., Chun, K. M., Nelson, F. E., Ekchian, J. A., and Heywood, J. B. (1984). Heat release analysis of engine pressure data, Society of Automotive Engineers, paper no. 841359.
3. Chun, K. M. and Heywood, J. B. (1987). Estimating heat-release and mass-of-mixture burned from spark-ignition engine pressure data, *Combustion Science and Technology*, **54**, 133–143.
4. Cheung, H. M. and Heywood, J. B. (1993). Evaluation of a one-zone burn-rate analysis procedure using production SI engine pressure data, Society of Automotive Engineers, paper no. 932749.
5. Sandoval, D. and Heywood, J. B. (2003). An improved friction model for spark-ignition engines, Society of Automotive Engineers, SAE paper no. 2003–01–0725.
6. Woschni, G. (1968). A universally applicable equation for the instantaneous heat transfer coefficient in the internal combustion engine, *SAE Transactions*, SAE paper no. 670931, **76**, 3065–3083.
7. Wiebe, J. J. (1970). *Brennverlauf und Kreisprozess von Verbrennungsmotoren*, VEB-Verlag Technik Berlin.
8. Caton, J. A. (2000). The effect of burn rate parameters on the operating attributes of a spark-ignition engine as determined from the second law of thermodynamics, in *Fuel Injection, Combustion, and Engine Emissions*, Vol. 2, Proceedings of the 2000 Spring Technical Conference, V. W. Wong (ed.), ICE-Vol. 34-2, pp. 49–62, the ASME Internal Combustion Engine Division, American Society of Mechanical Engineers, San Antonio, TX, 09-12 April.
9. Zhu, G. G., Daniels, C. F., and Winkelman, J. (2003). MBT timing detection and its closed-loop control using in-cylinder pressure signal, Society of Automotive Engineers, SAE Paper No. 2003–01–3266.
10. Ma, F., Wang, Y., Wang, J., Ding, S., Wang, Y., and Zhao, S. (2008). Effects of combustion phasing, combustion duration, and their cyclic variations on spark-ignition (SI) engine efficiency, *Energy & Fuels*, **22**, 3022–28.

11. Ayala, F. A., Gerty, M. D., and Heywood, J. B. (2006). Effects of combustion phasing, relative air-fuel ratio, compression ratio, and load on SI engine efficiency, Society of Automotive Engineers, SAE Paper No. 2006–01–0229.

12. Carvalho, L., de Melo, T., and Neto, R. (2012). Investigation on the fuel and engine parameters that affect the half mass fraction burned (CA50) optimum crank angle, Society of Automotive Engineers, SAE Paper No. 2012–36–0498.

13. Lavoie, G. A., Ortiz-Soto, E., Babajimopoulos, A., Martz, J. B., and Assanis, D. N. (2013). Thermodynamic sweet spot for high-efficiency, dilute, boosted gasoline engines, *International Journal of Engine Research*, **14**, 260–278.

14. Ponti, F., Ravaglioli, V., Serra, G., and Stola, F. (2009). Instantaneous engine speed measurement and processing for MFB50 evaluation, Society of Automotive Engineers, SAE Paper No. 2009–01–2747.

15. Ravaglioli, V., Morro, D., Serra, G., and Ponti, F. (2011). MFB50 on-board evaluation based on a zero-dimensional ROHR model, Society of Automotive Engineers, SAE Paper No. 2011–01–1420.

16. Caton, J. A. (2014). Combustion phasing for maximum efficiency for conventional and high efficiency engines, *Journal of Energy Conversion and Management*, **77** (1), 564–576.

17. Caton, J. A. (2011). Comparisons of global heat transfer correlations for conventional and high efficiency reciprocating engines, in proceedings of the 2011 Fall Technical Conference of the ASME Internal Combustion Division, paper no. ICEF2011-60017, Morgantown, WV, 02-05 October.

18. Hohenberg, G. F. (1979). Advanced approaches for heat transfer calculations, Society of Automotive Engineers, SAE Paper No. 790825.

19. Caton, J. A. (2010). An assessment of the thermodynamics associated with high-efficiency engines, in Proceedings of the ASME 2010 Internal Combustion Engine Division Fall Technical Conference, paper no. ICEF2010–35037, San Antonio, TX, 12–15 September.

20. Caton, J. A. (2012). The thermodynamic characteristics of high efficiency, internal-combustion engines, *Journal of Energy Conversion and Management*, **58**, 84-93.

21. Chang, J., Guralp, O., Filipi, Z., and Assanis, D. (2004). New heat transfer correlation for an HCCI engine derived from measurements of instantaneous surface heat flux, Society of Automotive Engineers, SAE Paper No. 2004–01–2996.

13

Cylinder Heat Transfer

13.1 Introduction

This case study concerns engine cylinder heat transfer. The results will include detailed time-resolved results for temperatures, heat transfer coefficients, and heat transfer rates. Engine performance results for the base case conditions will be described for a number of heat transfer correlations. These results will provide information on the sensitivity of the engine performance results to the heat transfer, and quantify the importance of the heat transfer component. After the basic results are described, this case study provides some results for engines employing low heat rejection (LHR) concepts. Finally, this case study ends with a consideration of the heat transfer as a function of exhaust gas recirculation (EGR). This last set of results illustrates some of the subtle thermodynamic aspects of engine cylinder heat transfer.

Cylinder heat transfer is one of the more important processes of internal combustion engines, and yet, after decades of research, comprehensive, universal convective heat transfer correlations are not available. What are available are global correlations that are based on approximate experimental energy surveys of specific engines operating at a few conditions. These global correlations have been acceptable for general understandings and, as shown in this chapter, for certain conditions engine performance results are not strongly dependent on the specific correlation used. Nevertheless, the next advances in engine technologies will be related to better understanding and managing of the energy streams, including the cylinder heat transfer. Furthermore, predictions of thermal nitric oxides (which are highly dependent on accurate temperature assessments) will be improved with more precise characterizations of the cylinder heat transfer. For these reasons, more thorough examinations of these global correlations are needed. These examinations should include results for a wide variety of engines and operating conditions.

Engine cylinder heat transfer exists because the components and lubricating oil in the cylinder must be maintained at temperatures significantly lower than the cylinder gas temperatures especially during the high temperature portions of the cycle. To accomplish this, the engine components need to be cooled. For most of the time, the heat transfer moves energy from the gases to the cylinder walls, piston, and cylinder head. For short periods of time, however, the cylinder gases may be at temperatures lower than the components and the heat transfer will be from the components to the gases. These items are illustrated in more detail in the results presented below.

An Introduction to Thermodynamic Cycle Simulations for Internal Combustion Engines, First Edition.
Jerald A. Caton © 2016 John Wiley & Sons, Ltd. Published 2016 by John Wiley & Sons, Ltd.

The cylinder heat transfer is largely convective, but some radiation from the flame(s) and any luminous particles (e.g., soot) may also exist. The energy that reaches the cylinder walls and other components is conducted through the metal to the lower temperature cooling media (liquid coolant or ambient air). The heat transfer from the metal to the cooling media is largely convective. Note that for the thermodynamic system defined for these studies, energy that reaches the cylinder walls and components is outside the control system.

An important distinction that should be emphasized at this point is that the correlations for global heat transfer convective coefficients are a major engineering simplification. To capture all the details of the actual engine cylinder heat transfer process is a gigantic task, and although progress has been attained over the last several decades, a definitive solution is lacking. The engine heat transfer process is complex due to a number of items such as the unsteady, compressible and turbulent flow field, the spatial and temporal variations of boundary conditions, the varying fluid properties, the surface details, and the interactions with the equally complex combustion process.

The global correlations for the cylinder heat transfer are often single expressions that are applied for the complete cylinder and are used on an instantaneous basis. The actual engine cylinder heat transfer is a complex process, and has extreme spatial and temporal variations. To expect a single, global correlation to capture all of these details is unrealistic. On the other hand, engineers often need practical, approximate correlations that can help complete evaluations that could not be done otherwise. For these reasons, the pursuit of appropriate, approximate correlations continues.

The following parts of this chapter include descriptions of the basic heat transfer relations, a brief summary of previous literature (with special emphasis on the heat transfer correlations used in this work), results from the simulation, and a brief summary of this chapter. The results will include those from conventional engines and from engines with low heat rejection concepts.

13.2 Basic Relations

Before reviewing some examples of correlations from the literature, a brief summary of the basic type of relations that are typically used to represent the global cylinder heat transfer is presented. In general, as mentioned above, engine cylinder heat transfer is a set of complex processes that involves turbulent flow, time and spatial temperature differences, chemically reacting flow, and variable (and often unknown) properties. Many studies have been conducted over the years to provide better understandings of the cylinder heat transfer, but no general consensus has ever evolved. Often new complicated correlations for the heat transfer are found to be difficult to use, and the improvements are just as suspect as from the simpler correlations.

Forced convective heat transfer is probably the dominant mechanism of cylinder heat transfer, especially for SI engines. The general form of the cylinder heat transfer is

$$\dot{Q} = htm\, h_{\mathrm{c}} A \left(T - T_{\mathrm{wall}} \right) \tag{13.1}$$

where $\dot{Q}$ is the overall heat transfer rate, h_{c} is the convective heat transfer coefficient, A is the surface area, $(T - T_{\mathrm{wall}})$ is the temperature difference, and htm is a "*heat transfer multiplier.*"

The value of *htm* will be 1.0 except for those cases where the sensitivity of engine performance on heat transfer levels is explored. In the following discussion, heat transfer is assumed positive from the gases to the wall.[1]

The heat transfer coefficient is generally given as a function of a Reynolds number

$$Nu = \frac{h_c L}{k_{gas}} = a Re^b = a\left(\frac{VL}{v}\right)^b \qquad (13.2)$$

where *Nu* is the Nusselt[2] number, h_c is the convective heat transfer coefficient, k_{gas} is the thermal conductivity of the gas, Re is the Reynolds number, and "*a*" and "*b*" are parameters selected to provide agreement with the experimental values. "*L*" and "*V*" are the characteristic length and velocity, respectively, and *v* is the kinematic viscosity. The characteristic length is often the engine bore, and the characteristic cylinder gas velocity is often the piston speed. The piston speed is not a great choice since during processes such as combustion and exhaust blow down, the piston speed would not be expected to be a good representation of the bulk gas velocity. The choice of the piston speed, however, is often the only convenient choice due to the lack of more detail information of gas velocities.

13.3 Previous Literature

The literature on internal combustion engine cylinder heat transfer is immense. This brief subsection will only provide a few examples of this literature. Many overviews of engine cylinder heat transfer have been published over the years. Two good summaries of the information up to the 1980s are provided by Heywood [1] and by Borman and Nishiwaki [2]. A more recent review of empirical correlations for engine cylinder heat transfer was published in 2006 [3]. In addition, most engine text books describe some of the background and previous literature concerning cylinder heat transfer. Some of the literature described in these references will be used in the current study and is mentioned next.

Some of the earliest works were related to providing global correlations for the convective heat transfer correlation. For example, Nusselt in 1923 [4] described one of the first heat transfer correlations for a diesel engine. The data used by Nusselt for this correlation were obtained from constant volume combustion facilities. Although this expression is not often used any more, the basic approach of Nusselt has continued throughout the history of developing correlations for engine cylinder heat transfer. Eichelberg [5] in 1939 provided a "lecture" on engine cylinder heat transfer for large two-stroke and four-stroke diesel engines. He presented some of the first instantaneous heat flux measurements, and proposed a correlation for the global heat transfer based on instantaneous cylinder gas temperatures and pressures.

---

[1] Note that the energy balance equations presented in earlier parts of this book have used the conventional definition of heat transfer as positive into the thermodynamic system. The alternative approach taken in this chapter has been adopted for convenience to provide a more direct examination of the cylinder heat transfer. In practice, the formulations in the simulation are consistent with the definition of heat transfer as positive into the thermodynamic system.

[2] The Nusselt number is named after the German physicist W. Nusselt (1882–1957). The first engine cylinder heat transfer correlation [4] presented in this chapter was due to W. Nusselt in 1923.

Annand [6] in 1963 presented a summary of eight previous global correlations (including the ones from Nusselt and Eichelberg), and from this information proposed a new correlation based on a reevaluation of this previous data. At the time of this work, this correlation [6] was considered one of the best ways to compute the cylinder heat transfer, and has been (and continues to be) used in various engine cycle simulations.

Probably one of the most widely used global correlations is due to Woschni [7] in 1967. His correlation was based on diesel engine data. One of the more novel items of the Woschni correlation is that it uses different constants and velocities for different portions of the engine cycle. For example, for the combustion period, he recommended a characteristic velocity that included a part based on the pressure increase relative to the motored pressure at the same instant. In this way, he hoped to capture the additional velocity due to combustion. Although this approach has some merit, no evidence is available to substantiate this aspect. Nevertheless, this correlation is widely used, and will serve as a reference point in the current work.

Hohenberg [8] in 1979 reexamined much of the previous cylinder heat transfer information including that from Woschni [7], and proposed a new correlation. This is often referred to as an update to the Woschni correlation, but in fact, this is simply another attempt to correlate data from one-dimensional, heat flux probes. Hohenberg based his work on both previous data, and on new data from his experiments. He utilized four different direct injection diesel engines, and completed extensive experiments that included measurements of heat flux, heat balance, and component temperatures.

13.3.1   Woschni Correlation

Since the Woschni [7] correlation is used in some of the studies presented in this book, and since it contains some unique aspects, this correlation will be described in some detail. In general, the other correlations are similar. He proposed the following correlation:[3]

$$Nu = \frac{h_c B}{k} = 0.035 \text{Re}^m \tag{13.3}$$

As mentioned above, the Reynolds number is given by

$$\text{Re} = \frac{wB}{v} \tag{13.4}$$

where w is an average characteristic cylinder gas velocity, and v is the kinematic viscosity of the boundary layer gases. Woschni [7] assumed the following dependencies:

$$k \propto T^{0.75} \tag{13.5}$$

$$\mu \propto T^{0.62} \tag{13.6}$$

$$p = \rho RT \tag{13.7}$$

$$v = \mu / \rho \tag{13.8}$$

---

[3] The value of "m" is given below ($m = 0.8$).

Combining the above relations results in the following expression for the instantaneous heat transfer coefficient:

$$h_c = CB^{m-1} p^m w^m T^{0.75-1.62m} \tag{13.9}$$

Woschni [7] suggested several different characteristic velocities (w) depending on the portion of the cycle. During intake, compression, and exhaust, he assumed that the characteristic velocity would be proportional to the mean piston speed. During combustion and expansion, he assumed the velocity would scale with some measure of the combustion intensity. Therefore, he proposed that the velocity for this period would be proportional to the increase in cylinder pressure above the motoring pressure. The pressure increase due to combustion is given by

$$(p - p_m) \tag{13.10}$$

where p_m is the cylinder pressure resulting from motoring (no combustion) for the same crank angle as "p." Finally, then, the average characteristic cylinder gas velocity for use in the Reynolds number is

$$w = \left[C_1 V_p + C_2 \frac{V_d T_r}{p_r V_r} (p - p_m) \right] \tag{13.11}$$

where V_p is the mean piston speed, V_d is the displaced volume, and p_r, T_r, and V_r are cylinder pressure, temperature and volume, respectively, at some reference condition (such as at intake valve close crank angle). The constants in eq. (13.11) are given by the following:

For the gas exchange period:	$C_1 = 6.18, C_2 = 0.0$
For the compression period:	$C_1 = 2.28, C_2 = 0.0$
For the combustion period:	$C_1 = 2.28, C_2 = 0.00324$

When the exponent (m) in eq. (13.3) is set to 0.8, the final (dimensional) correlation is:

$$h_c \left(\frac{W}{m^2 K} \right) = 3.26 \, B(m)^{-0.2} \, p(kPa)^{0.8} \times w \left(\frac{m}{s} \right)^{0.8} T(K)^{-0.55} \tag{13.12}$$

13.3.2 Summary of Correlations

Table 13.1 lists the four global engine cylinder heat transfer correlations considered in this chapter. The correlations are given in a *dimensional* form to better illustrate the dependencies on temperature, pressure, velocity, and so forth. The symbols are defined as follows: h_i is the convective heat transfer coefficient for correlation "i", p is the cylinder pressure, T is the cylinder gas temperature, u_p is the piston velocity, u_i is a specially defined velocity, L is the characteristic length, and V_{inst} is the instantaneous cylinder volume. Further details and values for the constants are given in the cited references.

Table 13.1 Heat transfer correlations[a] (dimensional form)

First author	Date	Correlation
Eichelberg [5]	1939	$h_1 = C_1 p^{0.5} T^{0.5} u_p^{0.33}$
Annand [6]	1963	$h_2 = C_2 L^{-0.3} p^{0.7} T^{-1.0} u_p^{0.7}$
Woschni [7]	1967	$h_3 = C_3 L^{-0.2} p^{0.8} T^{-0.35} u_4^{0.8}$
Hohenberg [8]	1979	$h_4 = C_4 L^{-0.2} p^{0.8} T^{-0.4} (u_p + 1.4)^{0.8} V_{inst}^{-0.06}$

[a]See cited references for details.

Listing the correlations in a *dimensional* form provides a fairly direct way to illustrate the different dependences of the correlations. For example, the Eichelberg [5] correlation depends on the gas temperature to a positive power of "0.5," whereas the other correlations depend on the gas temperature to various negative powers. Similar statements can be made about several of the other variables. In general, this demonstrates a serious weakness in the understanding of engine cylinder heat transfer. Hopefully, future work will strive to seek a more thorough understanding of engine cylinder heat transfer.

One of the parts of this study is to compare the cylinder heat transfer results from the simulation using each of these correlations for the convective coefficient. Similar studies have been completed in the past that have compared results using these and other engine cylinder heat transfer correlations (e.g., [3, 9]).

13.4   Results and Discussion

The approach of this study is to, first, complete a general overview of heat transfer results for one correlation, and, then, complete a comparison of the use of typical existing global heat transfer correlations. The following are results for the base case condition (1400 rpm, *bmep* = 325 kPa) for a compression ratio of 8:1, an inlet pressure of 50 kPa, and a start of combustion at −22°aTDC. The previous chapters have listed some of the operating conditions and output results for this case. A similar study concerning engine heat transfer for other conditions is available elsewhere [10].

Some related results on cylinder heat transfer are presented in Chapter 8 in relation to engine performance. The results of this case study are presented for the conventional engine, for an engine utilizing low heat rejection (LHR) concepts, and for an engine using exhaust gas recirculation.

13.4.1   Conventional Engine

Each of the components of the heat transfer expression is described first, and then the overall heat transfer rate is provided. Finally, the sensitivities of the results are shown as functions of the various heat transfer parameters including the core expressions for the heat transfer coefficients.

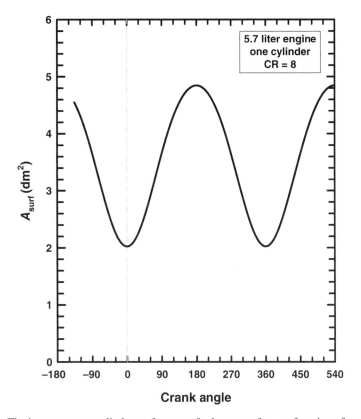

Figure 13.1   The instantaneous cylinder surface area for heat transfer as a function of crank angle

As shown above (eq. 13.1), the heat transfer rate is a function of surface area, heat transfer coefficient, and the temperature difference between the gas and the cylinder wall. Figure 13.1 shows the instantaneous surface area as a function of crank angle for the base case conditions. The surface area varies in a sinusoidal fashion as the piston moves between TDC and BDC. The minimum surface area is at TDC and the maximum area is at BDC. The surface area is a direct multiplier of the heat transfer rate and is significant. From TDC, the surface area increases by over a factor of 2.0 as the piston moves to BDC for a compression ratio of 8. This, alone, increases the instantaneous heat transfer rate by the same factor.

Figure 13.2 shows the instantaneous heat transfer coefficient from Woschni [7] as a function of crank angle for the base case conditions. The coefficient increases rapidly during combustion to attain its highest value slightly after TDC, and then it decreases during the expansion process. The high values during combustion are at least partly due to the high cylinder pressures and the high estimated cylinder gas velocity (eq. 13.11). The coefficient increases in a "step" fashion at EVO since the correlation switches to a higher constant for exhaust flow. In reality, this would be a more gradual change as the exhaust valve opens and this could be accommodated. The complexity of this accommodation, however, would not be warranted since the difference is modest. Other slight changes of the coefficient are directly related to the other valve events. In general, the heat transfer coefficient has a major effect on the heat transfer rate—especially, during the combustion process.

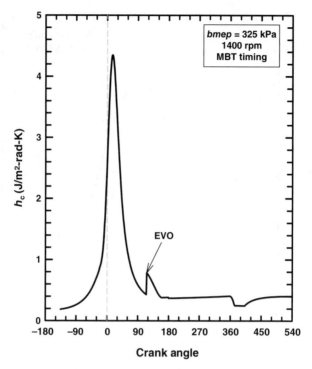

Figure 13.2   The instantaneous heat transfer coefficient using the Woschni expression as functions of crank angle for the base case conditions

Figure 13.3 shows the final item for determining the heat transfer rate—the temperature difference between the gases and the cylinder wall. This figure shows this temperature difference as a function of the crank angle for the base case conditions. The temperature difference increases rapidly during the combustion process, attains a maximum value at 27.8°aTDC, and then decreases during the expansion process. The high temperature values during combustion contribute to the high heat transfer rates.

Finally, Figure 13.4 shows the instantaneous heat transfer rate as a function of crank angle for the base case conditions. This heat transfer rate, of course, is the product of the three previous items: surface area, heat transfer coefficient, and temperature difference. The heat transfer rate increases rapidly during the combustion process and attains a maximum value of about 24.3 kJ/s at 20°aTDC, and then decreases during the expansion process. During the compression stroke until 90°bTDC, the heat transfer is from the cylinder walls to the gases since the wall temperature is greater than the gas temperature. After 90°bTDC, the gas temperature is greater than the wall temperature and the heat transfer is from the gases to the wall (positive heat transfer rates as defined in this chapter). During the intake process after 445°aTDC, the wall temperatures are again greater than the gas temperatures and the heat transfer is again from the wall to the gases.

As may be obvious, the heat transfer rate, heat transfer coefficient and gas temperatures are interconnected. For the conditions and correlation used, as the temperature increases or

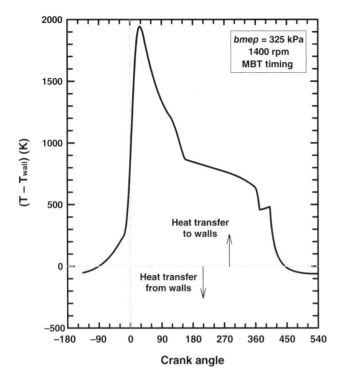

Figure 13.3 The instantaneous temperature difference as functions of crank angle for the base case conditions

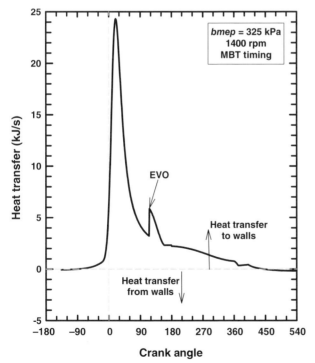

Figure 13.4 The instantaneous heat transfer rate as functions of crank angle using the Woschni correlation for the base case conditions

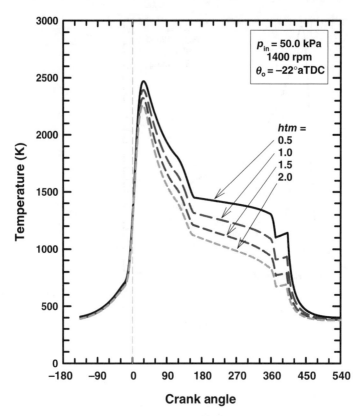

Figure 13.5   The instantaneous one-zone gas temperature as functions of crank angle using the Woschni correlation for four multiplying constants

pressure decreases, the heat transfer coefficient decreases (see Table 13.1). As the heat transfer coefficient decreases, the cylinder heat transfer rate decreases and temperatures decrease less. The net effect of these various actions is reflected in the final values of the gas temperatures and heat transfer rates.

The following results will now examine the sensitivity of the results to various heat transfer assumptions and approximations. For example, Figures 13.5 and 13.6 show the gas tempera-ture and heat transfer rate, respectively, as functions of crank angle for four multiplying values (*htm*) of the heat transfer: 0.5, 1.0 (base), 1.5, and 2.0. For these computations, the inlet pres-sure (50 kPa) and start of combustion ($-22°$aTDC) were maintained at their base case values. The higher the multiplying constant, the higher the heat transfer rate and therefore the lower the gas temperature. The effect is modest for the gas temperature, but more significant for the heat transfer rate for these conditions. The *maximum instantaneous* heat transfer rate is a factor of about 3.7 higher between the lowest and highest value of *htm*.

Figure 13.7 shows the brake thermal efficiency and brake mean effective pressure (*bmep*) as functions of the *htm* for the base input conditions. As the *htm* increases, the brake efficiency and the *bmep* decrease due to the increasing heat transfer. The effect is somewhat modest. For example, by doubling the cylinder heat transfer rate (*htm* = 2.0), the brake efficiency decreases

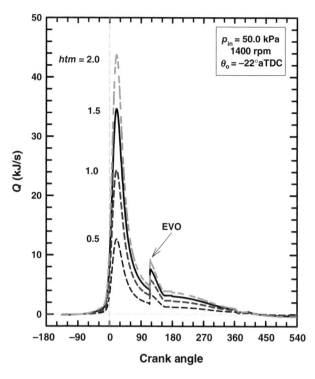

Figure 13.6 The instantaneous heat transfer rate as functions of crank angle using the Woschni correlation for four multiplying constants

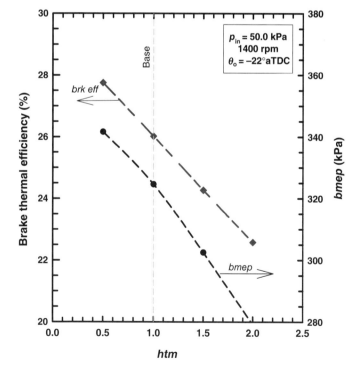

Figure 13.7 Brake thermal efficiency and brake mean effective pressure (*bmep*) as functions of the *htm* (the fraction of the heat transfer) for the base input conditions

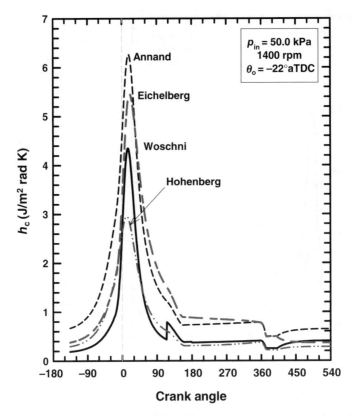

Figure 13.8   The instantaneous convective heat transfer coefficient as functions of crank angle for the four heat transfer coefficient expressions

from about 26.0% to 22.6%. In other words, large changes in the cylinder heat transfer rate result in modest changes in the efficiency and *bmep*. The reason for this is explained next.

The following explains the relation between changes in cylinder heat transfer and work output. For a decrease of the heat transfer, more thermal energy is retained in the cylinder. The conversion of this thermal energy to work is difficult (this aspect is a feature of the second law of thermodynamics). Therefore, only a fraction of this retained thermal energy is converted to work, and the remainder of the retained energy increases the exhaust gas energy. This conversion, by the way, is highly dependent on the values of the specific heats (for lower values of the specific heats, the conversion of the thermal energy to work is higher).

This insensitivity of the engine work output to the heat transfer implies that precise values of the convective coefficient may not be as critical as other items. A related conclusion is that significant reductions of the heat transfer will be needed to accomplish any performance gains. This is quantified later in this subsection.

Figures 13.8 and 13.9 compare the results using the four different convective heat transfer coefficients discussed above. Figure 13.8 shows the heat transfer coefficient as functions of crank angle for the four expressions for the same conditions. This comparison is based on the substitution of the individual correlations for the heat transfer coefficient. This means that the

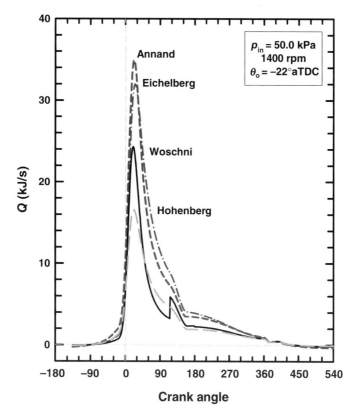

Figure 13.9   The instantaneous heat transfer rate as functions of crank angle for the four heat transfer coefficient expressions

cylinder gas temperatures and pressures will be different even though the operating and design parameters are the same. The heat transfer coefficient from Annand [6] provided the highest peak value and the coefficient from Hohenberg [8] provided the lowest peak value while the values from Eichelberg [5] and Woshni [7] were in between the other two. Figure 13.9 shows the resulting heat transfer rates as functions of crank angle using the four expressions. Again, the expression from Annand [6] resulted in the highest peak heat transfer rate while that from Hohenberg [8] resulted in the lowest peak rate. The resulting Hohenberg peak heat transfer rate would need to increase by 215% to attain the peak rate provided from the Annand correlation.

Several authors (e.g., [8]) have commented that the Woschni [7] correlation may overestimate the heat transfer during combustion, and underestimate the heat transfer during other portions of the cycle—at least for certain types of engines. Again, this is mentioned so as to emphasize the uncertainty with any of these correlations. For different operating conditions and design parameters, the use of these correlations, at best, provides estimates of the cylinder heat transfer.

Figure 13.10 shows the brake thermal efficiencies and *bmep* for each of the four heat transfer correlations for a constant inlet pressure of 50.0 kPa. The range between the highest and

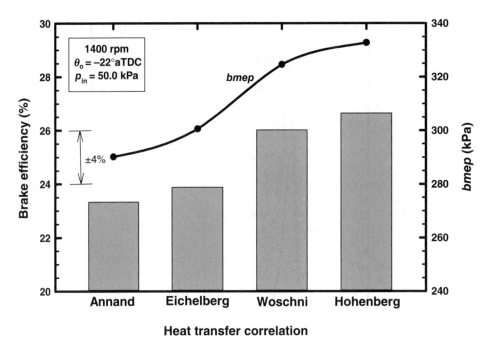

Figure 13.10 The brake thermal efficiency as determined using the four convective heat transfer coefficients for the same inlet conditions

lowest thermal efficiencies is about 13% (relative). Note the much smaller impact on engine performance even though the peak heat transfer was different by 215%. This reflects a number of items. First, the 215% difference is only for the peak heat transfer rates—for the majority of the time, the rates are not that different. Second, the lower heat transfer rates result in more thermal energy being retained in the cylinder. Only a portion of the retained energy, however, is converted to work. The remainder of the retained energy increases the energy content and gas temperature of the exhaust mass.

In other words, for a fairly wide range of convective heat transfer correlations, the changes of the net result are somewhat modest for the thermal efficiencies and performance parameters. This is because although the heat transfer is important, the lack of complete conversion of retained thermal energy mitigates the final result on engine work output. So, as mentioned above, the lack of a universally accepted global heat transfer correlation for many investigations is not a severe disadvantage, and studies of the thermodynamics of engines using these correlations remain useful.

The next consideration is to examine the details of the heat transfer for changes in an engine operating parameter. For this, the heat transfer as a function of engine speed will be examined. Some of this information extends the comments found in Chapter 8 concerning cylinder heat transfer. These results are for the base case (*bmep* = 325 kPa) using the heat transfer correlation from Woschni [7]. Figure 13.11 shows the ratio of the heat transfer rate and the heat transfer rate for 1400 rpm as functions of engine speed. The heat transfer rate increases as speed increases almost in a linear fashion. The reasons for this increase are examined next.

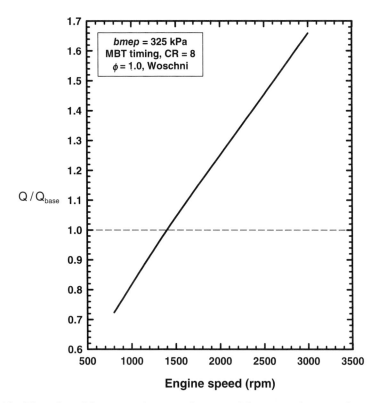

Figure 13.11   The ratios of the average heat transfer rate and the average heat transfer rate at 1400 rpm as functions of the engine speed

The items that result in the heat transfer rate are the surface area (which does not change for these results), the heat transfer coefficient, and the temperature difference. Figure 13.12 shows the ratios of the heat transfer coefficient and the temperature difference of the peak (one-zone) gas temperature[4] and the wall temperature (450 K). The temperature difference increases slightly and the heat transfer coefficient increases much more significantly as engine speed increases. The heat transfer coefficient [7] is proportional to the engine speed when expressed with units of "kW/m²-K." This increase of the heat transfer coefficient with increases of engine speed, therefore, is largely the reason for the higher heat transfer rates.

In terms of engine thermal efficiency, the item that is of interest is the relative heat transfer, or in other words, the ratio of the energy of the heat transfer relative to the energy of the fuel burned. Figure 13.13 shows the ratios of the relative heat transfer and the heat transfer energy (kJ) as functions of engine speed. Both parameters decrease as engine speed increases. This is in contrast to the results in Figure 13.11 which show the heat transfer rate increasing with increasing engine speed. This means that as engine speed increases, the heat transfer

---

[4] The use of the peak gas temperature is not unique, but provides a representative gas temperature. Other choices for this would be some average gas temperature. The results would be qualitatively the same since only the temperature trends are important for this discussion.

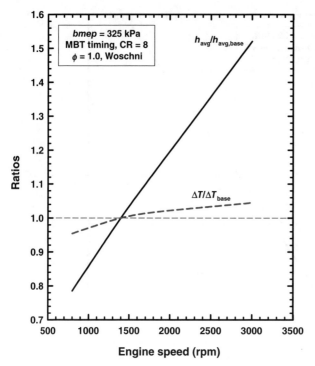

Figure 13.12 The ratios of the average heat transfer coefficient and the temperature difference of the peak gas temperature and the wall temperature as functions of the engine speed

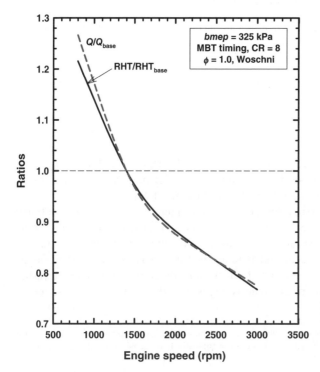

Figure 13.13 The ratios of the relative heat transfer (RHT) and the heat transfer energy (Q) as functions of the engine speed

rate increases but the energy fueling rate increases more. The net result is that the thermal efficiency will increase as engine speed increases since the relative heat transfer decreases.

As shown above, a significant portion of the fuel energy escapes the cylinder as heat transfer. This fact suggests that there may be some benefit of increased work (and efficiency) if at least some of the heat transfer energy was retained in the cylinder. This recognition leads to the next subsection which discusses the potential of low heat rejection engines.

13.4.2   Engines Utilizing Low Heat Rejection Concepts

The use of low heat rejection (LHR) concepts[5] for engine designs has a long history, and has probably been a consideration of engine developers from the very beginning. For example, a patent from 1930 was aimed at reducing the heat losses to increase the thermal efficiency and performance of the engine [11]. The idea of increasing the thermal efficiency and engine performance by reducing energy losses to the coolant and environment is an attractive consideration. The difficulty is reducing the energy losses and converting the retained energy to work. The normal approach for reducing the heat losses and retaining the thermal energy has been to insulate the cylinder. In effect, this means designing "barriers" for the heat transfer.

Since higher gas temperatures would result from any LHR concepts, the spark-ignition engine has not been considered seriously for this application. The higher gas temperatures would result in an increase of the probability of spark knock. This would mean an unfavorable adjustment of the combustion timing or a reduction of the compression ratio to avoid the occurrence of knock. This means that essentially all applications of LHR concepts have been directed toward compression ignition (diesel) engines.

In addition to increases of the thermal efficiency, the successful implementation of LHR concepts has several other advantages. The reduction or elimination of the cooling system not only reduces the engine weight and size, but reduces or eliminates a major maintenance item. Furthermore, removing cooling system components such as water pumps and fans would reduce the mechanical friction (parasitic losses) and add to the increase of the brake efficiency. In addition, the higher operating temperatures may allow lower grade fuels to be used successfully. For these reasons, the military has been especially interested in developing this concept [12]. Also, with any LHR concept the exhaust gas energy would be expected to increase. This additional energy in the exhaust gases would be beneficial for turbocharging and for alternative power production (e.g., turbo-compounding). From a second law perspective, energy in the exhaust gas is more useful than the same energy in the coolant.

Although the use of LHR concepts appears attractive, several disadvantages have been recognized from even the earliest attempts. First, the implementation of the required insulation technologies is difficult. The advanced materials required are generally more brittle and less durable. Successful materials and coatings for these applications must have high temperature strength, expansion coefficients similar to mating metals, low friction characteristics, good thermal shock resistance, light weight, and durability. Further, the higher gas temperatures could result in lower volumetric efficiencies which may tend to offset at least some of the potential gains, and these higher gas temperatures would result in higher nitric oxide emissions.

---

[5] Some of the literature describes LHR engines as "*adiabatic*" engines. This is, of course, a misuse of the term *adiabatic* since no actual device can ever be 100% adiabatic.

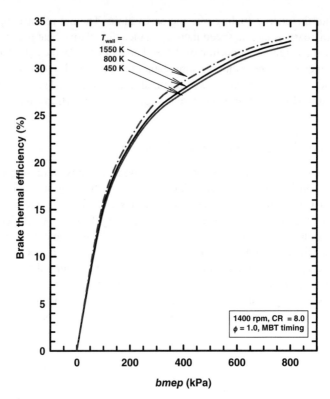

Figure 13.14   The brake thermal efficiency as functions of load (*bmep*) for three wall temperatures

Most early attempts at using LHR concepts were unsuccessful due to the lack of advanced materials (e.g., ceramics) and lubricants which are needed to survive the higher in-cylinder gas temperatures associated with these LHR concepts. Beginning in the 1970s, such advanced materials and lubricants were becoming more available. This provided the opportunity, at least in part, for major research and development programs to advance the LHR engine. This work was largely based on reducing the cylinder heat losses with the goal of obtaining higher efficiencies. Although much was learned from these programs, the general consensus was that thermal efficiency gains, if any, are small. Several overviews on developments related to LHR engines are available (e.g., [13–17]).

The engine cycle simulation (using the heat transfer correlation from Woschni [7]) was used to explore the use of LHR concepts for the base engine. Figures 13.14 and 13.15 show the brake thermal efficiency and Figure 13.16 shows the net indicated thermal efficiency as functions of load (*bmep*) for three wall temperatures: 450 K (standard case), 800 K, and 1550 K. The wall temperature of 1550 K provided near-zero net cylinder heat transfer. Zero net cylinder heat transfer means that the heat transfer from the walls to the gases is equal to the heat transfer from the gases to the walls for the complete event (two revolutions). This is essentially the maximum wall temperature for these conditions. In practice, the actual maximum possible wall temperature will probably be somewhat lower than this value.

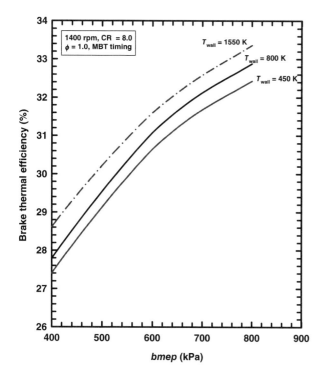

Figure 13.15 The brake thermal efficiency as functions of load (*bmep*) for three wall temperatures emphasizing the higher load conditions

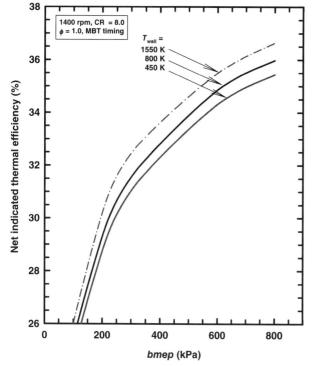

Figure 13.16 The net indicated thermal efficiency as functions of load (*bmep*) for three wall temperatures

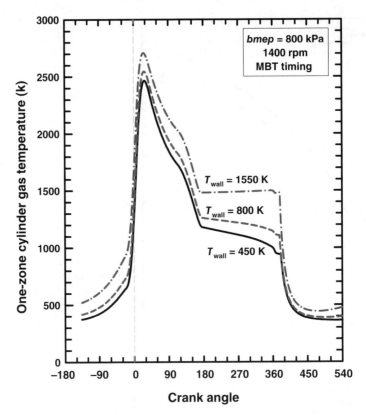

Figure 13.17   The instantaneous one zone cylinder gas temperature as functions of crank angle for two wall temperatures

The thermal efficiencies are slightly higher for the high wall temperature cases, and this is most evident at the higher loads (see Figure 13.15). The cases with the high wall temperatures result in a reduction of the cylinder heat transfer since the temperature difference between the hot gases and the cylinder walls is significantly decreased. These high wall temperatures would require advanced materials (probably ceramics or high temperature coatings), and special designs for alternative lubrication.

Figure 13.17 shows the instantaneous one-zone cylinder gas temperatures as functions of crank angle for the three cases. The peak temperatures are 2465 K, 2546 K, and 2707 K for the wall temperatures of 450 K, 800 K, and 1550 K, respectively. The higher gas temperatures for the higher wall temperature have several consequences. In this case, to maintain the constant load (*bmep* = 800 kPa), the inlet pressure was higher for the high wall temperature cases, and this countered the higher temperatures during the inlet process such that the volumetric efficiency was about the same for all three cases. For a naturally aspirated diesel engine (no throttle), the higher inlet temperatures would reduce the volumetric efficiency and no benefit would be possible from additional inlet pressure increases. To compensate, however, a diesel engine could use a supercharger or turbocharger to increase the inlet pressure.

The higher gas temperatures would result in higher exhaust gas temperatures for the high wall temperature cases. For these conditions, the energy average exhaust gas temperatures

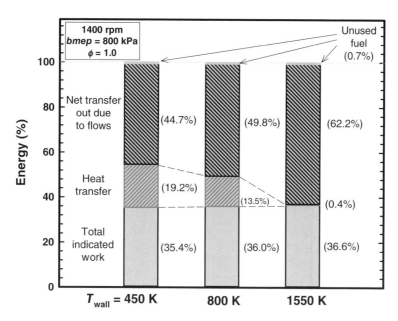

Figure 13.18 The energy distribution for the three wall temperatures

were 1340 K, 1443 K and 1679 K for the 450 K, 800 K, and 1550 K wall temperature cases, respectively. As noted above, these higher exhaust gas temperatures would be useful for turbocharging and turbo-compounding applications. Also, the higher gas temperatures toward the end of the compression stroke, as mentioned above, would tend to worsen any spark knock concerns for SI engines.

The results show a fairly modest increase of the thermal efficiencies with the highest gains for the highest loads. Figure 13.18 shows the distribution of the fuel energy for these three cases for a *bmep* of 800 kPa. For these conditions, the maximum net indicated thermal efficiency increase was about 1.2% (absolute) as the wall temperature increased from 450 K to 1550 K. The heat transfer was reduced by about 18.8% (absolute), and the exhaust gas energy was increased by about 17.5% (absolute). The relatively small amount of increase of the thermal efficiency is due to several items. First, the energy retained by reductions of heat loss late in the expansion stroke contributes little or nothing to work increases. Additionally, these results are indicative of the difficulty of converting thermal energy to work. In other words, although heat transfer is reduced, only a portion of the energy results in increased work. In this case, only about 6% (relative) of the energy associated with the reduced heat transfer is actually converted to work. This is further explained in terms of the thermodynamics next.

As explained elsewhere in this book (Appendix A of Chapter 3), the conversion of thermal energy into work is a strong function of the specific heats (or ratio of the specific heats). Low values of the specific heats (or higher values of the ratio of specific heats) results in higher conversions. Figure 13.19 shows the average of the ratio of specific heats (gamma) during combustion as functions of load (*bmep*) for the three wall temperature cases. The higher wall temperature cases (with the associated higher cylinder gas temperatures) result in significantly lower values for gamma. The impact of gamma on the conversion is as an exponent so the changes do not have to be large to be significant.

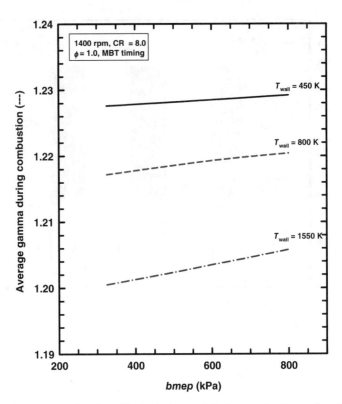

Figure 13.19 The average ratio of specific heats (gamma) during combustion as functions of load (*bmep*) for three wall temperatures

A major observation of the above results for the LHR concepts that has not been emphasized in the literature is the importance of the specific heats (or ratio of specific heats). As the various reviews [13–17] of LHR concepts have outlined, many experimental studies have reported low to zero increase of the thermal efficiencies when employing LHR concepts. This has often been attributed to changes in the combustion process, poor implementation of engine component insulation, or inconsistent experimental methodologies. As the above results indicate, however, another major aspect of the lack of increase of the thermal efficiencies may be due to the inherent thermodynamics associated with the resulting higher temperatures.

As a final comment regarding LHR concepts, at least one application has been proposed which may be a better match to these ideas. Thermodynamic simulation results [18] indicate that combining LHR concepts with low temperature combustion (LTC) engines may be attractive for achieving higher efficiencies. The reason that this combination may be successful is that the LTC engines result in low gas temperatures and this in turn results in lower values of the specific heats (higher values of the ratio of specific heats). Such a situation appears to be able to more effectively convert any retained thermal energy to work and thus provide more substantial thermal efficiency increases. In addition, for net zero heat transfer, the wall temperatures can be significantly lower than for the cases with conventional combustion. More details may be found in Reference 18.

13.4.3 Engines Utilizing Adiabatic EGR

As an example of some of the subtle thermodynamic aspects associated with the cylinder heat transfer, the efficiency and related parameters for an engine using exhaust gas recirculation (EGR) will be examined. The use of EGR was described in Chapter 8 and the use of EGR is a part of some of the other case studies. As described earlier, two EGR configurations may be considered that bracket the range of cooling that may be possible. The *cooled* EGR configuration assumed the exhaust gases were cooled to the nominal inlet mixture temperature (typically 319.3 K for the work reported in this book). This would represent the maximum cooling. The *adiabatic* EGR configuration assumed no cooling such that the exhaust gases entered the inlet stream at the engine-out temperature. This would represent zero cooling.

For this example, the adiabatic EGR configuration is examined since this provides results which highlight some of the subtle aspects of cylinder heat transfer. The engine operating condition is a *bmep* of 325 kPa, 1400 rpm, a compression ratio of 8, and an equivalence ratio of 1.0. The heat transfer correlation is from Woschni [7] with an *htm* of 1.0. Figure 13.20 shows the net indicated and brake thermal efficiency, and the relative heat transfer as functions of the EGR level. To achieve the constant *bmep*, as the EGR level increases, the inlet pressure increases. Since the inlet pressure increases, the cylinder gas pressures increase as

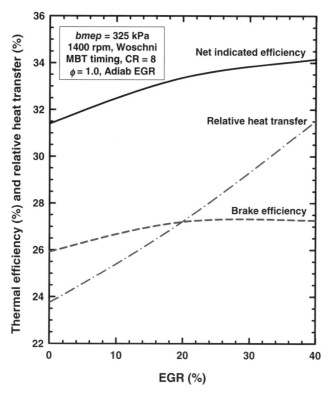

Figure 13.20   The thermal efficiencies and relative heat transfer as functions of the EGR level for the adiabatic EGR configuration

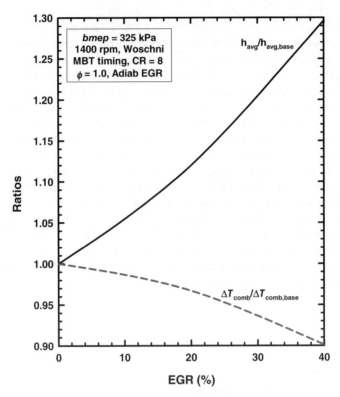

Figure 13.21   The ratio of the heat transfer coefficient (h_{avg}) and the ratio of the temperature difference during combustion (ΔT_{comb}) as functions of the EGR level for the adiabatic EGR configuration

the EGR level increases. The net indicated and brake thermal efficiencies increase as the EGR level increases. These trends will be explained by examining some of the fundamental aspects related to these conditions. As shown in Figure 13.20, the relative heat transfer increases as the EGR level increases. Although not shown, the actual heat transfer also increases as the EGR level increases. The reasons for this heat transfer increase are described next.

Since the surface area is constant, the two major aspects of the cylinder heat transfer are the temperature difference between the gases and the cylinder walls, and the heat transfer coefficient (see eq. 13.1). Figure 13.21 shows the ratio of the average temperature difference for each case during the combustion period and the average temperature difference for the 0% EGR case as functions of the EGR level. This temperature metric decreases with increases of the EGR level largely due to the reduced combustion temperatures due to the dilution effect of the EGR.

Figure 13.21 also shows the ratio of the average heat transfer coefficient for each case and average heat transfer coefficient for the 0% EGR case. This ratio increases with increases of the EGR level. This is largely due to the increase of cylinder pressures and the decrease of cylinder gas temperatures. As provided in Table 13.1, the heat transfer coefficient from Woschni [7] is proportional to the cylinder pressure to a fractional power and inversely proportional

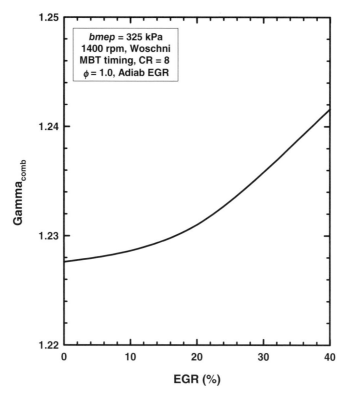

Figure 13.22   The ratio of the specific heats during combustion (Gamma$_{comb}$) as a function of the EGR level for the adiabatic EGR configuration

to the cylinder temperature to a fractional power. Since pressure increases and temperature decreases as the EGR level increases, both these parameters will result in an increasing heat transfer coefficient as the EGR level increases. The net result of the temperature and heat transfer coefficient changes is that the heat transfer coefficient changes dominate, and the relative (and actual) heat transfer increases as the EGR level increases (in spite of the gas temperature decreases).

Returning to Figure 13.20, the net indicated thermal efficiency increases even though the relative heat transfer increases with increases of the EGR level. The reason for this is important and it is not always clearly demonstrated. The use of EGR (or other dilution techniques) results in lower cylinder temperatures, and these lower temperatures result in increases of the ratio of specific heats as the EGR level increases. Figure 13.22 shows the average ratio of specific heats during the combustion period (gamma$_{comb}$) as a function of the EGR level for these conditions. Since the ratio of specific heats is important relative to the conversion of thermal energy into work, even modest increases of the ratio provides significant increases of the thermal efficiency. So, in spite of the increases of the cylinder heat transfer, the net indicated thermal efficiency increases largely due to the increases of the ratio of specific heats.

Finally, the brake thermal efficiency increases but, for the higher EGR levels, the increases are somewhat mitigated. This is due to the increasing mechanical friction. Figure 13.23 shows

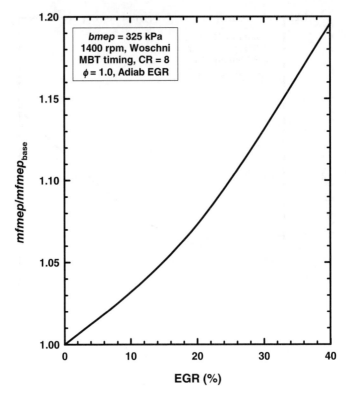

Figure 13.23 The ratio of the mechanical friction mean effective pressure and the mechanical friction mean effective pressure for 0% EGR as a function of the EGR level for the adiabatic EGR configuration

the ratio of the mechanical friction mean effective pressure and the mechanical friction mean effective pressure for 0% EGR as a function of the EGR level. This parameter increases as the EGR level increases largely due to the increasing cylinder pressures.

The above example of the cylinder heat transfer associated with the use of adiabatic EGR provides several important results to highlight some of the details of the cylinder heat transfer and the importance of the thermodynamics related to IC engines. The use of adiabatic EGR, however, is not the preferred application approach. As explained elsewhere in this book, the more favorable application of EGR is to cool the exhaust gases as much as practical. Cooled EGR configurations contribute the most to the thermal efficiencies and are the most effective for the reduction of nitric oxides.

13.5   Summary and Conclusions

Engine cylinder heat transfer is one of the more important processes of internal combustion engines, and yet after decades of research, an acceptable understanding is still lacking. Many global correlations for the convective heat transfer coefficient have been proposed over the years. The current work has compared and contrasted the use of four different engine cylinder heat transfer correlations. Another similar study has examined these and other correlations for a higher load and speed condition [10]. In addition, this case study has examined low heat rejection engines and the use of adiabatic EGR. The major results from the current work may be summarized as follows:

- In general, the cylinder heat transfer is from the gases to the cylinder walls, piston top and cylinder head bottom. For a short period of time during the compression and intake processes, the heat transfer is from the hot walls to the cool entering gases.
- For the four correlations examined, the *maximum* heat transfer coefficient varied by 215% for the same engine operating condition.
- As expected, using correlations with lower heat transfer resulted in higher thermal efficiencies. For the correlations examined, the overall difference for the brake efficiency was about 13% (relative). This much lower impact on the efficiencies was due to the fact that not all the energy retained due to the lower heat transfer was converted to work.
- Other results are available elsewhere [10] for other conditions. In addition to the *bmep* and brake efficiency, these results show the effects of the different heat transfer correlations on exhaust gas temperatures and exhaust energy, and on the exergy destruction during combustion.
- Although the difference between the various engine cylinder heat transfer correlations may be large, the final results in terms of engine efficiency and performance are not affected as much.
- Examples were provided that showed that the important aspect of cylinder heat transfer with respect to thermal efficiency is the relative heat transfer. For increases of engine speed, the heat transfer rate increases, but the relative heat transfer decreases.
- Low heat reduction (LHR) concepts have been proposed for IC engines to recover some of the energy associated with the normal cylinder heat transfer for additional work (and efficiency). Thermodynamic results indicate that only a small portion of the retained energy can be converted to work.
- One proposal for using LHR concepts that appears to have the potential for some success is to couple LHR concepts with LTC engines where the adverse effects of the change of the specific heats are mitigated.
- The use of adiabatic EGR was examined to illustrate some subtle characteristics associated with the cylinder heat transfer. For the conditions examined, cylinder heat transfer increased even though the gas temperatures decreased due to increases of the heat transfer coefficient.
- For the use of adiabatic EGR, the net thermal efficiency increased slightly even though the relative heat transfer increased. This was a result of the increase of the ratio of specific heats. The brake thermal efficiency increased less due to increases of the mechanical friction which was because of the higher cylinder pressures.

References

1. Heywood, J. B (1988) *Internal Combustion Engine Fundamentals*, McGraw-Hill Book Company, New York.
2. Borman G. and Nishiwaki, K. (1987) Internal-combustion engine heat transfer, *Progress in Energy and Combustion Science*, **13**, 1–46.
3. Finol C. A. and Robinson, K. (2006) Thermal modelling of modern engines: a review of empirical correlations to estimate the in-cylinder heat transfer coefficient, *Proceedings of the Institution of Mechanical Engineers – Part D – Journal of Automotive Engineering*, **220** (12) 1765–81.

4. Nusselt, W. (1923) Der warmeubergang in der verbrennungskraftmaschine (the heat transfer in the internal combustion engine), V. D. I.Forschungsheft, 264.
5. Eichelberg, G. (1939) Some new investigations on old combustion engine problems, *Engineering*, **148**, 463–466 and 547–550.
6. Annand, W. J. D. (1963) Heat transfer in the cylinders of reciprocating internal combustion engines, *Proceedings of the Institution of Mechanical Engineers*, **177** (36), 973–996.
7. Woschni, G. (1968) A universally applicable equation for the instantaneous heat transfer coefficient in the internal combustion engine, *SAE Transactions*, SAE paper no. 670931, **76**, 3065–83.
8. Hohenberg, G. F. (1979) Advanced approaches for heat transfer calculations, Society of Automotive Engineers, SAE Paper No. 790825, 1979.
9. Han, S. B., Chung, Y. J., Kwon, Y. J., and Lee, S. (1979) Empirical formula for instantaneous heat transfer coefficient in spark ignition engine, Society of Automotive Engineers, SAE Paper No. 972995.
10. Caton, J. A. (2011) Comparisons of global heat transfer correlations for conventional and high efficiency reciprocating engines, in Proceedings of the ASME Internal Combustion Engine Division 2011 Fall Technical Conference, paper no. ICES2011–60017, Morgantown, WV, 02–05 October.
11. Christensen, N. C. (1930) Internal-combustion engine, patent no. US1755501A.
12. Schwarz, E., Reid, M., Bryzik, W., and Danielson, E. (1993) Combustion and performance characteristics of a low heat rejection engine, Society of Automotive Engineers, SAE Paper No. 930988.
13. Zucchetto, J., Myers, P., Johnson, J., and Miller, D. (1988) An assessment of the performance and requirements for 'adiabatic' engines, *Science*, **240** (5), 1157–62.
14. Siegla, D. C. and Alkidas, A. C. (1989) Evaluation of the potential of a low-heat-rejection diesel engine to meet future epa heavy-duty emission standards, Society of Automotive Engineers, SAE paper no. 890291.
15. Reddy, C. S., Domingo, N., and Graves, R. L. (1990) Low heat rejection engine research status: where do we go from here?, Society of Automotive Engineers, SAE paper no. 900620.
16. Jaichandar, S. and Tamilporai, P. (2003) Low heat rejection engines – an overview, Society of Automotive Engineers, SAE paper no. 2003-01-0405.
17. Smith, J. E. and Churchill, R. (1989) A concept review of low-heat-rejection engines, *Applied Mechanics Review*, **42** (3), 71–90.
18. Caton, J. A. (2011) Thermodynamic advantages of low temperature combustion (LTC) engines using low heat rejection (LHR) concepts, Society of Automotive Engineers, SAE paper no. 2011–01–0312.

14

Fuels

14.1 Introduction

Most of the work reported in this book has examined engine operation using isooctane as the fuel. This chapter will examine engine operation as a function of different pure fuels. The use of various fuels for internal combustion engines began as early as the late 1800s during the development of the first IC engines [1–6]. Over the years, these fuels have included gasoline, diesel, natural gas, liquefied petroleum gas (LPG), other fuel gases, alcohols, and numerous other gaseous and liquid fuels. Each of these fuels has advantages and disadvantages. Often a fuel is considered for IC engines when it is a more economical choice than the current fuel. Also, in recent years, fuels that possess naturally lower emissions are often considered as replacements for the more conventional petroleum fuels.

Gasoline and diesel have been the conventional fuels for spark-ignited and compression-ignited engines since the 1900s. Over the years these two fuels have been and continue to be modified to provide the performance and emission characteristics that are required [1–2]. As stated elsewhere, the success of IC engines has been at least in part due to the excellent match of the conventional fuels and engines [1–3].

The literature on fuels for engines is vast and has included economic assessments, technical evaluations, and "well-to-wheel" considerations. This previous work has included commercial fuels, special blends and pure fuels. No single reference was found which included the eight fuels of the current work, and only a few previous studies appear to have completed a consistent thermodynamic evaluation of a set of fuels.

The objectives of the current work (using the engine cycle simulation) are to compare the performance and second law parameters[1] of a conventional spark-ignition engine operated on each of eight fuels: methane, propane, hexane, isooctane, methanol, ethanol, carbon monoxide, and hydrogen. The comparisons are completed for a base case part load condition, and as functions of speed and load. Complete details of this and related work are available elsewhere [7–9].

---

[1] The second law results may be more understandable after reading Chapter 9 which reports the results from the second law study.

An Introduction to Thermodynamic Cycle Simulations for Internal Combustion Engines, First Edition.
Jerald A. Caton © 2016 John Wiley & Sons, Ltd. Published 2016 by John Wiley & Sons, Ltd.

14.2   Fuel Specifications

Eight fuels were selected for this investigation. Four of these are normal alkanes: methane, propane, hexane and isooctane. Two are alcohols: methanol and ethanol. The final two fuels are carbon monoxide and hydrogen. Although carbon monoxide is not typically considered a fuel, it possesses some unique features and was included for completeness.

Table 14.1, a list of these eight fuels and some of their major specifications, includes the name of the fuel, the molecular formula, the molecular mass, the stoichiometric mass air–fuel ratio (AF), the lower heating value (LHV), the fuel exergy, and the ratio of the fuel exergy and LHV. Note that the ratio of the fuel exergy and LHV is near 1.0, and ranges from a low of 0.909 (for carbon monoxide) to a high of 1.056 (for ethanol).

All liquid fuels are considered vapor as they enter the thermodynamic control system. By not including the vaporization energy of the liquid fuels, the following comparisons are some-what more direct. Since the fuel vaporization process is a complex process of heat transfer, mass transfer, fluid flow and fuel properties, it is difficult to establish a consistent and impartial technique for including this process.

For completeness, however, the following comments are intended to provide some indications of the role that the fuel vaporization could potentially play regarding the engine thermodynamics. As an example, methanol and isooctane are compared. Methanol (an alcohol fuel) is known to possess a high level of vaporization energy compared to isooctane. Methanol and isooctane have vaporization energies of 1103 and 308 kJ/kg$_{fuel}$, respectively. In a simple air stream, 100% vaporization of a stoichiometric mixture of these liquid fuels and air could result in a temperature decrease of about 128 K and 19 K, respectively. Some of the difficulties of this simple analysis are the amount of heating and the source of the energy for this heating. Most often the vaporization is not complete and some liquid droplets remain in the air stream. For these and other reasons, the vaporization process has been omitted in the current work.

Table 14.1   Fuel specifications [1,2,3,4, 12]

Fuel	Molecular mass (kg/kmol)	AF (stoich)	LHV (MJ/kg$_f$)	Fuel exergy (MJ/kg$_f$)	Fuel exergy/ LHV
Hydrogen, H_2	2.016	34.15	120.0	113.5	0.946
Methane, CH_4	16.04	17.17	50.0	49.9	0.998
Propane, C_3H_8	44.10	15.61	46.4	47.1	1.015
Hexane, C_6H_{14}	86.18	15.16	45.1	46.1	1.022
Isooctane, C_8H_{18}	114.2	15.07	44.4	45.5	1.025
Methanol, CH_3OH	32.04	6.45	20.0	21.1	1.055
Ethanol, C_2H_5OH	46.07	8.97	26.9	28.4	1.056
Carbon monoxide	28.01	2.46	10.1	9.18	0.909

14.3   Engine and Operating Conditions

The engine selected for this study is the automotive, 5.7 liter, V–8 configuration described elsewhere. The heat transfer correlation is from Woschni [10] with a heat transfer multiplier (*htm*) of 1.33. The operating condition examined is a moderate load (*bmep* = 325 kPa), moderate speed (2000 rpm) condition. The compression ratio is 8.1:1, and most of the results are for MBT combustion timing.

14.4   Results and Discussion

First, the assumptions and constraints of the study are described. Then the results from this study are presented in the following three subsections: (i) basic results, (ii) engine performance results, and (iii) second law results.

14.4.1   Assumptions and Constraints

As with most of the results presented in this book, the combustion process is considered successful and proceeds as dictated by the Wiebe function for all fuels. Again, the focus of the work is to compare and contrast the fuels from a thermodynamic perspective and not to describe the details of the combustion process. The Wiebe constants were the same for all fuels: $a = 5.0$ and $m = 2.0$.

14.4.2   Basic Results

The use of different fuels alters several aspects of the engine thermodynamics. One difference is that for part load operation, the inlet pressure must be adjusted for equal power (all else the same). Also, for each fuel and for each operating condition, the start of combustion is adjusted so as to obtain the maximum brake torque (MBT timing). These results are for premixed mixtures of fuel vapor and air.

For the base case conditions, three of the fuels are compared in Tables 14.2–14.4. The three fuels selected represent the different types of fuels: normal alkanes (isooctane), alcohol (methanol), and hydrogen. Table 14.2 lists some of the basic engine inputs and some of the basic engine results for the part load base case operating condition for these three fuels. Due to the higher stoichiometric air–fuel ratio for hydrogen relative to isooctane (34.3 vs. 15.1), to produce an equal engine output (say, *bmep*), the hydrogen-fueled engine requires a higher inlet manifold pressure (all else equal). Methanol, on the other hand, has a much lower heating value, and to produce an equal engine output, the inlet pressure must also be higher. Table 14.2 lists inlet pressures of 49.9, 54.0, and 59.8 kPa for isooctane, methanol, and hydrogen, respectively.

Table 14.3 lists the distribution of the fuel *energy* among the major energy items for the three fuels. The brake power percentage is the brake thermal efficiency. The brake efficiency is slightly higher for the methanol case primarily because of the lower exergy destruction during combustion (more effective conversion of thermal energy to work). In general, however, the differences between the fuels are modest. In summary, for all fuels for these conditions, the percentage of the fuel energy for the brake work values are about 24–26%, the heat losses are

Table 14.2   Base case engine inputs and results (*bmep* = 325 kPa, 2000 rpm, MBT timing, ϕ = 1.0) (Woschni [10] heat transfer correlation, *htm* = 1.33)

Item	C_8H_{18}	CH_3OH	H_2
Displaced volume (dm^3)	5.733	5.733	5.733
Engine speed (rpm)	2000	2000	2000
bmep (kPa)	325	325	325
Compression ratio	8.1	8.1	8.1
Comb duration (°CA)	60	60	60
Inlet temperature (K)	319	319	319
Cylinder wall temp (K)	450	450	450
Inlet pressure (kPa)	49.9	54.0	59.8
Start of comb (°bTDC)	22.0	22.0	20.0
AF_{stoich}	15.13	6.45	34.32
Fuel LHV (kJ/kg)	44,400	20,000	120,000
Mech friction mep (kPa)	74.8	75.3	76.1
Exhaust pressure (kPa)	104.4	104.9	105.7
Residual fraction	0.1027	0.0999	0.1016

Table 14.3   Base case engine *energy* results (*bmep* = 325 kPa, 2000 rpm, MBT timing, ϕ = 1.0) (Woschni [10] heat transfer correlation, *htm* = 1.33)

Item	C_8H_{18}	CH_3OH	H_2
		(Percentage of fuel energy)	
Brake power	25.51	26.37	24.55
Mechanical friction	5.84	6.07	5.73
Heat loss	25.49	25.82	25.46
Exhaust	42.50	41.07	43.58
Unburned	0.67	0.67	0.67

25.4–25.8%, the energy values of the exhaust are about 41–44%, and the mechanical friction losses are about 5–6%.

Table 14.4 lists the distribution of the fuel *exergy* among six major items. The percentage of the fuel exergy that is used as indicated power (similar to the indicated thermal efficiency) is about 30.7% for the isooctane case and about 30.8% and 32.2% for the methanol and hydrogen cases, respectively. The percentage exergy that leaves due to heat transfer is about 20.7% for the isooctane case, and about 20.3% and 22.6% for the methanol and hydrogen cases, respectively. The exergy retained in the exhaust gases is about 25.6% for the isooctane case, and about 30.6% and 31.6% for the methanol and hydrogen cases, respectively. Finally, the percentage exergy destroyed during the combustion process is about 21% for the isooctane case, and about 16.1% and 11.3% for the methanol and hydrogen cases, respectively. This latter item, the exergy destroyed during the combustion process, will be discussed in more detail in the following.

Table 14.4 Base case engine *exergy* results (*bmep* = 325 kPa, 2000 rpm, MBT timing, ϕ = 1.0) (Woschni [10] heat transfer correlation, *htm* = 1.33)

Item	C_8H_{18}	CH_3OH	H_2
	(Percentage of fuel exergy)		
Indicated power	30.72	30.83	32.20
Heat loss exergy	20.67	20.25	22.61
Comb exergy destruction	20.99	16.12	11.26
Exhaust exergy	25.59	30.63	31.59
Intake mixing destruction	1.36	1.51	1.66
Unburned	0.67	0.67	0.68

Comparing the values in Tables 14.3 and 14.4 provides some insight into the usefulness of the second law analyses. For example, for these fuels and this condition, the energy associated with the exhaust flow is about 42% of the fuel energy, but the exergy associated with the exhaust is only about 26–32% (depending on the fuel). This indicates that only a fraction of the energy leaving as exhaust is recoverable as work. Of course, any actual device will only be able to convert a portion of the exhaust exergy to work due to the conversion irreversibilities.

Figure 14.1 shows the gross indicated, net indicated and brake thermal efficiency for each fuel for the part load (*bmep* = 325 kPa) base engine condition. The fuels are located from left to right from lowest gross indicated efficiency to highest gross indicated efficiency. The relative order of the fuels is (from lowest efficiency to highest efficiency): carbon monoxide, hydrogen, methane, propane, hexane, isooctane, ethanol, and methanol. This order is consistent with other work reported in the literature for some of these fuels (e.g., [11]). The differences between the fuels with respect to their efficiencies are modest, but some trends can be identified.

For methanol and ethanol compared to isooctane, the slightly higher efficiencies are due at least in part to the lower heat loss, lower exergy destruction during combustion, and lower pumping work. The lower heat loss is due to slightly lower gas temperatures which are due to the slightly lower heating value *per mixture mass*. The low exergy destruction permits a more effective conversion of the thermal energy to work relative to the other fuels.

The n-alkanes (isooctane, hexane, propane, and methane) have about the same thermal efficiencies. The slightly lower efficiencies (relative to the alcohol fuels) are due to the higher heat loss, higher exergy destruction, and the higher pumping work.

For hydrogen compared to isooctane, the thermal efficiencies are slightly lower due to higher heat loss and a lower ratio of the product moles and reactant moles. The gas temperatures are higher for the hydrogen case compared to the isooctane case. The lower efficiencies for hydrogen exist even though the exergy destruction and pumping work was lower for hydrogen. The lower efficiencies result in slightly higher exhaust gas energy and temperatures.

Finally, carbon monoxide has the lowest efficiency largely due to high heat loss (due to high temperatures), a low average ratio of specific heats during the combustion process, and a low ratio of product moles and reactant moles. This low efficiency for carbon monoxide exists even though this fuel had lower pumping work and the lowest exergy destruction.

Figure 14.2 shows results similar to those in Figure 14.1, but for a wide open throttle (WOT) condition. For this work WOT was defined as having an inlet pressure of 95 kPa. The pumping work is, of course, lower for the WOT condition compared to the part load case (Figure 14.1).

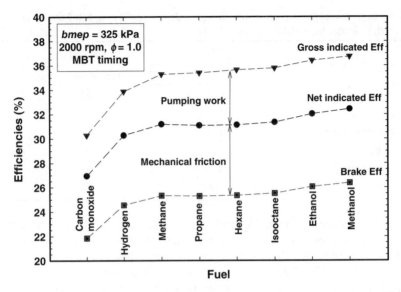

Figure 14.1   The gross indicated, net indicated, and brake thermal efficiency for each fuel for the part load base engine conditions. *Source:* Caton 2010. Reproduced with permission from Elsevier

Each fuel results in a different *bmep* for these conditions. Nevertheless, the ordering of the fuels from lowest to highest efficiency remains nearly the same as for the part load condition. The reasons for these efficiency numbers are similar to those described above for the part load case.

The energy and exergy distributions for the base case conditions for all eight fuels are presented in Appendix 14.A for completeness.

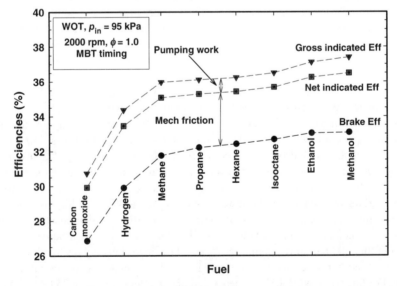

Figure 14.2   The gross indicated, net indicated, and brake thermal efficiency for each fuel for the WOT base engine conditions. *Source:* Caton 2010. Reproduced with permission from Elsevier

14.4.3   Engine Performance Results

The next set of results shows engine performance as functions of the engine speed and load. Figure 14.3 shows the inlet pressure as functions of engine speed for the eight fuels. The variation of inlet pressure is dictated by the constraint of constant brake load (in this case, a *bmep* of 325 kPa). Both the shape of these curves and their relative position can be related to the differences of the brake thermal efficiencies (shown below), heating value, fuel vapor density, displaced air by the fuel vapor, and other such features. As engine speed increases, the brake efficiency first increases (due to the decreasing relative heat transfer), reaches a maximum and then decreases (as mechanical friction becomes more important). The inlet pressure curve is the opposite of the efficiency curve which is a result of the constant load condition.

In Figure 14.3, the relative position of the various curves reflects the individual fuel and the associated incoming fuel energy. The incoming fuel energy is related to the fuel density (fuel molecular mass), the displaced air due to the fuel vapor volume, the heating value, and the brake efficiency. For example, hydrogen requires the highest inlet pressures due largely to its low molecular mass, and isooctane requires the lowest inlet pressure due largely to its high molecular mass. The other fuels are distributed between these two roughly in proportion to their molecular masses.

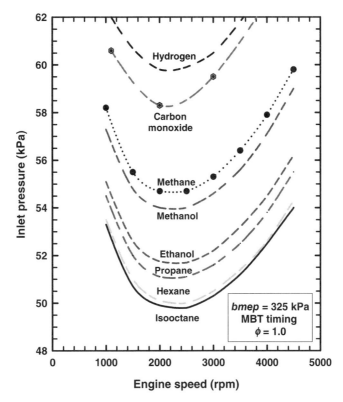

Figure 14.3   Inlet pressures as functions of engine speed for the base case for the eight fuels. *Source:* Caton 2010. Reproduced with permission from Elsevier

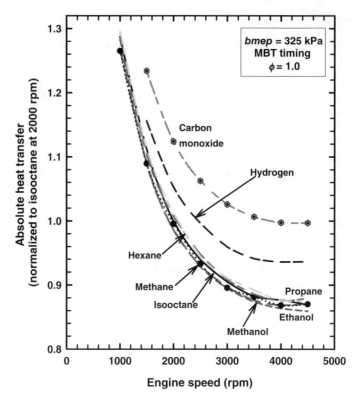

Figure 14.4   Absolute heat transfer (normalized with isooctane at 2000 rpm) as functions of engine speed for the base case for the eight fuels. *Source:* Caton 2010. Reproduced with permission from Elsevier

Figure 14.4 shows the absolute heat transfer as functions of engine speed for the base case conditions for each of the eight fuels for a *bmep* of 325 kPa. The heat transfer has been normalized by the value of the heat transfer for isooctane at 2000 rpm. This was found to be a more concise way to illustrate the heat transfer results. Other ways include reporting the relative heat transfer, but this was found to be somewhat misleading since the fuel energy consumed for each fuel was different. For example, carbon monoxide and hydrogen have lower relative heat transfer compared to the other fuels, but higher actual values (Figure 14.4). This was due to the higher fuel consumption for carbon monoxide and hydrogen.

The heat transfer decreases as the engine speed increases due largely to the decreasing available time for the heat transfer as engine speed increases. Carbon monoxide and hydrogen have the highest heat transfer, and methanol and ethanol have the lowest. These features will contribute to interpreting the following results.

Figure 14.5 shows the net indicated thermal efficiencies as functions of engine speed for the eight fuels for a *bmep* of 325 kPa. The indicated thermal efficiency increases as engine speed increases for all cases due largely to the decreasing importance of the heat transfer (Figure 14.4) for all fuels. The indicated thermal efficiencies for the alcohols are the highest, and for carbon monoxide and hydrogen are the lowest for these conditions. The relative order of the fuels with respect to the efficiencies is the net result of the pumping losses (Figure 14.3), heat transfer (Figure 14.4), thermodynamic properties (e.g., the ratio of specific heats) and fuel energy input

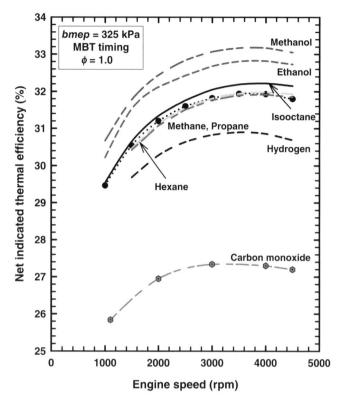

Figure 14.5 The net indicated thermal efficiency as functions of engine speed for the eight fuels for a *bmep* of 325 kPa, an equivalence ratio of 1.0, and MBT timing. *Source:* Caton 2010. Reproduced with permission from Elsevier

density. These various items are different for the various fuels, and no single explanation is adequate to explain all of the results. Some of these explanations are presented above regarding the base case. One of the major contributors, however, is the lower heat transfer associated with the alcohols and the higher heat transfer associated with carbon monoxide and hydrogen.

Figure 14.6 shows the brake thermal efficiency as functions of engine speed for the eight fuels for a *bmep* of 325 kPa. These values are lower compared to the net indicated thermal efficiencies due to the mechanical friction. The mechanical friction increases with increases of speed and this is what causes the decrease in the brake efficiencies for engine speeds above about 2000 rpm. For any engine speed, the fuels are ranked in the same order as discussed above for the net indicated thermal efficiencies.

Related to the above results, Figure 14.7 shows the enthalpy-averaged exhaust gas tempera-ture as functions of engine speed for eight fuels for a *bmep* of 325 kPa, an equivalence ratio of 1.0, and MBT timing. The enthalpy-averaged exhaust gas temperature is based on the energy of the flowing exhaust gases, and accounts for the exhaust gas temperatures, energy, and flow rates. The enthalpy-averaged exhaust gas temperature relates to the energy of the exhaust better than a simple time-average of the exhaust temperatures. For all fuels, as engine speed increases, the exhaust gas temperature increases. This is largely due to the decreasing relative heat transfer as engine speed increases. For the hydrocarbons and alcohols, the exhaust gas

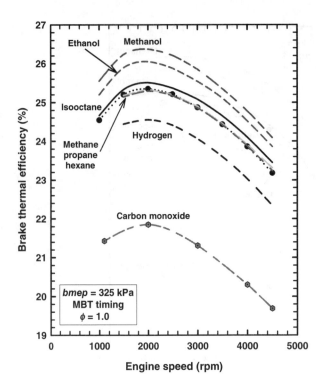

Figure 14.6   The brake thermal efficiency as functions of engine speed for the eight fuels for a *bmep* of 325 kPa, an equivalence ratio of 1.0, and MBT timing. *Source:* Caton 2010. Reproduced with permission from Elsevier

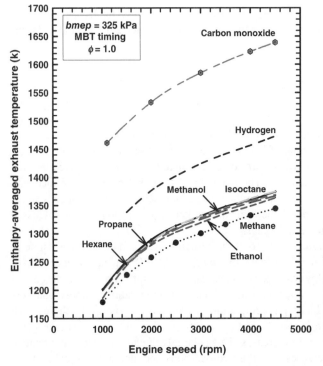

Figure 14.7   The enthalpy-averaged exhaust gas temperature as functions of engine speed for eight fuels for a *bmep* of 325 kPa, an equivalence ratio of 1.0, and MBT timing. *Source:* Caton 2010. Reproduced with permission from Elsevier

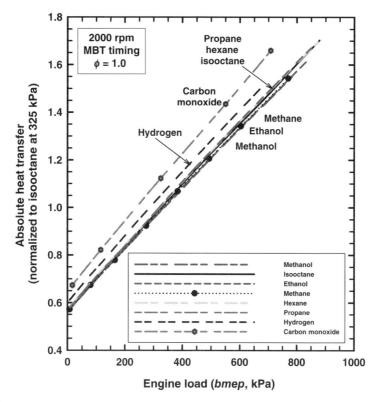

Figure 14.8   The absolute heat transfer (normalized with isooctane at 325 kPa) as functions of engine load (*bmep*) for the eight fuels for an engine speed of 2000 rpm, an equivalence ratio of 1.0, and MBT timing. *Source:* Caton 2010. Reproduced with permission from Elsevier

temperatures range between about 1180 K and 1370 K for these conditions. The exhaust gas temperatures are higher for hydrogen and highest for carbon monoxide. These temperatures are consistent with the results in Figures 14.2–14.6.

Figure 14.8 shows the absolute heat transfer (normalized with heat transfer for isooctane with a *bmep* of 325 kPa) as functions of engine load (*bmep*) for the eight fuels for an engine speed of 2000 rpm, an equivalence ratio of 1.0, and MBT timing. For all fuels, the heat transfer increases with increases of load. The absolute magnitude of the heat transfer increases with load due mainly to the increasing gas temperatures; but, as a percentage of the fuel energy, the relative heat transfer decreases since the fuel input energy is increasing faster as load increases. Carbon monoxide and hydrogen resulted in the highest heat transfer, and methanol and ethanol the lowest heat transfer. These results are consistent with the temperatures associated with each of the fuels.

Figures 14.9 and 14.10 show the net indicated and the brake thermal efficiencies, respectively, as functions of engine load (*bmep*) for the eight fuels for 2000 rpm. In general, over the range of loads, the alcohols resulted in the highest efficiencies, and the four alkanes resulted in the next highest efficiencies. The lowest thermal efficiencies were for hydrogen and carbon monoxide. The four alkanes resulted in intermediate results. These efficiencies are consistent with the above explanations.

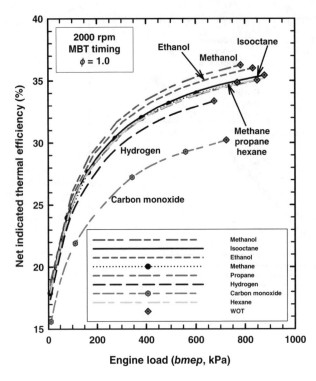

Figure 14.9 The net indicated thermal efficiency as functions of engine load (*bmep*) for the eight fuels for an engine speed of 2000 rpm, an equivalence ratio of 1.0, and MBT timing. *Source:* Caton 2010. Reproduced with permission from Elsevier

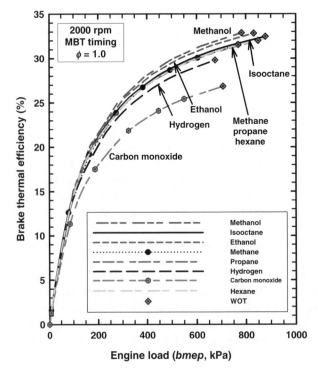

Figure 14.10 The brake thermal efficiency as functions of engine load (*bmep*) for the eight fuels for an engine speed of 2000 rpm, an equivalence ratio of 1.0, and MBT timing. *Source:* Caton 2010. Reproduced with permission from Elsevier

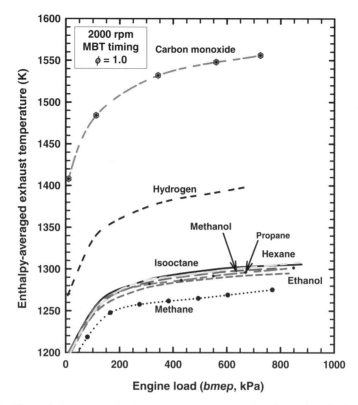

Figure 14.11 The enthalpy-averaged exhaust gas temperature as functions of engine load (*bmep*) for eight fuels for an engine speed of 2000 rpm, an equivalence ratio of 1.0, and MBT timing. *Source:* Caton 2010. Reproduced with permission from Elsevier

Another feature illustrated in Figures 14.9 and 14.10 is the highest loads possible (shown with diamond shape symbols and labeled "WOT") for each fuel with an inlet pressure of 95 kPa. The use of isooctane results in the highest maximum load (maximum *bmep* of about 886 kPa) for these conditions relative to the other fuels. The use of hydrogen, on the other hand, resulted in the lowest maximum load (maximum *bmep* of about 684 kPa) due to the higher volume of fuel vapor for stoichiometric operation. The higher volume of fuel vapor displaces air and reduces the fuel that can be oxidized. Finally, carbon monoxide also was limited in maximum output—about 707 kPa maximum *bmep* for these conditions.

Figure 14.11 shows the enthalpy-averaged exhaust gas temperature as functions of engine load (*bmep*) for eight fuels for an engine speed of 2000 rpm, an equivalence ratio of 1.0, and MBT timing. In general, as load increases the exhaust gas temperatures increase for all fuels. This is consistent with the previous results and largely reflects the higher temperatures associated with the higher loads.

The "order" of the fuels with respect to exhaust temperature level is roughly related to their individual heat transfer, gas temperature, and thermal efficiency values. For example, carbon monoxide and hydrogen have the highest exhaust gas temperatures, and they have the lowest thermal efficiencies, and the highest one-zone peak gas temperatures. Carbon monoxide has the lowest relative heat transfer and highest total (air plus fuel) flow rate of all the fuels, and these items contribute to the higher exhaust gas temperatures. Methane, on the other hand, has the lowest exhaust gas temperatures of all the fuels. This is largely a result of the mid-level thermal efficiency and a mid-level relative heat transfer for methane compared to the other fuels. The other fuels are closely grouped in terms of their exhaust gas temperature, and the differences between these fuels is largely related to the fuels' thermal efficiencies, heat transfer and gas temperatures. For example, methanol and ethanol have the highest heat transfer levels of all the fuels which contribute to their lower exhaust gas temperatures.

14.4.4   Second Law Results

As shown above, the combustion process is responsible for the majority of the exergy destruction during engine operation. Exergy destruction during the combustion process is the result of a number of irreversible molecular processes. These include the mixing of reactant species, chemical reactions, heat transfer among the molecules at different temperatures, and mixing of product species. Of these, heat transfer among the molecules is expected to be one of the largest contributors to the exergy destruction. Other items that are important are the number of moles produced during the reaction.

Figure 14.12 shows the percentage of the fuel exergy destroyed during the combustion process for the eight fuels for a *bmep* of 325 kPa, an equivalence ratio of 1.0, and 2000 rpm. The four normal alkanes are shown in the right portion of the figure and hydrogen, carbon monoxide, and the two alcohols are shown in the left portion of the figure. The exergy destruction percentage increases slightly as the n-alkanes increase in size (more carbon; more hydrogen).

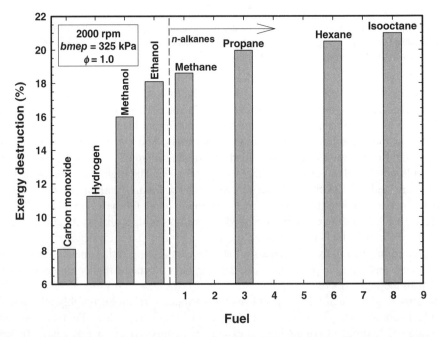

Figure 14.12   The percentage of the fuel exergy destroyed during the combustion process for the eight fuels for a *bmep* of 325 kPa, an equivalence ratio of 1.0, and 2000 rpm. *Source:* Caton 2010. Reproduced with permission from Elsevier

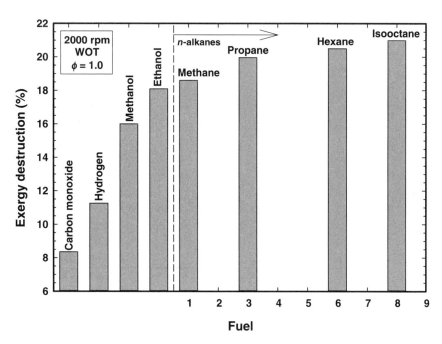

Figure 14.13 The percentage of the fuel exergy destroyed during the combustion process for the eight fuels for wide open throttle (WOT), an equivalence ratio of 1.0, and 2000 rpm. *Source:* Caton 2010. Reproduced with permission from Elsevier

The alcohol fuels result in slightly lower exergy destruction. This appears to be related to the oxygen atom contained in both molecules.

Hydrogen has the lowest destruction for the traditional fuels, and this appears to be due to the fact that it is the simplest molecule (in addition, hydrogen combustion is at higher temperatures which is discussed below). Finally, carbon monoxide has the lowest destruction for all fuels. Not only is carbon monoxide a simple molecule, but it also contains an oxygen atom. In addition, carbon monoxide and hydrogen possess the characteristic that the total number of moles is reduced during the reaction (which is not true of the other fuels). The decrease of the number of moles helps in mitigating the increase of entropy.

For completeness, Figure 14.13 shows results similar to Figure 14.12, but for WOT conditions. The values for the exergy destruction during combustion are similar to the values for the part load case.

The exergy destruction is expected to correlate with the gas temperatures. As shown elsewhere [12], the exergy destruction during combustion for a range of engine operating and design parameters correlated best with the burned gas temperature. Figure 14.14 shows the exergy destruction as functions of the maximum burned zone gas temperature. As shown, for higher temperatures during the combustion process, the exergy destruction decreases. The temperature differences for all the fuels except carbon monoxide and hydrogen are within about ±1% for this correlation. Due to the relatively small temperature differences for the various fuels, the exergy destruction during the combustion process does not have a high correlation with temperature for these conditions. The effects of temperature during combustion are still important, but the fuel molecular characteristics appear to also play an important role.

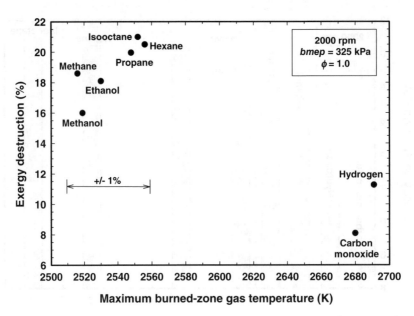

Figure 14.14   The percentage of the fuel exergy destroyed during the combustion process for the eight fuels as functions of the maximum burned zone temperature for a *bmep* of 325 kPa, an equivalence ratio of 1.0, and 2000 rpm. *Source:* Caton 2010. Reproduced with permission from Elsevier

14.5   Summary and Conclusions

A thermodynamic engine cycle simulation was used to examine the performance of eight fuels used in a spark-ignition engine. The fuels examined were methane, propane, hexane, isooctane, methanol, ethanol, carbon monoxide, and hydrogen. The fuels were examined as vapor, and no energy for vaporization was included in this assessment. The base case was a part load condition which included a *bmep* of 325 kPa at 2000 rpm. Inlet pressure was adjusted to attain the same load. This resulted in slightly lower pumping losses for fuels such as hydrogen and carbon monoxide. Even so, engine performance was similar for the eight fuels for the conditions examined.

The greatest differences were for the values of the exergy destruction during the combustion process for the various fuels. For the conditions examined, carbon monoxide had the lowest destruction (about 8% of the fuel exergy) while isooctane had the highest value (about 21% of the fuel exergy). The other fuels had exergy destruction values in between these two values and the actual amount of exergy destruction seems to correlate best with the burned gas temperature, the complexity of the fuel molecule, and the presence (or absence) of oxygen atoms.

References

1. Guibet, J.-C. (1999). *Fuels and Engines*, Volume 1, Editions Technip, Paris.
2. Guibet, J.-C. (1999). *Fuels and Engines*, Volume 2, Editions Technip, Paris.
3. Owen, K. and Coley, T. (1995). *Automotive Fuels Reference Book*, second edition, Society of Automotive Engineers, Inc.

4. Weaver, C. S. (1990). *Gaseous Fuels and Other Alternative Fuels*, Society of Automotive Engineers, Inc., SP–832.
5. Szybist, J. P., Chakravathy, K., and Daw, C. S. (2012). Analysis of the impact of selected fuel thermochemical properties on internal combustion engine efficiency, *Energy & Fuels*, **26**, 2798–2710.
6. Heywood, J. B., (1988). *Internal Combustion Engine Fundamentals*, McGraw-Hill Book Company, New York.
7. Caton, J. A. (2010). Implications of fuel selection for an SI engine: results from the first and second laws of thermodynamics, *Fuel*, **89**, 3157–3166.
8. Caton, J. A. (2009). A thermodynamic evaluation of the use of alcohol fuels in a spark-ignition engine, presented at the 2009 SAE Powertrains, Fuels and Lubricants Meeting, Society of Automotive Engineers, SAE paper no. 2009–01–2621, Grand Hyatt San Antonio, San Antonio, TX, 02–04 November.
9. Caton, J. A. (2006). First and second law analyses of a spark-ignition engine using either isooctane or hydrogen, in Proceedings of the 2006 Fall Conference of the ASME Internal Combustion Engine Division, Sacramento, CA, 5–8 November 2006.
10. Woschni, G. (1968). A universally applicable equation for the instantaneous heat transfer coefficient in the internal combustion engine, *SAE Transactions*, SAE paper no. 670931, **76**, 3065–3083.
11. Farrell, J. T., Stevens, J. G., and Weissman, W. (2006). A second law analysis of high efficiency low emission gasoline engine concepts," Society of Automotive Engineers, SAE paper no. 2006–01–0491.
12. Caton, J. A. (2015). Correlations of exergy destruction during the combustion process for internal combustion engines, *International Journal of Exergy*, **16** (2) 183–213.
13. Caton, J. A. (2010). The destruction of exergy during the combustion process for a spark-ignition engine, in Proceedings of the ASME Internal Combustion Engine Division 2010 Fall Technical Conference of the, paper no. ICES2010–35036, San Antonio, TX, 12–14 September.

Appendix 14.A: Energy and Exergy Distributions for the Eight Fuels at the Base Case Conditions (*bmep* = 325 kPa, 2000 rpm, ϕ = 1.0 and MBT timing)

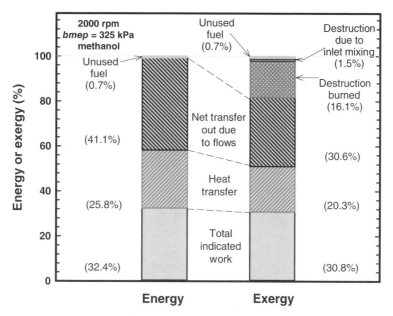

Figure 14.A.1 Results for methanol[2]. *Source:* Caton 2010. Reproduced with permission from ASME

---

[2] Versions of these figures are available in Reference 13.

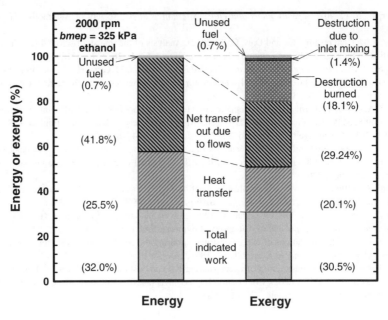

Figure 14.A.2   Results for ethanol. *Source:* Caton 2010. Reproduced with permission from ASME

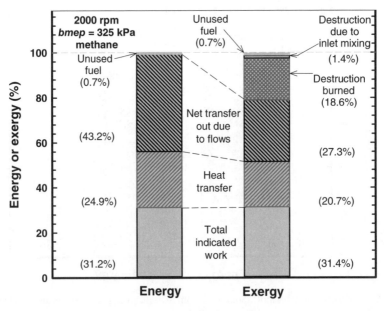

Figure 14.A.3   Results for methane. *Source:* Caton 2010. Reproduced with permission from ASME

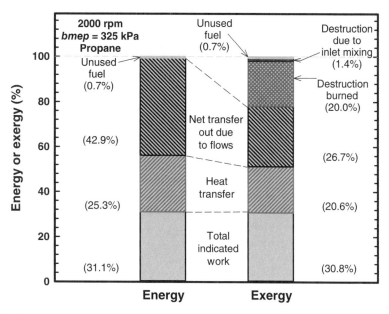

Figure 14.A.4   Results for propane. *Source:* Caton 2010. Reproduced with permission from ASME

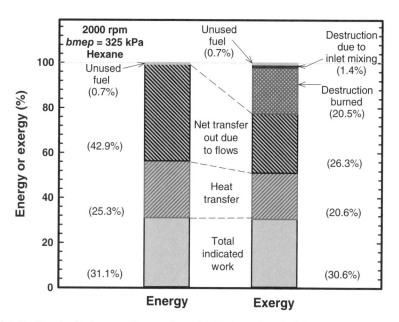

Figure 14.A.5   Results for hexane. *Source:* Caton 2010. Reproduced with permission from ASME

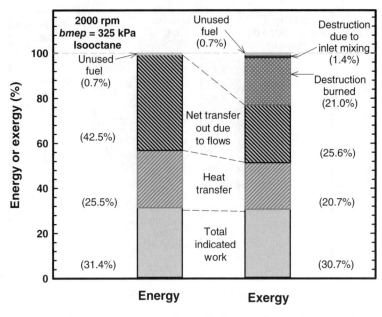

Figure 14.A.6   Results for isooctane. *Source:* Caton 2010. Reproduced with permission from ASME

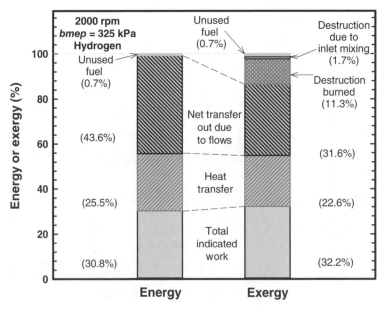

Figure 14.A.7   Results for hydrogen. *Source:* Caton 2010. Reproduced with permission from ASME

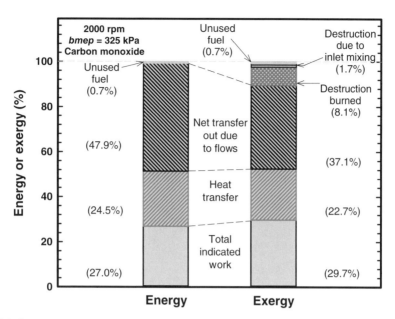

Figure 14.A.8 Results for carbon monoxide. *Source:* Caton 2010. Reproduced with permission from ASME

15

Oxygen-Enriched Air

15.1 Introduction

The results presented throughout this book have been for engine operation using "normal" air with an oxygen concentration of 21%[1] (by volume). The current chapter is aimed at extending the results to include inlet air with greater amounts of oxygen.

Some of the motivations for developing engines to operate with oxygen-enriched air include increasing the power density, reducing CO and HC emissions, improving cold start emissions, reducing cyclic variations, and extending the lean limits. In addition, for potential future engines operating on hydrogen, oxygen may be a by-product of the hydrogen production. This oxygen may then be available at relatively low cost. The expected disadvantages include increased nitric oxide emissions, increased tendencies to knock, and the lack of practical and economical systems to provide the oxygen-enriched air.

This study is an examination of the effects of oxygen enrichment of the inlet air on engine performance and nitric oxide emissions for a spark-ignition engine. Oxygen-enriched inlet air is expected to affect the combustion process, and thus result in changes in engine performance. The objective of this work is to quantify these expectations for a range of operating conditions.

Another related topic is the effect of oxygen enrichment on the destruction of exergy during combustion. This topic is included in a related technical paper [1].

The following sections of this chapter include a summary of previous literature, a summary of the engines and operating conditions for the current study, and a presentation of the results. The results will include overall and detail engine and thermodynamic parameters, and results for nitric oxide production. Finally, this chapter will end with a brief summary and conclusions.

---

[1] The calculations used an oxygen concentration of 20.95% for normal air; however, for simplicity, the text uses the number 21%.

An Introduction to Thermodynamic Cycle Simulations for Internal Combustion Engines, First Edition.
Jerald A. Caton © 2016 John Wiley & Sons, Ltd. Published 2016 by John Wiley & Sons, Ltd.

15.2　Previous Literature

This section will provide a brief summary of the background information concerning the use of oxygen-enriched combustion air in spark-ignition engines. The following citations, then, are examples of the literature on this topic, but are not intended to be comprehensive. Considerable work also has been completed on oxygen-enriched air for compression-ignition (diesel) engines, but these references are not reviewed here.

Engine operation using oxygen-enriched inlet air was reported as early as 1946. Hawthorne [2] (1946) described the use of oxygen-enriched air to increase the performance of aircraft engines. In the current era, Wartinbee [3] (1971) described an experiment using oxygen-enriched air to investigate the potential of reducing emissions. Although he reported reduced carbon monoxide and hydrocarbon emissions, he also noted increases in nitric oxide emissions.

Quader [4] (1978) published a paper on the results of experiments using a single-cylinder, spark-ignition engine operating with oxygen-enriched inlet air. He used oxygen from a high pressure cylinder to produce inlet air with up to 32% oxygen. The engine was operated at 1200 rpm with a 6.8:1 compression ratio and at MBT spark timing for a range of equivalence ratios. He found a significant increase in nitric oxide emissions as oxygen concentration increased particularly for lean mixtures. Hydrocarbon emissions decreased with increasing oxygen concentrations, and the lean limit was lower for increases in oxygen enrichment. Quader noted that the combustion period (from ignition to end of combustion) decreased for increases in oxygen enrichment for all equivalence ratios. The indicated thermal efficiency and the volumetric efficiency decreased as oxygen enrichment increased for all equivalence ratios.

Quader [4] (1978) stated that many interacting factors were responsible for the above observations. The use of oxygen enrichment causes higher combustion temperatures and flame propagation velocities. These items, in turn, cause shorter burn durations. The higher combustion temperatures may result in more heat transfer to the coolant and the earlier combustion may result in longer times for heat transfer (due to higher gas temperatures earlier in the expansion stroke). Another effect may be the reduction of the ratio of heat capacities for the product gases which causes a reduction in the thermal efficiency.

Detuncq et al. [5] (1988) reported on an experimental study of the use of oxygen-enriched air for a single cylinder, natural gas fueled, spark-ignition engine. A major motivation of this work was related to the observation that natural gas fueled engines typically have lower specific power output compared to liquid-fueled engines. The use of oxygen-enriched air was anticipated to improve the specific power output of natural gas fueled engines. They examined air mixtures with up to 28% oxygen by volume. Relative to the conventional gasoline engine, the use of oxygen-enriched inlet air resulted in higher specific power, higher rates of pressure rise, higher exhaust temperatures, higher energy losses to the cooling water, and lower thermal efficiencies.

Kajitani et al. [6] (1992) examined modest levels of oxygen enrichment (up to 23% oxygen by volume) on engine performance and in-cylinder reactions. They used two different single-cylinder, spark-ignition engines in two laboratories for different aspects of the study. In contrast to Quader (1978), they found an increase in the brake thermal efficiency as the oxygen enrichment increased for fixed spark timing. Also, the brake mean effective pressure increased for increases in the oxygen enrichment for any fixed spark timing. The exhaust gas temperature increased slightly for increases in the oxygen enrichment for any fixed spark timing. They also found that the combustion duration decreased and the cyclic variation reduced for increases in the oxygen enrichment. Kajitani *et al.* [7] (1993) reported on an extension of the above work to include studies at part load operation and during the warm-up period.

Ng et al. [8] (1993) presented results for a vehicle using a 3.1 liter V–6, spark-ignition engine. The engine was a variable fuel version (VFV), and was tested with inlet air with up to 30% (by volume) oxygen. They noted that for oxygen enrichment above 30%, the engine would knock. The vehicle was mounted on a chassis dynamometer and was tested over the FTP emissions cycle. They found reductions of carbon monoxide and hydrocarbon emissions particularly for the cold phase of the FTP as oxygen enrichment increased. On the other hand, they noted increases in nitric oxide emissions as oxygen enrichment increased.

Maxwell et al. [9] (1993) reported on the performance and emissions at wide-open throttle for a single-cylinder, spark-ignition engine for oxygen-enriched inlet air with oxygen concentration up to 25%. They used both gasoline and natural gas. They found significant increases in power output, efficiency, and exhaust gas temperatures; and substantial reductions of hydrocarbons and carbon monoxide emissions for either fuel.

Poola et al. [10,11] (1995, 1996) published two papers describing their work using a spark-ignition engine equipped vehicle. They reported both engine out and converter out emissions from the federal test procedure (FTP). They used oxygen-enriched air with up to 25% oxygen concentration. They found carbon monoxide and hydrocarbon decreases of up to 50% using the oxygen-enriched air compared to using normal air. The corresponding nitric oxide emissions, however, were up to 79% higher. Time-resolved results indicated that most of the hydrocarbon and carbon monoxide reductions came during the cold-start portion of the cycle.

As the above indicates, considerable interest exists concerning the use of oxygen-enriched inlet air for internal combustion engines. A majority of this previous work has been experimental. In contrast to experimental work, new and unique insights on the use of enriched inlet air are possible by the use of engine cycle simulations.

15.3   Engine and Operating Conditions

As with most of the work reported in this book, the base engine for this study was the 5.7 liter, V–8 engine. As described below, additional engines were used in this study. In brief, for oxygen-enriched inlet air, engines with smaller displacements could be used and still generate the same power. The operating conditions examined include wide-open throttle with an inlet pressure of 95 kPa. The equivalence ratio was stoichiometric, the compression ratio was 8.1, the combustion duration was 50°CA, and the heat transfer correlation was from Woschni [12] with an *htm* of 1.33. Other details may be found in Reference 13.

15.4   Results and Discussion[2]

Results from this study are presented in five sections. First, the strategy used in this study is described. Then, examples of the thermodynamic properties for different levels of oxygen-enriched air are provided. This is followed by results for the base case conditions. Finally, the last two sections present engine performance and nitric oxide emissions as functions of engine parameters for a range of oxygen-enriched inlet air.

---

[2] Another version of the material in this chapter is available in Reference 13.

15.4.1   Strategy for This Study

A comparison of engine operation using oxygen-enriched inlet air may be conducted in a number of ways. Most experimental studies have reported engine performance as functions of the inlet oxygen concentration using one engine with a constant displaced cylinder volume. For different levels of inlet oxygen concentration, this results in different power outputs for the different oxygen concentrations (all else constant). Another approach is to compare the operation with different sized engines. For this approach, for each inlet air oxygen concentration, the engine size (displaced volume) is adjusted so that power output remains constant. This latter approach provides the capability to examine the fundamental differences due to the thermodynamics without the complication of different power outputs. From a practical point of view, this type of comparison provides insight into the potential "downsizing" of engines operating with oxygen enriched inlet air.

Knock has not been considered in the following study. For the conditions examined, "end gases" may be at higher temperatures and pressures for some oxygen concentrations. This, therefore, would tend to enhance the occurrence of knock. Higher octane fuels or the use of anti-knock fuel additives could be used to avoid knock for these conditions. In any case, the detail consideration of knock was beyond the scope of this study.

Also, note that the mechanical friction mean effective pressure has not been varied as a function of engine displacement. In general, as engine size decreases, the relative importance of friction increases. For the relatively modest engine size change proposed in this study and due to the relatively large variation of friction correlations, the friction mean effective pressure was assumed to remain unchanged. This assumption could easily be removed, but this would not change the general nature of the results of this study.

15.4.2   Basic Thermodynamic Properties

The use of inlet air with oxygen enrichment alters several aspects of the thermodynamics compared to the use of traditional air. These aspects include the increased oxygen concentration in the inlet air, the different equilibrium species concentrations in the product gases, and the different thermodynamic properties in both the reactant and product gas mixtures. One consequence of these differences in properties is reflected in different gas temperatures. These results for the thermodynamic properties (and for the results from the engine computations) are for isooctane and "air" mixtures.

The oxygen concentration in the reactant mixture increases as the oxygen concentration in the inlet "air" increases. Figure 15.1 shows the equilibrium oxygen concentrations (with a log scale) in the product gases as functions of the equivalence ratio for normal air (with 21% oxygen) and air enriched to 42% oxygen. These results are for typical combustion gases at 2750 K and 3000 kPa. As expected, the product oxygen concentration is higher for the 42% oxygen inlet air for all equivalence ratios. In addition, the oxygen concentration in the product mixture decreases rapidly as equivalence ratio increases. For the rich equivalence ratios, the difference is modest.

Figure 15.2 shows the mixture enthalpy values for the reactant and product gases as functions of temperature for both normal air (with 21% oxygen) and oxygen-enriched air with 42% oxygen for an equivalence ratio of 1.0 and for 3000 kPa. On this scale, the enthalpy values for the reactant mixtures are similar with the case of the oxygen-enriched air having

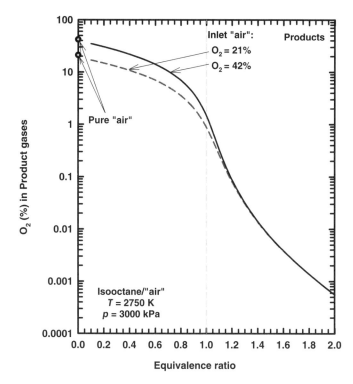

Figure 15.1 Oxygen concentration in the product mixture as functions of equivalence ratio for normal air (21% oxygen) and oxygen-enriched air (42% oxygen). *Source:* Caton 2005. Reproduced with permission from ASME

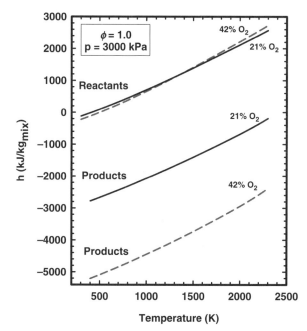

Figure 15.2 Enthalpy in the reactant product mixtures as functions of temperature for an equivalence ratio of 1.0 and a pressure of 3000 kPa for normal air (21% oxygen) and oxygen-enriched air (42% oxygen). *Source:* Caton 2005. Reproduced with permission from ASME

slightly lower enthalpy values for temperatures below about 1300 K, and slightly higher for temperatures above about 1300 K. The enthalpy values for the products are markedly different with the case of the oxygen-enriched air possessing much lower enthalpy values. This is due to the greater relative importance of species such as CO_2 and H_2O as the nitrogen component decreases.

15.4.3   Base Engine Performance

The base engine selected for this study is the standard automotive 5.7 dm^3, V–8, spark-ignition engine. A wide-open throttle (WOT) operating condition at 2500 rpm was selected for the base operating case. A wide-open throttle operating condition was considered relevant since a portion of the current study involved selecting the engine size (displaced volume), and this is largely dictated by WOT performance.

With the above input parameters, the simulation was used for the case of normal (21% oxygen) and oxygen-enriched inlet air (32%) for the base case condition: 2500 rpm and 95.0 kPa (inlet pressure). Table 15.1 lists some of the engine parameters and some of the major output parameters for these two cases. For normal air (21% oxygen), the brake power output was 110 kW. For the case of 32% oxygen, to obtain 110 kW, the engine displacement could be decreased to 4.18 dm^3 liter. Further details about the engine scaling are presented below.

The two engines have different mean effective pressures due to the different displaced volumes with the smaller engine (higher oxygen concentration) having the greater mean effective pressures. The specific fuel consumption is higher (and the thermal efficiency is lower) for the smaller engine, and this is largely due to the higher heat loss and higher exhaust temperature of the smaller engine. This higher heat loss of the small engine is due to the higher gas temperatures which are a result of the higher oxygen concentration (less nitrogen dilution).

Table 15.1 also lists the distribution of the energy values for the two engine cases. This distribution is another way to examine the differences between the two cases, and these results are consistent with those described above. The exhaust gas and heat loss energies are higher and the brake work energy is lower for the smaller (higher oxygen concentration) engine compared to the larger engine (normal air). These results are more fully explained below.

Detailed thermodynamic results from the cycle simulation are presented next. Figure 15.3 shows the cylinder pressure and average (one-zone) gas temperature as functions of crank angle for the base case operating conditions for inlet air with 21% (solid lines) and 40% (dashed lines) oxygen concentrations. The higher oxygen concentration of 40% was selected to emphasize the differences (and may not be a practical oxygen level). The engine size for this higher oxygen concentration was 3.65 dm^3 to obtain the same power output of 110 kW. The cylinder pressures and gas temperatures have a similar change with respect to crank angle, but are proportionally higher for the higher oxygen concentration (smaller engine). The case with the higher oxygen concentration (lower nitrogen dilution) results in higher average (one-zone) temperatures and higher cylinder pressures.

Figure 15.4 shows the average (one-zone) temperatures as functions of crank angle for normal air (21% oxygen) and for four levels of oxygen-enriched inlet air. The gas temperatures proportionally increase as the oxygen concentration of the inlet air increase. This is largely a result of the reduced level of diluents (i.e., nitrogen molecules) for the higher oxygen inlet air concentrations, and the resulting higher combustion temperatures.

Table 15.1 Comparison of performance results for the same power and combustion duration

Item	32% oxygen (4.18 dm³)		21% oxygen (5.73 dm³)	
Inputs:				
p_{inlet} (kPa)	–	95.0	–	95.0
Equivalence ratio	–	1.0	–	1.0
Spark timing	–	MBT	–	MBT
Combustion start	–	−22.0	–	−20.0
Comb duration	–	50°	–	50°
Heat transfer	–	Woschni	–	Woschni
htm	–	1.33	–	1.33
fmep (kPa)	–	98.0	–	98.0
Results:				
	Indicated[a]	Brake	Indicated[a]	Brake
mep (kPa)	1353	1255	1017	920
sfc (g/kW hr)	250.3	269.7	219.4	242.6
η (%)	32.4	30.1	37.0	33.4
Torque (N•m)	449.9	417.5	464.2	419.8
Power (kW)	117.8	109.3	121.6	109.9
max one-zone temp	–	2963	–	2554
max T_b (K)	–	3075	–	2685
CA of max T_b	–	9.2	–	10.5
p_{peak} (kPa)	–	5686	–	4525
CA of p_{peak}	–	13.2	–	15.0
$\dot{m}_{fuel}$(g/s)	–	8.19	–	7.41
$\dot{m}_{air}$(g/s)	–	82.3	–	112.0
T_{exh} (K)	–	1741	–	1312
Residual fraction	–	0.043	–	0.053
Energy dist:				
Brake work (%)	–	30.07	–	33.42
Friction (%)	–	2.33	–	3.53
Heat loss (%)	–	21.02	–	19.04
Exhaust (%)	–	45.91	–	43.28
Unused (%)	–	0.67	–	0.67
Total (%)	–	100.0	–	100.0

[a]Net indicated is for all four strokes.

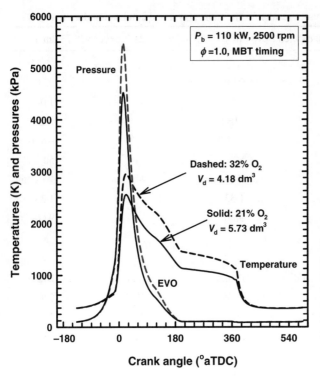

Figure 15.3   Cylinder pressures and temperatures as functions of crank angle for 21% and 40% oxygen concentrations in the inlet air for the base case. *Source:* Caton 2005. Reproduced with permission from ASME

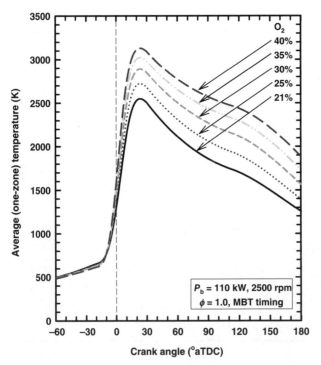

Figure 15.4   Average (one-zone) cylinder gas temperature as functions of crank angle for five oxygen concentrations in the inlet air for the base case. *Source:* Caton 2005. Reproduced with permission from ASME

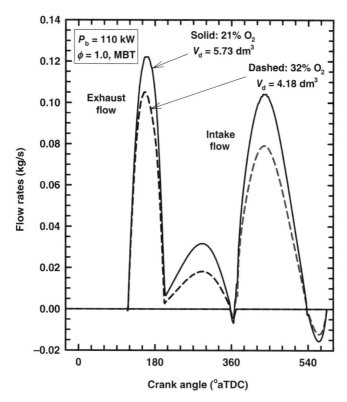

Figure 15.5 Exhaust and intake flow rates as functions of crank angle for 21% and 40% oxygen concentration in the inlet air. *Source:* Caton 2005. Reproduced with permission from ASME

Figure 15.5 shows the mass flow rates through the engine for the base case operating conditions for inlet air with 21% and 40% oxygen concentrations. The case for the higher oxygen concentration requires lower mass flow rates to obtain the same power output of 110 kW. Again, these reduced flow rates for the higher oxygen inlet air are largely due to the reduced level of diluents.

15.4.4   Parametric Engine Performance

Next, the engine performance will be described for different levels of oxygen enrichment for the *same power output*. As described above, this is a unique comparison that was first reported in the literature in 2005 [13]. Again, as the oxygen concentration increases, the engine displacement must be reduced to obtain the same power level (all else the same). Figure 15.6 shows the engine displacement and a scaling factor as functions of oxygen concentrations in the inlet air from 21% to 40%. The scaling factor was used to reduce all the geometric dimensions (bore, stroke, valve diameters, …) in a uniform fashion. For example, the base engine has a 5.73 dm^3 displacement and represents a scaling factor of 1.0. As oxygen concentration increases, the scaling factor and displaced volume decreased to maintain the 110 kW power

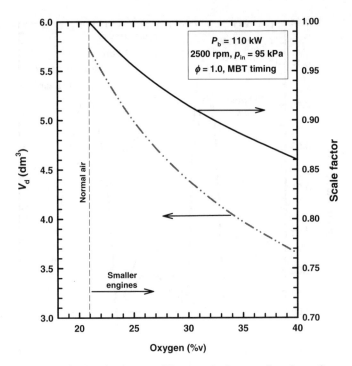

Figure 15.6 Engine total displaced volume and linear scale factor as functions of oxygen concentration in the inlet air for three spark timings for the base case. *Source:* Caton 2005. Reproduced with permission from ASME

output. As oxygen concentration increased from 21% to 40%, the scaling factor decreased from 1.0 to about 0.86 and the displaced volume decreased from 5.73 dm^3 to about 3.65 dm^3.

Figure 15.7 shows the brake mean effective pressure (*bmep*) as functions of the oxygen concentration in the inlet air for three spark timings: MBT timing, 10°CA advanced timing, and 10°CA retarded timing. Also shown is the scale factor used for each oxygen concentration. The brake mean effective pressure increases with oxygen concentration which reflects the decreasing diluents present. For this range of oxygen concentration change, the *bmep* increased about 58%. For non-MBT spark timing (both the advanced and the retarded timing cases), the *bmep* decreases.

Figure 15.8 shows the brake thermal efficiency as functions of oxygen concentration in the inlet air for three spark timings: MBT timing, 10°CA advanced timing, and 10°CA retarded timing. The thermal efficiency decreases with increasing oxygen concentration for all timings. For the 10°CA of advance or retarded timing, the efficiency is reduced by about 4% (a factor of 0.96). The reasons of the decrease of efficiency with increasing oxygen concentration are examined in the following.

Figure 15.9 shows the specific brake power (kW/dm^3) and brake thermal efficiency as functions of oxygen concentration in the inlet air for MBT spark timing for the base case operating conditions. The specific brake power increases as the oxygen concentration in the inlet air increases even though the brake thermal efficiency decreases. The specific brake power increases about 58% as the oxygen concentration increases from 21% to 40%. This result

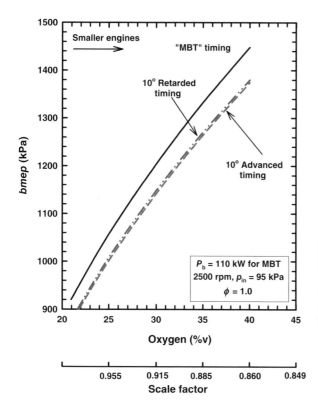

Figure 15.7 Brake mean effective pressure as functions of oxygen concentration in the inlet air for three spark timings for the base case. *Source:* Caton 2005. Reproduced with permission from ASME

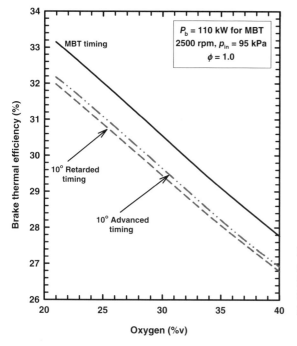

Figure 15.8 Brake thermal efficiency as functions of oxygen concentration in the inlet air for three spark timings for the base case. *Source:* Caton 2005. Reproduced with permission from ASME

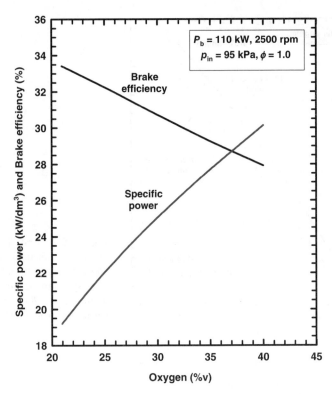

Figure 15.9 Specific brake power and brake thermal efficiency as functions of oxygen concentration in the inlet air for MBT spark timing for the base case. *Source:* Caton 2005. Reproduced with permission from ASME

reflects the net gain of specific power with the use of oxygen-enriched inlet air. Although the higher oxygen concentrations reduce the thermal efficiency, the gains from the reduced diluents (reduced nitrogen concentration) dominate for the conditions examined.

The next several figures illustrate the results that help explain the reduction in the thermal efficiency with increasing oxygen concentration in the inlet air. Figure 15.10 shows the heat loss as a percentage of the fuel energy (relative heat transfer) as functions of the oxygen concentration of the inlet air for the three spark timings. The relative heat transfer increases slightly as the oxygen concentration increases due largely to the higher gas temperatures. For the MBT timing case, the increase is about 7.2% for the range of oxygen concentration examined. This effect is most obvious for the initial increases of oxygen concentration. For oxygen concentrations above about 35%, this effect is much less. For cases with retarded and advanced timing, the relative heat transfer is lower and higher, respectively. For the retarded timing, the peak cylinder pressures and temperatures are lower which results in lower relative heat transfer. For the advanced timing, the peak cylinder pressures and temperatures are higher, resulting in higher relative heat transfer.

Figure 15.11 shows the exhaust gas temperature as functions of the oxygen concentration in the inlet air for three spark timings. The exhaust gas temperature increases almost linearly with increases in the oxygen concentration for all three timings. The exhaust gas temperature

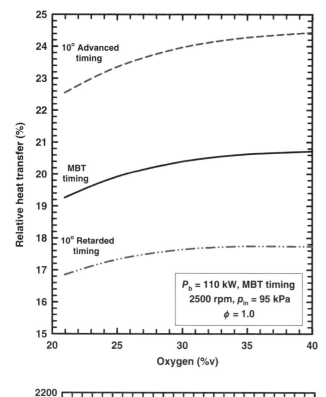

Figure 15.10 Heat loss as a percentage of the fuel energy as functions of oxygen concentration in the inlet air for three spark timings for the base case. *Source:* Caton 2005. Reproduced with permission from ASME

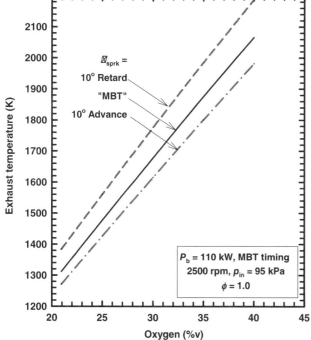

Figure 15.11 Exhaust gas temperature as functions of oxygen concentration in the inlet air for three spark timings for the base case. *Source:* Caton 2005. Reproduced with permission from ASME

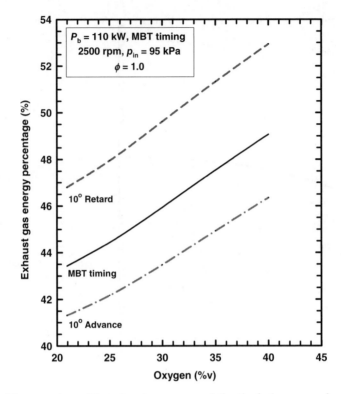

Figure 15.12   The percentage of the exhaust gas energy relative the fuel energy as functions of oxygen concentration in the inlet air for three spark timings for the base case. *Source:* Caton 2005. Reproduced with permission from ASME

is highest for the retarded timing since combustion is completed later in the cycle and there is less time for heat transfer. For the opposite reason (early combustion and more time), the advanced timing case has the lowest exhaust gas temperatures.

Figure 15.12 shows the exhaust gas energy as a percentage of the fuel energy as functions of the oxygen concentration of the inlet air for three spark timings. The exhaust gas energy increases roughly in a linear fashion with respect to the oxygen concentration. For the MBT spark timing case, the exhaust gas energy increases by about a factor of 1.13 for the range of oxygen examined. The cases for retarded timing have the highest exhaust gas energy and the cases for the advanced timing have the lowest exhaust gas energy. These results are consistent with the exhaust gas temperatures shown in Figure 15.11.

Figure 15.13 shows the brake thermal efficiency as functions of the oxygen concentration in the inlet air for three different burn durations. The burn duration is expected to change with increasing amounts of oxygen concentration, but only a limited amount of information is available. As the burn duration decreases, the efficiency increases but at diminishing amounts. For burn durations less than about 30°CA, the gains in efficiency are negligible.

In summary, the decrease of thermal efficiency as the inlet air oxygen concentrations increases is largely due to the increase of the heat transfer due to the higher combustion and expansion gas temperatures. Although not shown here (but covered in a subsequent case

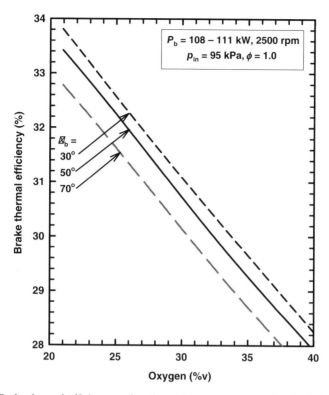

Figure 15.13   Brake thermal efficiency as functions of oxygen concentration in the inlet air for three burn durations for the base case. *Source:* Caton 2005. Reproduced with permission from ASME

study), another key reason for the decrease of thermal efficiencies is the decrease of the ratio of specific heats (due to the higher temperatures) for the cases with the higher inlet oxygen concentration. The lower value of the ratio of specific heats results in less effective conversion of thermal energy into work.

15.4.5   Nitric Oxide Emissions

A subsequent case study (Chapter 17) in this book describes the methodology for the computation of the nitric oxide concentrations. These calculations are based on the extended Zeldovich mechanism using the kinetics recommended by Dean and Bozzelli [14]. Figures 15.14 and 15.15 show the instantaneous nitric oxide concentrations as functions of crank angle for three spark timings for the base case conditions for 21% and 25% oxygen in the inlet air, respectively. The nitric oxide formation rate is greatest for the advanced spark timings and for the higher oxygen concentration levels. The nitric oxide forms rapidly, reaches a maximum, and then decreases during the latter portions of the expansion stroke. The decrease varies in magnitude depending on the specific conditions. For the 21% oxygen case, the nitric oxide decreasing process is complete before 90°aTDC. For the 25% oxygen case, the nitric oxide decreasing process continues up to the exhaust valve open time. For higher oxygen

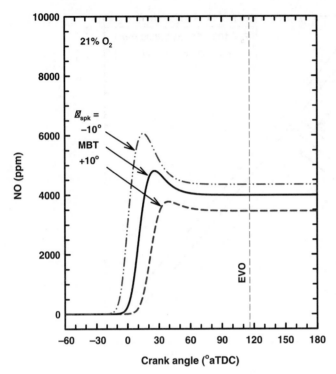

Figure 15.14 Instantaneous nitric oxide concentrations as functions of crank angle for the 21% O_2 case for three spark timings for the base case conditions. *Source:* Caton 2005. Reproduced with permission from ASME

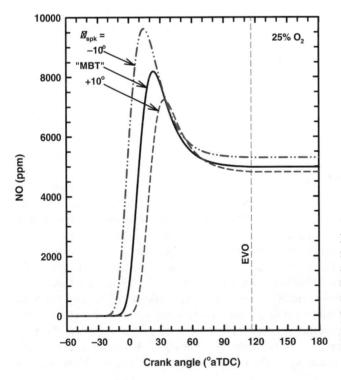

Figure 15.15 Instantaneous nitric oxide concentrations as functions of crank angle for the 25% O_2 case for three spark timings for the base case conditions. *Source:* Caton 2005. Reproduced with permission from ASME

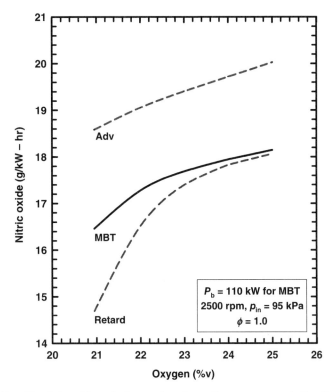

Figure 15.16   Specific nitric oxide emissions as functions of oxygen concentration for three spark timings for the base case conditions. *Source:* Caton 2005. Reproduced with permission from ASME

concentrations, this decreasing process had not ended when the exhaust valve was opened. For simplicity, these cases have not been examined at this time. The process once the exhaust valve opens would be different than simple closed system expansion.

Figure 15.16 shows the specific nitric oxide emission as functions of oxygen concentration for three spark timings. In general, as the oxygen concentration increase the nitric oxide emission increases. For the MBT spark timing case, the nitric oxide emissions increase by about 11% as the oxygen concentration increases from 21% to 25%. These increases are due largely to the higher gas temperatures and to a lesser degree due to the higher oxygen concentrations as the oxygen concentration of the inlet air is increased. The change of the nitric oxide emission as oxygen concentration increases is different for the three spark timings. This is explained by examining the results presented in Figures 15.14 and 15.15. The nitric oxide emissions are suppressed more at the advanced and MBT spark timings than at the retarded spark timing. This is a result of the trade-off between the temperatures and time available as shown in Figures 15.14 and 15.15.

15.5   Summary and Conclusions

This study has examined the effects of oxygen-enriched inlet air on the performance and emissions of a spark-ignition engine using a thermodynamic engine cycle simulation. One unique feature of these computations is that the comparisons were completed for the same power output by allowing engine size (volume displacement) to change with oxygen concentration.

The overall objective of this study was to quantify the effects of higher oxygen concentrations on engine performance and nitric oxide emissions. Such results are useful for better evaluating the advantages and disadvantages of this concept. Results from this study have indicated the following:

- Thermodynamic properties are a function of the oxygen concentration of the inlet air. Properties for the reactant and product mixtures are affected.
- In general, as the oxygen concentration of the inlet air increases, the cylinder gas pressure and temperature increases.
- For the same power output, as the oxygen concentration of the inlet air increases, the displaced volume (engine size) decreases. In other words, the specific power (kW/dm^3) increases. For an increase of oxygen concentration from 21% to 40%, the specific power increases about 58%.
- For the same power output, as the oxygen concentration of the inlet air increases, the thermal efficiency decreases due to greater heat losses and lower values for the ratio of specific heats.
- For the same power output, as the oxygen concentration of the inlet air increases, the nitric oxide emissions increase. For an increase in the oxygen concentration from 21% to 25%, the specific nitric oxide emissions increased about 11%.

References

1. Caton, J. A. (2005). Use of a cycle simulation incorporating the second law of thermodynamics: results for spark-ignition engines using oxygen enriched combustion air, 2005 SAE International Congress and Exposition, Society of Automotive Engineers, Cobo Hall, Detroit, MI, 11–14 April.
2. Hawthorne, E. P. (1946). Oxygen injection as a means of increasing aero-engine performance, *Aircraft Engineering and Aerospace Technology*, **18**(10), 330–335.
3. Wartinbee, Jr., W. J. (1971). Emissions study of oxygen enriched air, Society of Automotive Engineers, SAE paper number 710606.
4. Quader, A. A. (1978). Exhaust emissions and performance of a spark ignition engine using oxygen enriched intake air, *Combustion Science and Technology*, **19**, 81–86.
5. Detuncq, B., Williams, J., Guernier, C., Gou, M., and Fraser, Y. (1988). Performance of a Spark ignition engine fueled by natural gas using oxygen enriched air, Society of Automotive Engineers, SAE paper number 881658.
6. Kajitani, S., Sawa, N., McComiskey, T., and Rhee, K. T. (1992). A spark ignition engine operated by oxygen enriched air, Society of Automotive Engineers, SAE paper number 922174.
7. Kajitani, S., Clasen, E., Campbell, S., and Rhee, K. T. (1993). Partial-load and start-up operations of a spark-ignition engine with oxygen enriched air, Society of Automotive Engineers, SAE paper number 932802.
8. Ng, H. K., Sekar, R. R., Kraft, S. W., and Stamper, K. R., (1993). The potential benefits of intake air oxygen enrichment in spark ignition engine powered vehicle, Society of Automotive Engineers, SAE paper number 932803.
9. Maxwell, T. T., Setty, V., Jones, J. C., and Narayan, R. (1993). The effect of oxygen enriched air on the performance and emissions of an internal combustion engine, Society of Automotive Engineers, SAE paper number 932804.
10. Poola, R. B., Ng, H. K., Sekar, R. R., Baudino, J. H., and Colucci, C. P. (1995). Utilizing intake-air oxygen-enrichment technology to reduce cold-phase emissions, Society of Automotive Engineers, SAE paper number 952420.
11. Poola, R. B., Sekar, R. R., Ng, H. K., Baudino, J. H., and Colucci, C. P. (1996). The effects of oxygen-enriched intake air on FFV exhaust emissions using M85, Society of Automotive Engineers, SAE paper number 961171.

12. Woschni, G. (1968). A universally applicable equation for the instantaneous heat transfer coefficient in the internal combustion engine, *SAE Transactions*, SAE paper no. 670931, **76**, 3065–3083.

13. Caton, J. A. (2005). The effects of oxygen enrichment of combustion air for spark-ignition engines using a thermodynamic cycle simulation," in proceedings of the ASME Internal Combustion Division 2005 Spring Technical Conference, Chicago, IL, 05–07 April.

14. Dean, A. M., and Bozzelli, J. W. (2000). Combustion chemistry of nitrogen, in *Gas-Phase Combustion Chemistry*, Gardiner, W. C., Jr. (ed.) pp. 125–342, Springer-Verlag, New York.

16

Overexpanded Engine

16.1 Introduction

For conventional reciprocating engines, the compression and expansion strokes are equal. The gases at the end of a conventional expansion process, however, may possess the potential to produce more work if allowed to expand to a greater volume. By the use of special linkages, the expansion ratio could be greater than the compression ratio and thereby provides the potential for greater overall thermodynamic gains. This concept, originally patented [1] in 1887 by James Atkinson, a British engineer, is known as the Atkinson cycle or (in a generic sense) an overexpanded cycle.

Another concept that is closely related is often called the Miller cycle [2], and involves using either early or late intake valve closing to reduce the effective compression ratio relative to the expansion ratio. This concept has been relatively successful and exists in a number of commercial engines (e.g., Reference 3). Although the Atkinson and Miller cycles are related, they possess unique features. For example, the Miller cycle with late intake valve closing will move gases into the intake system during the initial period of the compression stroke. Also, some studies have shown that certain timings of the intake valve closing cause poor combustion [4]. The Atkinson cycle, on the other hand, can implement optimized valve events independent of the compression and expansion strokes. The Atkinson cycle, however, will require a more complex mechanism and incur additional frictional losses.

Previous studies of the Atkinson and Miller cycles are numerous (e.g., References 5–8). Most of these studies are based on the Miller cycle; only a few are based on mechanical features which provide a longer expansion stroke relative to the compression stroke (e.g., Reference 9).

The purpose of this case study is to quantify the effects of expansion ratio on engine performance. In particular, engine configurations with a wide range of expansion ratios for a fixed compression ratio will be examined. Additional information on this topic may be found in Reference 10.

An Introduction to Thermodynamic Cycle Simulations for Internal Combustion Engines, First Edition.
Jerald A. Caton © 2016 John Wiley & Sons, Ltd. Published 2016 by John Wiley & Sons, Ltd.

16.2   Engine, Constraints, and Approach[1]

16.2.1   Engine and Operating Conditions

The selected engine for this study was the automotive, 5.7 liter V–8 configuration operating at 1400 rpm. The operating conditions included a part load case (*bmep* = 325 kPa) and a wide-open throttle (WOT) case using the Woschni [11] heat transfer correlation with an *htm* of 1.33.

16.2.2   Constraints

As with the majority of the other case studies, the combustion process was considered successful for all conditions. In addition, mechanical friction was based on conventional engines (further comments on friction are described in Section 16.3.2). Due to these assumptions and approximations, the following results may be slightly optimistic.

16.2.3   Approach

In all the other parts of this book, a conventional engine was examined which had the same compression ratio and expansion ratio. Since a greater expansion ratio is expected to be important in terms of thermodynamic work production, the Atkinson cycle was examined. To study the Atkinson cycle, a number of approaches may be considered. The current study adopted an approach which was based on the following set of assumptions and approximations:

- The piston motion will be assumed to be approximated by two superimposed motions based on the classic "slider-crank" mechanism.
- The clearance volume will be based on the compression ratio (CR) and the stroke dimension (S_{comp}) of the standard engine.
- By selecting an expansion ratio (ER) and using the clearance volume from above, the expansion stroke (S_{exp}) can be determined.
- The transition between the compression and expansion strokes will be at 0.0°CA (TDC during combustion) and 360°aTDC.
- The engine mechanical friction will be based on the standard engine using the compression ratio. The slight increase of friction due to the longer expansion stroke will be neglected. (The effects of other friction assumptions are considered at the end of the results section.)
- The engine performance parameters (such as *bmep*) will be based on the standard engine configuration using the compression stroke to determine the displaced volume.
- As mentioned above, items such as knock, preignition, roughness, and cyclic variability are ignored for this study.

---

[1] A version of the material in this chapter is available in Reference 10.

First, some of the required variables will be defined. The compression ratio follows the normal definition

$$CR = \frac{V_{comp}}{V_{cl}}$$ (16.1)

where V_{comp} is the total cylinder volume based on the compression stroke length (S_{comp}) and V_{cl} is the clearance volume. The expansion ratio is

$$ER = \frac{V_{exp}}{V_{cl}}$$ (16.2)

where V_{exp} is the total cylinder volume based on the expansion stroke length (S_{exp}). The "displaced" volumes are

$$V_{disp,compression} = \frac{\pi B^2}{4} S_{comp}$$ (16.3)

$$V_{disp,expansion} = \frac{\pi B^2}{4} S_{expansion}$$ (16.4)

A multiplier factor for the expansion stroke relative to the compression stroke may also be defined

$$\frac{S_{exp}}{S_{comp}} = \frac{ER}{CR} = \text{Stroke Multiplier}$$ (16.5)

For example, if an expansion ratio of 15 is desired and the compression ratio is 10, then the expansion stroke must be increased by a factor of 1.5 relative to the compression stroke.

16.3   Results and Discussion

16.3.1   Part Load

The first set of results is for the part load operating condition with a *bmep* of 325 kPa and at 1400 rpm. The first example of results for the Atkinson cycle is based on a compression ratio of 15 and an expansion ratio of 20. Figure 16.1 shows the instantaneous cylinder volume as a function of crank angle for this configuration, and for reference, also shows the cylinder volume for the conventional case where the compression ratio and expansion ratio are the same (and equal to 15). To the right of the figure is a schematic of the cylinder arrangement with the two stroke dimensions indicated. The cylinder volume decreases during the compression stroke from intake valve close (IVC) to TDC (0.0°CA), then increases to a maximum volume during the expansion stroke from TDC to 180°aTDC. Although two expressions were used for the volume computation, the transition appears smooth on this scale.

Figure 16.2 shows the related change of volume as functions of crank angle for the above case for a compression ratio of 15 and an expansion ratio of 20. Again, for reference, the change of volume is also shown for the conventional case where the compression ratio and expansion ratio are the same (and equal to 15). Also again, the transition between the two expressions appears smooth. For other cases with greater differences between the compression ratio and expansion ratio (say, CR = 6 and ER = 30), the transitions were not as smooth.

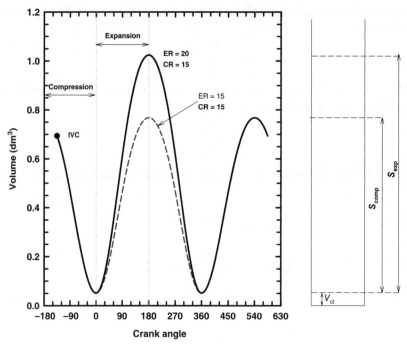

Figure 16.1   Cylinder volume as a function of crank angle for a compression ratio of 15 and for expansion ratios of 15 (dashed lines) and 20 (solid lines). Schematic of engine cylinder shown on the right side. *Source:* Caton 2007. Reproduced with permission from ASME

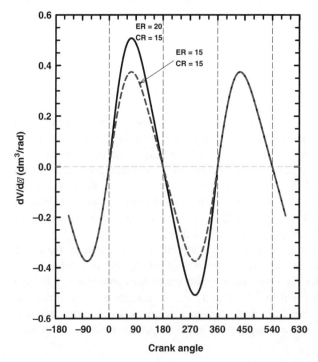

Figure 16.2   Rate of change of volume as functions of crank angle for a compression ratio of 15 and for expansion ratios of 15 (dashed lines) and 20 (solid lines). *Source:* Caton 2007. Reproduced with permission from ASME

Figure 16.3 shows the computed brake thermal efficiency as functions of expansion ratio for three compression ratios (6, 10, and 15) for the base case part load engine conditions. For these computations, the *bmep* was constant (= 325 kPa)—so as the brake efficiency increased, the inlet pressure was reduced. For each compression ratio case, as the expansion ratio increases the thermal efficiency first increases, reaches a maximum, and then decreases. The expansion ratio for the highest efficiency appears to be equal to about the compression ratio plus three. The use of this expansion ratio provides an absolute thermal efficiency increase of about 1% (e.g., from 24% to 25%). These results are examined more fully in the following.

First, the case with a compression ratio of 15 and an expansion ratio of 20 will be examined. This case represents the highest efficiencies of the cases examined. Figure 16.4 shows the log of the cylinder pressure as functions of the log of the cylinder volume. The conventional case (dashed lines) with CR = ER = 15 is compared with the case with an expansion ratio of 20 (solid lines). The longer expansion stroke results in a maximum cylinder volume of about 0.973 dm^3, while the conventional case has a maximum cylinder volume of 0.717 dm^3. The inlet pressure is slightly lower for the overexpanded case, and this results in generally lower pressures throughout most of the cycle.

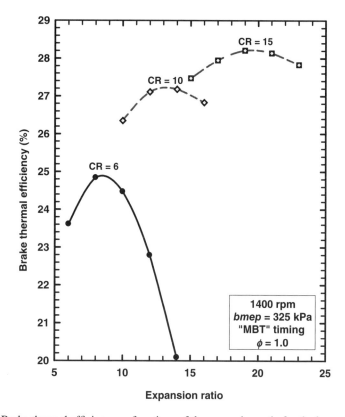

Figure 16.3   Brake thermal efficiency as functions of the expansion ratio for the base case for three compression ratios. *Source:* Caton 2007. Reproduced with permission from ASME

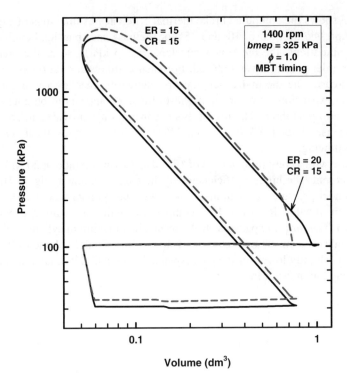

Figure 16.4   Log of the cylinder gas pressure as functions of the log of the cylinder volume for the base case for a compression ratio of 15 and for expansion ratios of 15 (dashed lines) and 20 (solid lines). *Source:* Caton 2007. Reproduced with permission from ASME

Figure 16.5 shows the log of the one-zone, cylinder gas temperature as functions of the log of the cylinder volume for the base case conditions for a compression ratio of 15 and an expansion ratio of 20 (solid lines), and for the conventional case (CR = ER = 15; dashed lines). Denoted in this figure by circles are the various events (such as EVO, EVC, ...). Starting with the intake valve closing (IVC), the temperature increases during the compression stroke and during the combustion process. After the majority of the expansion stroke, the exhaust valve opens (EVO) at 116°aTDC, and after BDC, the temperature decreases during the exhaust stroke. The two cases have similar temperatures, but the overexpanded case has slightly higher temperatures for most of the cycle.

Figure 16.6 shows the intake and exhaust flow rates as functions of crank angle for the two cases. The most significant difference is the lower flow rates during the blow-down portion of the exhaust flow for the overexpanded case. This is because of the greater expansion which results in a lower cylinder pressure at the time of exhaust valve opening. This is compensated by the greater flow rates during the exhaust displacement phase of the exhaust process. The effect of the exhaust valve open time is examined below.

The next items to discuss are the results in Figure 16.3 which indicated that for each compression ratio a specific expansion ratio provided the maximum efficiency. This will be examined

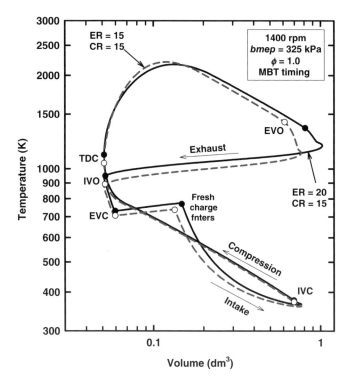

Figure 16.5 Log of the one-zone, cylinder gas temperature as functions of the log of the cylinder volume for the base case for a compression ratio of 15 and for expansion ratios of 15 (dashed lines) and 20 (solid lines). *Source:* Caton 2007. Reproduced with permission from ASME

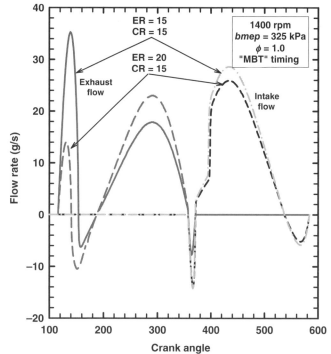

Figure 16.6 Exhaust and intake flow rates as functions of crank angle for the base case for a compression ratio of 15 and for expansion ratios of 15 (dashed lines) and 20 (solid lines). *Source:* Caton 2007. Reproduced with permission from ASME

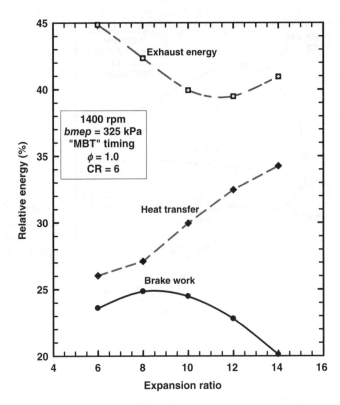

Figure 16.7   Relative energy as functions of the expansion ratio for the base case for a compression ratio of 6. *Source:* Caton 2007. Reproduced with permission from ASME

for the case with a compression ratio of 6.0. Figure 16.7 shows the relative energy distribution (based on the lower heating value of the fuel) for this compression ratio of 6.0 as functions of the expansion ratio. The brake work percentages are equivalent to the earlier shown brake thermal efficiencies. The relative heat transfer increases as the expansion ratio increases. This is largely due to the increase of surface area as the expansion ratio increases. The relative energy leaving in the exhaust first decreases and then increases slightly for the expansion ratios greater than about 12. This is a net result of the energy delivered as work and transferred out of the system via heat transfer. For the higher expansion ratios, the work is decreasing faster than the heat transfer is increasing, and this results in a net increase of the exhaust energy.

The cases which will be examined in some detail are for expansion ratios of 8 (near maximum efficiency) and 10 (both for a compression ratio of 6). The question is, why does the thermal efficiency begin to decrease even as the expansion ratio increases from 8 to 10. Figure 16.8 shows the log pressure as a function of the log volume for these two cases. These results are similar to those in Figure 16.4. Also shown in this figure is the location of the exhaust valve open (EVO) event. Due to the longer expansion stroke, this occurs at a lower pressure for the case with an expansion ratio of 10. Figure 16.9 shows the intake and exhaust flow rates as functions of crank angle for the two cases. As shown above (Figure 16.6), the most significant difference is the much lower flow rates during the blow-down portion for the case with

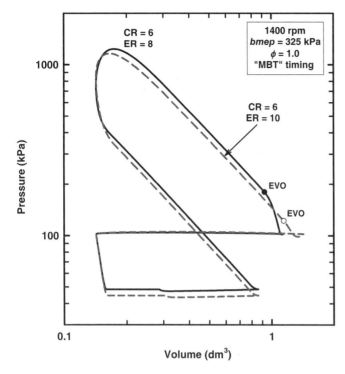

Figure 16.8 Log of the cylinder gas pressure as functions of the log of the cylinder volume for the base case for a compression ratio of 6 and expansion ratios of 8 and 10. *Source:* Caton 2007. Reproduced with permission from ASME

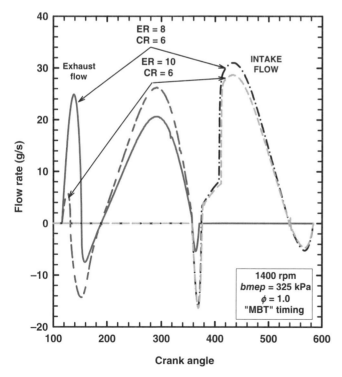

Figure 16.9 Exhaust and intake flow rates as functions of crank angle for the base case for a compression ratio of 6 for two expansion ratios: (i) optimum, ER = 8, and (ii) not optimum, ER = 10. *Source:* Caton 2007. Reproduced with permission from ASME

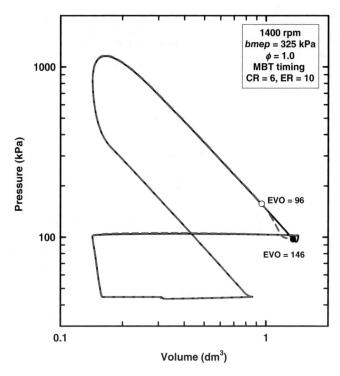

Figure 16.10 Log of the cylinder gas pressure as functions of the log of the cylinder volume for the base case for a compression ratio of 6 and an expansion ratio of 10 for two different exhaust valve open timings. *Source:* Caton 2007. Reproduced with permission from ASME

an expansion ratio of 10. In fact, for expansion ratios higher than 10, due to the low cylinder pressures, the initial blow-down is back into the cylinder (rather than out of the cylinder). The lack of an effective exhaust blow-down for higher expansion ratios is one of the reasons that the higher expansion ratios result in decreasing efficiencies.

Since the timing of the exhaust valve opening may be important, this was examined as well. Figure 16.10 shows the log of the cylinder pressure as a function of the log of the cylinder volume for the case with a compression ratio of 6 and an expansion ratio of 10 for two EVO timings: an early opening at 96°aTDC and a late opening at 146°aTDC. The cylinder pressures and thermal efficiencies are nearly the same for these changes of the EVO.

Finally, for completeness, the destruction of exergy during the combustion process as a function of expansion ratio was examined. For the case with a compression ratio of 6, the percentage of exergy destroyed was nearly constant, and ranged between 21.74% and 21.79% for expansion ratios between 6 and 14. One of the major reasons for this negligible difference is the modest changes of the combustion gas temperatures as the expansion ratio increased.

16.3.2   Wide-Open Throttle

In addition to the above results for a part load condition, results were obtained for the use of greater expansion ratios for the case of wide-open throttle (p_{inlet} = 95 kPa). Figure 16.11 shows

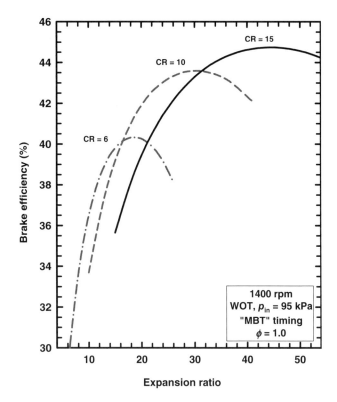

Figure 16.11   Brake thermal efficiency as a function of expansion ratio for compression ratios of 6, 10, and 15 for an engine speed of 1400 rpm and an inlet manifold pressure of 95 kPa (WOT). *Source:* Caton 2007. Reproduced with permission from ASME

the brake thermal efficiency as functions of the expansion ratio for three compression ratios (6, 10, and 15) for WOT. For each compression ratio, the thermal efficiency first increases, reaches a maximum, and then decreases. The expansion ratio for the highest thermal efficiency is about 19, 29, and 44 for the compression ratios of 6, 10, and 15, respectively. The use of these expansion ratios resulted in an increase of the thermal efficiency of about 10% (absolute). For example, for a compression ratio of 6, the brake thermal efficiency increased from about 30% to slightly over 40% as the expansion ratio increases from 6 to about 19. Compared to the previous part load cases, the WOT cases result in much greater improvements with greater values of expansion ratios. The following will discuss some aspects of these results.

Figure 16.12 shows the log of the cylinder pressure as a function of the log of the cylinder volume for the WOT case with the highest brake thermal efficiency (of the cases examined) which was for a compression ratio of 15 and an expansion ratio of 44 (solid line). For comparison, Figure 16.12 also shows (dashed line) the results for the WOT case for a compression ratio of 15 and an expansion ratio of 15 (standard engine configuration). The pressures during the compression stroke are the same for both configurations, but the pressures during the expansion are different, and of course, for the overexpanded configuration, the expansion proceeds to a much greater volume. The additional area within the pressure–volume diagram

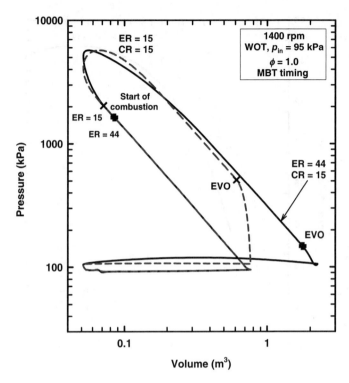

Figure 16.12 Log of cylinder pressure as a function of log of cylinder volume for a compression ratio of 15 and expansion ratios of 15 and 44 for an engine speed of 1400 rpm and an inlet manifold pressure of 95 kPa (WOT). *Source:* Caton 2007. Reproduced with permission from ASME

associated with the overexpanded configuration is indicative of the additional work (power) obtained with the greater expansion. The pressures for the exhaust stroke are generally higher for the overexpanded configuration, but the pressures during the intake stroke are nearly the same.

Figure 16.13 shows the relative increase in the brake thermal efficiency as a function of the increase of the length of the expansion stroke (relative to the length of the compression stroke) for the case with a compression ratio of 15. As an example, the expansion stroke length must increase about three times the compression stroke length to achieve the highest thermal efficiency for the conditions examined. This could be impractical for the modest gains.

Figure 16.14 shows the mass flows as functions of crank angle for two cases. For the standard configuration, the blow-down portion of the exhaust flow is significant whereas for the overexpanded case, the blow-down portion is much less significant. For the overexpanded case, the majority of the exhaust flow occurs during the displacement portion of the exhaust flow. In addition, for the overexpanded case, the backflow after the blow-down is much more significant. For expansion ratios greater than about 44 (not shown), the exhaust process is compromised, and when the exhaust valve first opens, the flow is back into the cylinder. Note that the intake flows are nearly the same for both configurations.

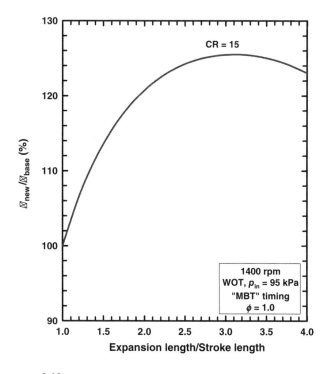

Figure 16.13 Ratio of brake thermal efficiencies as a function of the ratio of the expansion stroke length and compression stroke length for a compression ratio of 15 for an engine speed of 1400 rpm and an inlet manifold pressure of 95 kPa (WOT). *Source:* Caton 2007. Reproduced with permission from ASME

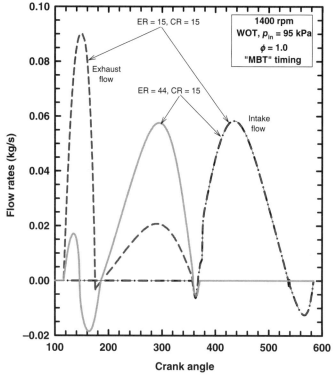

Figure 16.14 Mass flow rates as a function of crank angle for a compression ratio of 15 and two expansion ratios (15 and 44) for an engine speed of 1400 rpm and an inlet manifold pressure of 95 kPa (WOT). *Source:* Caton 2007. Reproduced with permission from ASME

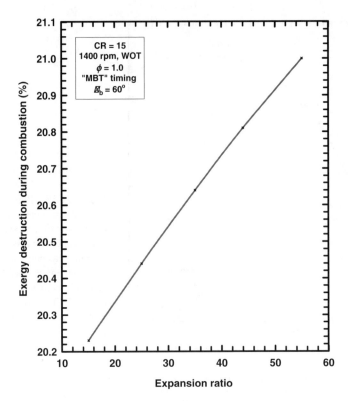

Figure 16.15   Exergy destruction during combustion as a function of the expansion ratio for a compression ratio of 15 for an engine speed of 1400 rpm and an inlet manifold pressure of 95 kPa (WOT). *Source:* Caton 2007. Reproduced with permission from ASME

For completeness, Figure 16.15 shows that the exergy destruction during combustion increases slightly (from about 20.2% to 21.0%) as the expansion ratio increases from 15 to 55 for a compression ratio of 15 for the WOT case. This slight increase is due to the lower combustion temperatures associated with the higher expansion cases. Since the work output increases for the higher expansion ratios, the gas temperatures decrease which is a major cause of exergy destruction during the combustion process [10].

One final consideration is the mechanical friction. For an overexpanded configuration, the mechanical friction would be expected to be higher due to the additional mechanism and the longer stroke. This cannot be quantified without some knowledge of the specific design and experimental information. To explore this issue, some estimates were completed. By assuming that the friction was due to an engine configuration possessing the expansion ratio for both the compression and expansion strokes (a higher estimate for the friction component), the maximum thermal efficiency decreased by about 2%, absolute (i.e., from 45% to 43%). In addition, the optimum expansion ratio for the higher friction case decreased. Again, a more detailed consideration of friction would be needed for a particular mechanism.

16.4   Summary and Conclusions

This case study provided results for the effects of expansion ratio on performance, efficiency, and second law parameters for an automotive, spark-ignition engine. The cases with greater expansion ratios than compression ratios are often called "Atkinson" cycles. For these cases, the expansion ratio was assumed independent of the compression ratio. The results of this study may be summarized as follows:

- The study examined expansion ratios up to four times the compression ratio. For most cases, the thermal efficiency first increased as expansion ratio increased, then attained a maximum efficiency, and then decreased. The decrease in efficiency after the maximum value was shown to be due to increased heat losses, increased friction, and ineffective exhaust processes (due to the reduced cylinder pressure at the time of exhaust valve opening).
- For the part load cases, the higher expansion ratio provided only modest gains due to increased pumping losses associated with the constant load requirement (which resulted in lower inlet pressures as the efficiency increased).
- For the wide-open throttle cases, however, the higher expansion ratios provided significant gains. For example, for a compression ratio of 10, expansion ratios of 10 and 30 provided brake thermal efficiencies of about 34% and 43%, respectively. Although the net thermodynamic gains are significant, large expansion ratios such as 30 may not be practical for most applications.
- For the wide-open throttle cases, for a compression ratio of 15, the exergy destroyed during the combustion process increased slightly from about 20.2% to 21.0% as the expansion ratio increased from 15 to 55.

References

1.   Atkinson, J. (1887). Gas engine, US Patent no. 367,496.
2.   Miller, R. H. (1947). Supercharging and internal cooling for high output, American Society of Mechanical Engineers, *SAE Transactions*, **69**, 453–464.
3.   Goto, T., Hatamura, K., Takizawa, S., Hayama, N., Abe, H., and Kanesaka, H. (1994). Development of V6 Miller cycle gasoline engine, Society of Automotive Engineers, SAE paper no. 940198.
4.   Hara, S., Nakajima, Y., and Nagumo, S. (1985). Effects of intake-valve closing timing on spark-ignition engine combustion, Society of Automotive Engineers, SAE paper no. 850074.
5.   Boggs, D. L., Hilbert, H. S., and Schechter, M. M. (1995). The Otto-Atkinson Cycle engine – fuel economy and emissions results and hardware design, Society of Automotive Engineers, SAE paper no. 950089.
6.   Chen, L., Lin, J., Sun, F., and Wu, C. (1998). Efficiency of an Atkinson engine at maximum power density, *Energy Conservation and Management*, **39** (3/4), 337–341.
7.   Shiga, S., Hirooka, Y., Miyashita, Y., Yagi, S., MacHacon, H. T., Karasawa, T., and Nakamura, H. (2001). Effect of over-expansion cycle in a spark-ignition engine using late-closing of intake valve and its thermodynamic consideration of the mechanism, *International Journal of Automotive Technology*, **2** (1), 1–7.
8.   Wu, C., Puzinauskas, P. V., and Tsai, J. S. (2003). Performance analysis and optimization of a supercharged Miller cycle Otto engine," *Applied Thermal Engineering*, **23**, 511–521.
9.   Raynes, S. H. (1998). An Atkinson cycle engine for low pollution, *proceedings of Combustion Engines and Hybrid Vehicles*, Institute of Mechanical Engineers, London, paper no. C529, 28 April.
10. Caton, J. A. (2007). The effects of compression ratio and expansion ratio on engine performance including the second law of thermodynamics: results from a cycle simulation, in proceedings of the ICEF2007 Fall Conference of the ASME Internal Combustion Engine Division, Charleston, SC, 14–17 October.
11. Woschni, G. (1968). A universally applicable equation for the instantaneous heat transfer coefficient in the internal combustion engine, *SAE Transactions*, SAE paper no. 670931, **76**, 3065–83.

17

Nitric Oxide Emissions

17.1 Introduction

Nitric oxide (NO) is one of the several regulated species from engines. In practice, most regulations limit the combination of nitric oxide and nitrogen dioxide (NO_2)—together these two nitrogen species are known as NO_x. For internal combustion engines, for most conditions, the emitted NO_x is dominated by nitric oxide. From at least the 1960s, engine cycle simulations have been used to estimate nitric oxide emissions. Since that time, the knowledge of nitric oxide formation and destruction mechanisms, and the associated chemical kinetics have continually improved, and the estimates from engine cycle simulations have reflected this improved knowledge. In addition, engine cycle simulation features such as the use of multiple zones has enabled these simulations to more accurately capture the high cylinder gas temperatures which is critical for the computation of nitric oxide emissions.

The thermodynamic engine cycle simulation with the three-zone combustion model is appropriate for investigating the formation and destruction of nitric oxides based on the Zeldovich mechanism (described below). This case study will illustrate the use of the engine cycle simulation using this submodel. The exhaust nitric oxide emissions will be determined as functions of engine design and operating parameters. Results will be provided from the basic chemical kinetics, and from the engine simulation. The results from the simulation will be shown as time-resolved quantities and as overall engine outputs. Some of these results have been compared to experimental values in a separate publication [1]. Other work which is similar to that reported in this chapter is available [2,3].

The purpose of the current work is to provide detailed information about the nitric oxide formation and destruction mechanisms for a range of engine design and operational parameters. Three of the specific goals of this work were (i) to provide detail descriptions of the thermal nitric oxide formation and destruction processes during combustion, (ii) to determine the sensitivity of the nitric oxide computations to the major assumptions and approximations typically used, and (iii) to ascertain the change of nitric oxide emissions with changes in engine design and operating conditions. The following sections of this chapter will include a brief discussion of the background of the nitric oxide kinetics, the engine specifications and operating conditions, and the results.

An Introduction to Thermodynamic Cycle Simulations for Internal Combustion Engines, First Edition.
Jerald A. Caton © 2016 John Wiley & Sons, Ltd. Published 2016 by John Wiley & Sons, Ltd.

17.2   Nitric Oxide Kinetics

A vast amount of work has been completed over the years on the kinetics of nitric oxide formation and destruction. This section will merely highlight some of the relevant literature as it pertains to the current work. Four major mechanisms have been proposed for the formation of nitric oxides in combustion processes: (i) thermal, (ii) prompt, (iii) nitrous oxide, and (iv) fuel nitrogen. In addition, this section will briefly summarize some of the kinetic rate expressions that have been proposed for nitric oxide.

17.2.1   Thermal Nitric Oxide Mechanism

As is well known, the Zeldovich mechanism has been successfully used to describe the elementary chemical steps responsible for the majority of the nitric oxide produced for engine conditions [4]. The original Zeldovich mechanism was extended to include a third reaction which has been shown to be important for some combustion conditions [5]. The extended Zeldovich mechanism is

$$O + N_2 \rightleftarrows NO + N \tag{17.1}$$

$$N + O_2 \rightleftarrows NO + O \tag{17.2}$$

$$N + OH \rightleftarrows NO + H \tag{17.3}$$

Since the rates of these reactions are generally slower than the combustion rates, the majority of the combustion reactions may be assumed to be unaffected by the above steps. Further, the concentrations of O, N_2, O_2, O, OH and H may be assumed to have reached equilibrium values. The nitrogen atoms, on the other hand, exist at low concentrations ($\sim 10^{-7}$) and a steady-state approximation is valid for these species [4]. Miller and Bowman [6] have indicated situations where the steady-state approximation may not be completely valid.

These reactions are known as the thermal nitric oxide rates since the strong triple bond of the N_2 molecule will only break apart at high temperatures. Furthermore, as illustrated in Section 17.5.1, reaction (17.1) has the slowest rates and therefore is often the limiting reaction. In fact, reaction (17.1) is so slow that for typical engine conditions, nitric oxide may not achieve equilibrium concentrations (this is shown in Section 17.5.2).

17.2.2   "Prompt" Nitric Oxide Mechanism

In 1971, Fenimore [7] observed nitric oxide production that far exceeded the values expected from the thermal mechanism. He found that for rich mixtures, another pathway existed for the production of nitric oxide. This pathway was due to a series of reaction steps at the flame front that involved the radical CH. Generally, the radical CH is considered a transient species in the overall hydrocarbon oxidation process. Fenimore [7], however, noted that the radical CH may be an important species in a pathway that forms nitric oxide.

This pathway is particularly important for rich mixtures. In general, this pathway is initiated by the rapid reaction of hydrocarbon radicals with molecular nitrogen, and leads to intermediates such as amines or cyano compounds (e.g., HCN) that then react to form nitric oxide. The most important of these reactions are

$$CH + N_2 \rightleftarrows HCN + N \tag{17.4}$$

$$C + N_2 \rightleftarrows CN + N \tag{17.5}$$

The HCN and CN then react through a series of subsequent fast reactions to form nitric oxide.

Lavoie and Blumberg [8] reported that prompt nitric oxide may be important for high dilution conditions such as with exhaust gas recirculation or lean operation when thermal nitric oxide production is low. In general, when the thermal nitric oxide production is below about 100 ppm, then the prompt nitric oxide mechanism could be significant.

17.2.3   Nitrous Oxide Route Mechanism

In 1970, Lavoie et al. [5] proposed that the following reactions could be responsible for significant NO.

$$N_2 + O + M \rightleftarrows N_2O + M \tag{17.6}$$

$$O + N_2O \rightleftarrows NO + NO \tag{17.7}$$

This pathway for nitric oxide production is named for the nitrous oxide molecule, N_2O. This mechanism may be most important for lean conditions (which suppresses the prompt nitric oxide), low temperatures (which suppresses the thermal nitric oxide) and high pressures (which promotes the above three body reactions).

17.2.4   Fuel Nitrogen Mechanism

For fuels with bound nitrogen, such as coal and some heavy distillate, another mechanism of nitric oxide production is possible. The nitrogen-containing compounds will vaporize during combustion, and the released nitrogen will participate in a series of reactions to form nitric oxide. Often these mechanisms proceed via the formation of NH_3 (ammonia) and HCN (hydrocyanic acid). For most current spark-ignition engine fuels, however, fuel-bound nitrogen is negligible or zero.

17.3   Nitric Oxide Computations

Of the four major mechanisms described above, the extended Zeldovich mechanism is often considered the dominate source of nitric oxides—particularly for spark-ignition engines. Therefore the computation of nitric oxides in this work will be based on the extended Zeldovich mechanism, and for completeness, the mechanism is summarized here. Most of this description follows the presentation by Heywood [4]. The extended Zeldovich mechanism for thermal nitric oxides is well accepted for computing nitric oxides in engines. From these chemical steps, the following formation rate for nitric oxide may be derived.

$$
\begin{aligned}
\frac{d[NO]}{dt} &= k_1^+[O]_e[N_2]_e + k_2^+[N]_{ss}[O_2]_e \\
&\quad + k_3^+[N]_{ss}[OH]_e - k_1^-[NO]_a[N]_{ss} \\
&\quad - k_2^-[NO]_a[O]_e - k_3^-[NO]_a[H]_e
\end{aligned}
\tag{17.8}
$$

where the rate coefficients (k_i^*) may be obtained from a number of sources, $[N]_{ss}$ is the steady-state concentration of nitrogen atoms, and $[NO]_a$ is the actual (and not the equilibrium) concentration of nitric oxide, and $[XX]_e$ is the equilibrium concentration of the remaining species. The k_i^+ represents the forward rate coefficient, and the k_i^- represents the reverse rate coefficient for reaction "i". The two rate coefficients are related to the equilibrium constant

$$
K_i^{\text{equil}} = \frac{k_i^+}{k_i^-}
\tag{17.9}
$$

A similar expression is used to determine the steady-state concentration of nitrogen atoms.

$$
\begin{aligned}
\frac{d[N]}{dt} &= k_1^+[O]_e[N_2]_e - k_2^+[N]_{ss}[O_2]_e \\
&\quad - k_3^+[N]_{ss}[OH]_e - k_1^-[NO]_a[N]_{ss} \\
&\quad + k_2^-[NO]_a[O]_e + k_3^-[NO]_a[H]_e
\end{aligned}
\tag{17.10}
$$

For the steady-state assumption,

$$
\frac{d[N]}{dt} = 0
\tag{17.11}
$$

this leads to

$$
[N]_{ss} = \frac{k_1^+[O]_e[N_2]_e + k_2^-[NO]_a[O]_e + k_3^-[NO]_a[H]_e}{k_1^-[NO]_a + k_2^+[O_2]_e + k_3^+[OH]_e}
\tag{17.12}
$$

This can be combined with the above expression to result in a simplified rate expression.

$$
\frac{d[NO]}{dt} = \frac{2R_1\left(1-\beta^2\right)}{(1+\beta K)}
\tag{17.13}
$$

where the terms are

$$
R_1 = k_1^+[O]_e[N_2]_e = k_1^-[NO]_e[N]_e
\tag{17.14}
$$

$$
R_2 = k_2^+[N]_e[O_2]_e = k_2^-[NO]_e[O]_e
\tag{17.15}
$$

$$R_3 = k_3^+ [N]_e [OH]_e = k_3^- [NO]_e [H]_e \tag{17.16}$$

$$\beta = \frac{[NO]_a}{[NO]_e} \tag{17.17}$$

$$K = \frac{R_1}{R_2 + R_3} \tag{17.18}$$

The use of eq. (17.13) for the production of nitric oxide has two specific implications. First, for each reaction, the rate coefficient of either the forward or the reverse reaction is sufficient. Second, the concentration of the steady-state nitrogen atom is incorporated into the expression.

17.3.1 Kinetic Rates

For many chemical reactions, the kinetic rate expressions are often not known in an exact fashion. Four references from 1973 through 2000 have included recommended values for the kinetic rate expressions for the Zeldovich mechanism [6,9–11]. As these references demonstrate, the recommended rate constants have differed through the years. Fortunately, the estimates for the reaction rates for the important first reaction have not changed too much over the last 20 years or so. For example, for 2300 K, the rate has ranged from a low of 5.08×10^6 to a high of 1.08×10^7 cc/gmol-sec. The values recommended by Dean and Bozzelli [11] have been used in the current work, and Table 17.1 lists these reaction rate constants. The forward rates are from Reference 11, and the reverse rates are from the use of eq. (17.9).

The nitric oxide emissions may be expressed in a number of forms. In the computations, the net mass of nitric oxide formed is obtained, but a fraction remains in the cylinder with the residual gases. The emitted mass of nitric oxide may be found from the mass of nitric oxide formed as follows:

$$\left(m_{NO} \right) = \left(m_{NO} \right)_{emitted} = \left(m_{NO} \right)_{formed} \left(1 - x_r \right) \tag{17.19}$$

Table 17.1   Rate constants from Dean and Bozzelli [11]

Reaction (listed in eqs. (17.1)–(17.3))	Rate constant as listed in Reference 11 (cc/gmol-s)	Rate constant as obtained from K_{eq} (cc/gmol-s)
1 (forward)	1.95×10^{14} exp (−38660/T)	
−1 (reverse)		4.11×10^{13} exp (−715/T)
2 (forward)	9.0×10^{09} T exp (−3270/T)	
−2 (reverse)		1.9×10^{09} T exp (−19410/T)
3 (forward)	1.1×10^{14} exp (−565/T)	
−3 (reverse)		2.92×10^{14} exp (−24675/T)

One measure of the emission level is the emission index (EI_{NO}):

$$EI_{NO} = \frac{m_{NO}}{m_f}$$ (17.20)

The specific nitric oxide emissions may be obtained from the following (for a four-stroke cycle engine):

$$spNO = \frac{m_{NO}}{\dot{W}_b}\left(\frac{60}{2}\right)N$$ (17.21)

or

$$spNO = m_{NO}\frac{bsfc}{m_f} = EI_{NO}(bsfc)$$ (17.22)

where spNO (g_{NO}/Kw hr) is the specific nitric oxide emissions, m_{NO} is the mass of nitric oxide emitted per cycle, N is the engine speed, $\dot{W}_b$ is the brake power, m_f is the mass of fuel per cycle, and x_r is the residual fraction.

In terms of ppm,

$$ppm_{NO} = m_{NO}\frac{MW_{exh}}{MW_{NO}}\frac{N}{\dot{m}_f}(AF+1)10^6$$ (17.23)

where MW_{exh} is the molecular mass (weight) of the exhaust gases, MW_{NO} is the molecular mass (weight) of nitric oxide, and AF is the mass air–fuel ratio.

As an internal consistency check,

$$spNO = ppm_{NO}\frac{MW_{NO}}{MW_{exh}}\frac{\dot{m}_{exh}}{\dot{m}_f}bsfc$$ (17.24)

where $\dot{m}_{exh}$ is the mass flow rate of exhaust gases.

Another internal consistency check,

$$ppm_{NO} = \frac{(m_{NO})_{formed} / MW_{NO}}{m_{cyl} / MW_{exh}}10^6$$ (17.25)

where m_{cyl} is the mass in the cylinder.

17.4 Engine and Operating Conditions

The following results are based on the 5.7 liter, V–8 engine. The base case operating conditions include a stoichiometric mixture, a *bmep* of 325 kPa, 1400 rpm, a compression ratio of 8.0, a burn duration of 60°CA, and using isooctane as the fuel. The simulation uses the heat transfer correlation from Woschni [12] with an *htm* of 1.0.

17.5   Results and Discussion

Results from the basic chemical kinetic relationships and from the engine cycle simulation are described in the following four subsections: (i) basic chemical kinetic results, (ii) time (crank angle) resolved nitric oxide results, (iii) nitric oxide engine results, and (iv) nitric oxide results for mass fraction burned variations.

17.5.1   Basic Chemical Kinetic Results

Before the engine results are presented, this subsection will include results which highlight the basic chemical kinetics[1] of nitric oxide formation and decomposition for representative engine conditions. The first set of results is for simple nitric oxide formation in high temperature and high pressure combustion products. The temperature, pressure, and concentrations of the other species remain constant. Figure 17.1 shows the nitric oxide concentrations[2] as functions of time for three temperatures for 1000 kPa and an equivalence ratio of 1.0. The horizontal, dashed lines represent the equilibrium nitric oxide concentration for the given conditions.

As shown in Figure 17.1, the nitric oxide concentration increases rapidly at first and then more slowly as it approaches its final equilibrium value. For the highest temperature (2600 K), the nitric oxide has reached about 90% of its equilibrium value in about 3 ms. For the next temperature (2500 K), 90% is reached in slightly more than 7 ms. For the case of 2400 K, to reach 90% of the final equilibrium nitric oxide concentration takes more than 16 ms. Note that at 1000 rpm, 60 crank angle degrees is 10 ms. This means that typically within the time available for nitric oxide formation, the nitric oxide concentration may not have reached equilibrium especially for the lower temperature periods. Furthermore, as the temperature decreases during the expansion stroke, the concentration of nitric oxide eventually may be kinetically limited or "frozen" (the concentration will no longer change).

In addition, Figure 17.1 shows that the nitric oxide production rate is highly temperature dependent. For these conditions, the final nitric oxide concentration may vary by a factor of two or more for a change in temperature of about 200 K.

Figure 17.2 shows the nitric oxide concentrations as functions of time for four equivalence ratios for 1000 kPa and 2600 K. These conditions are representative of engine conditions, but typically the temperatures are not independent of the equivalence ratio. As shown, for this relatively high temperature, the nitric oxide reaches its equilibrium value for all cases within about 8 ms. The final nitric oxide concentration increases from about 1500 ppm to about 11,500 ppm (1.15%) as the equivalence ratio decreases from 1.2 to 0.8. As the equivalence ratio decreases, the concentration of oxygen increases. For constant temperature and pressure, as the equivalence ratio decreases, equilibrium concentrations of O-atoms, OH, and N-atoms increase while concentrations of H-atoms decrease. Higher concentrations of O-atoms, OH, and N-atoms and lower concentrations of H-atoms increase equilibrium (and kinetic) NO. Other results for the more normal engine situation where gas temperatures change with changes in equivalence ratio are presented below.

---

[1] Unless otherwise mentioned, the chemical kinetics for the nitric oxide computations are from Dean and Bozzelli [11].

[2] All concentrations are "wet" values: i.e., the concentrations are based on the total moles including all water.

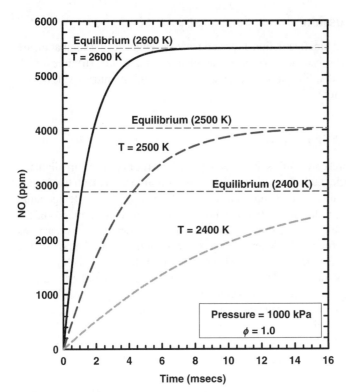

Figure 17.1 The nitric oxide concentration as a function of time for three temperatures for 1000 kPa and an equivalence ratio of 1.0. *Source:* Caton 2002. Reproduced with permission from ASME

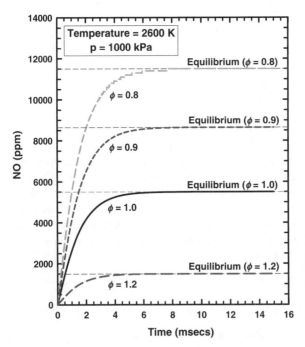

Figure 17.2 The nitric oxide concentration as a function of time for four equivalence ratios for 2600 K and 1000 kPa. *Source:* Caton 2002. Reproduced with permission from ASME

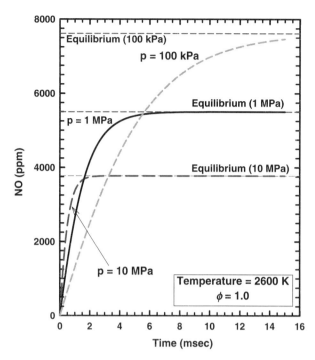

Figure 17.3 The nitric oxide concentration as a function of time for three pressures for 2600 K and an equivalence ratio of 1.0

Figure 17.3 shows the nitric oxide concentrations as functions of time for three pressures for 2600 K and an equivalence ratio of 1.0. As pressure decreases, the final equilibrium value increases, and the time to reach this final value increases. For the 100 kPa case, the final equilibrium concentration is not attained within the 16 ms of the figure. The nitric oxide production is not too sensitive to the pressure. For these conditions, decreasing the pressure from 10 MPa to 1 MPa causes nitric oxide to increase, and the final equilibrium concentrations increase from about 3800 to 5500 ppm. This is largely the result of less dissociation for the higher pressures.

The next figure (Figure 17.4) shows the nitric oxide concentration results as functions of time for different multiples of the reaction rate coefficients for the case with a temperature of 2600 K, a pressure of 1000 kPa, and an equivalence ratio of 1.0. These types of results are useful to understand and quantify the sensitivity of the computed nitric oxides to the values of the kinetic constants (which are often not known with a high degree of precision). The curve for the base case conditions is the same as shown in Figure 17.1. The curve labeled "$R_1 \times 2.0$" represents the results for the case where reaction (17.1) has been increased by a factor of 2.0. This resulted in the fastest increase of the nitric oxide concentration values. Similarly, the curve labeled "$R_1/2.0$" represents the results for the case where reaction (17.1) has been decreased by a factor of 2.0. This resulted in the slowest increase of the values for the nitric oxide concentration. As expected, reaction (17.1) is the most important for these conditions.

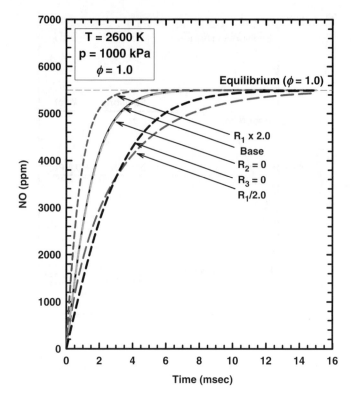

Figure 17.4 The nitric oxide concentration as a function of time for different reaction rate cases for 2600 K and 1000 kPa, and an equivalence ratio of 1.0

On the other hand, setting reaction (17.2) to zero resulted in only a modest difference. This suggests that reaction (17.2) is not controlling the overall rate. Setting reaction (17.3) to zero, however, reduced the nitric oxide concentrations. This suggests that the third reaction rate does contribute to the overall nitric oxide formation. Similar computations have been completed for actual engine conditions [2].

17.5.2 Time-Resolved Nitric Oxide Results

For most of the following results, the engine cycle simulation was executed for the part load base case condition: 1400 rpm and a *bmep* of 325 kPa as described above. For the base case, the brake-specific nitric oxide concentration is 18.9 g_{NO}/bkW-hr (3593 ppm).

Figure 17.5 shows the nitric oxide concentrations[3] as functions of crank angle for the adiabatic and boundary layer zones and the total nitric oxide for the base case. In general, the nitric oxide forms slowly at first, and then a little after TDC begins to increase rapidly. Near the end of combustion, the majority of the nitric oxide has formed. For these conditions, near

---

[3] For consistency, nitric oxide concentrations are reported based on the total moles of the exhaust even prior to exhaust valve opening.

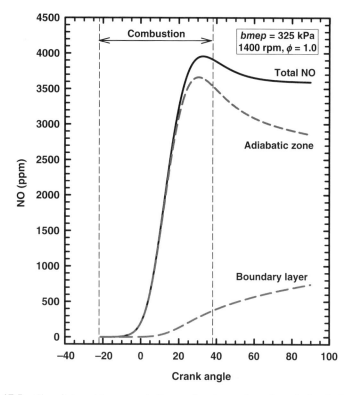

Figure 17.5   The nitric oxide concentration as functions of crank angle for the base case.
Source: Caton 2002. Reproduced with permission from ASME

the end of combustion, the nitric oxide concentration decreased slightly since the instantaneous values exceeded the equilibrium concentration and gas temperatures were high enough for continuing reactions.

The nitric oxide is formed in the adiabatic zone as described above, and then due to mass transfer some portion moves to the boundary layer as the boundary layer grows. The decrease of nitric oxide in the adiabatic zone at the end of combustion is largely due to the mass transfer to the boundary layer. A small decrease due to the kinetics also occurs.

More information on the total and individual zone masses as functions of crank angle for the base case is available in Chapter 7 on basic results. During the combustion period, the unburned mass decreases while the burned mass increases. The mass of the adiabatic zone closely follows the burned zone mass until near the end of combustion. The mass of the boundary layer increases steadily from the start of combustion until the computations were terminated at 90°aTDC.

The next set of results are aimed at illustrating the importance of selecting the appropriate gas temperature for computing nitric oxide concentrations. Figure 17.6 shows the burned gas temperatures (T_b), the adiabatic zone gas temperature (T_{ad}), and the one-zone average cylinder gas temperature (T_{avg}) as functions of crank angle for the base case conditions. The one-zone average gas temperature results in the same total energy that exists in the combined burned

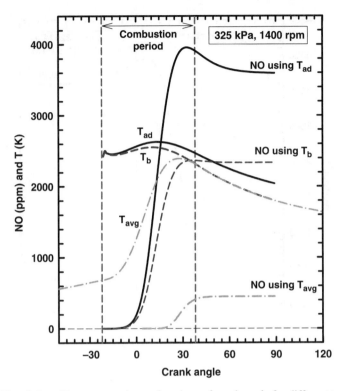

Figure 17.6   The nitric oxide concentration as functions of crank angle for different temperature assumptions for the base case

and unburned zones. The average gas temperature transitions from the unburned gas temperature to the burned gas temperature during the course of the combustion process.

Figure 17.6 also shows the computed instantaneous nitric oxide values based on the different temperatures: the one-zone gas temperature, the burned zone gas temperature, and the adiabatic zone gas temperature. For this comparison, the same procedures were used (same volumes, pressures, and so forth). This means, for example, that the volume of the adiabatic zone is used, but the temperature is changed to be one of the other two described above (the burned gas temperature or the one-zone gas temperature).

As expected, the highest nitric oxide values are obtained when using the adiabatic zone temperature. The values for the final nitric oxide concentration were about 3593 ppm, 2335 ppm and 450 ppm for the adiabatic zone gas temperature, the burned zone gas temperature, and the one-zone cylinder gas temperature, respectively. For these conditions, then, the final computed nitric oxide concentration based on the adiabatic zone temperature is about 50% higher than if the burned zone gas temperature was used. Clearly, the nitric oxide computations are highly dependent on the modeling of the in-cylinder gas temperatures, and the use of multiple zones to capture the high gas temperatures appears particularly necessary for nitric oxide computations. In addition, note the poor result for the nitric oxide concentrations when using the one-zone gas temperature.

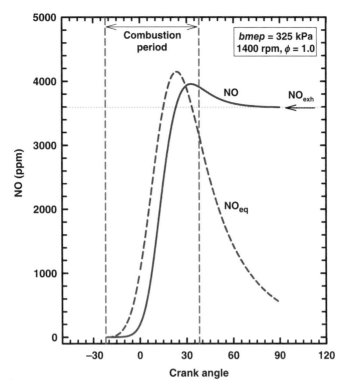

Figure 17.7   The instantaneous and equilibrium nitric oxide concentration as functions of crank angle for the base case

Figure 17.7 shows the instantaneous (rate-controlled) nitric oxide and the instantaneous equilibrium nitric oxide concentrations (NO_{eq}) as functions of crank angle for the base case. The equilibrium nitric oxide concentrations are based on the instantaneous gas temperatures, pressures, and composition. The actual rate-controlled nitric oxide concentrations are lower than the equilibrium values for most of the combustion period. After about 30°aTDC, the equilibrium values decrease as the temperatures and pressures decrease during the expansion period. The rate-controlled nitric oxide concentrations, however, remain close to the peak value and do not decrease very much during the expansion period. This is largely due to decreasing kinetic rates due to the decreasing temperatures during the expansion process. For other conditions, the rate-controlled nitric oxide concentrations were lower than the equilibrium values. In general, therefore, equilibrium nitric oxide concentrations are not particularly relevant to the instantaneous rate-controlled values for most conditions.

Figure 17.8 shows the instantaneous nitric oxide concentrations as functions of crank angle for three combustion timings: (i) 20°CA before (advanced) the MBT timing, (ii) the MBT timing, and (iii) 20°CA after (retarded) the MBT timing. The location of the combustion duration is shown for the MBT timing case. The advanced timing case results in the highest nitric oxide values due to the higher temperatures associated with advanced timing. Conversely, the retarded timing case results in the lowest nitric oxide values due to the lower temperatures associated with retarded timing. For the advanced timing case, the instantaneous nitric oxide

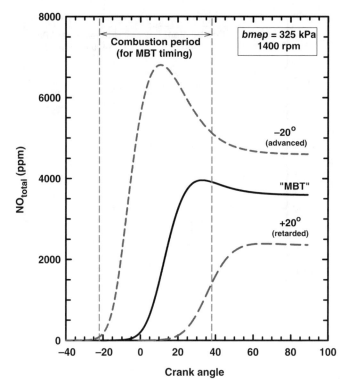

Figure 17.8   The nitric oxide concentration as functions of crank angle for three combustion timings for the base case

concentration reaches a maximum during combustion and then significantly decreases during the end of combustion and expansion process. All three cases reach a "frozen" concentration after about 70°aTDC—this will be the exhaust nitric oxide concentration level.

Figures 17.9 and 17.10 show nitric oxide concentrations as functions of crank angle for different equivalence ratios. Figure 17.9 shows fuel lean cases for three equivalence ratios: 0.93, 0.8, and 0.7. An equivalence ratio of 0.93 resulted in the highest nitric oxide concentration, and for lower equivalence ratios, the nitric oxide values were lower. This is largely a consequence of the lower combustion temperatures associated with lean mixtures. Figure 17.10 shows the results for equivalence ratios greater than 0.93. As the equivalence ratio increases, the nitric oxide concentrations decrease. Again, this is largely due to the lower temperatures associated with the richer equivalence ratios. As shown in the next subsection, for equivalence ratios lower than stoichiometric but greater than 0.93, the temperature effect is less than the oxygen effect (see Figure 17.12).

17.5.3   Engine Nitric Oxide Results

The next set of results are for nitric oxide exhaust concentrations as functions of engine load (*bmep*), engine speed, equivalence ratio, compression ratio, relative spark timing, and exhaust

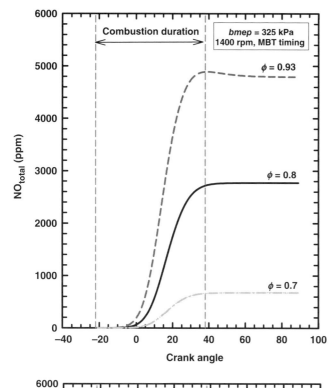

Figure 17.9 The nitric oxide concentration as functions of crank angle for three equivalence ratios for the base case

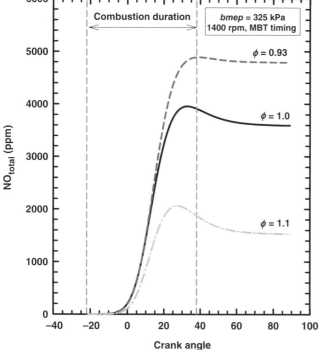

Figure 17.10 The nitric oxide concentration as functions of crank angle for three equivalence ratios for the base case

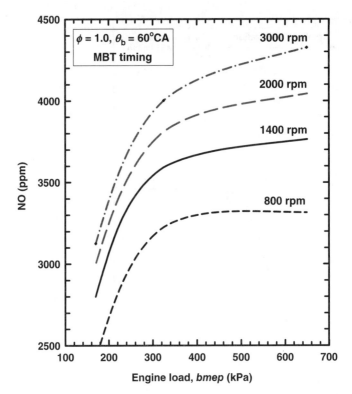

Figure 17.11   The nitric oxide concentration as functions of engine load (*bmep*) for four engine speeds

gas recirculation. The conditions for these computations, unless otherwise stated, include an equivalence ratio of 1.0, burn duration of 60°CA, and MBT spark timing. For the base case with an equivalence ratio of 1.0, the brake-specific nitric oxide concentration is 18.9 g/Kw-hr (3593 ppm).

Figure 17.11 shows the nitric oxide concentrations (ppm) as functions of engine load (as *bmep*) for four different engine speeds. Nitric oxide concentrations increase monotonically for each speed as engine load increases, and increase monotonically as engine speed increases. This is largely due to the higher temperatures associated with the higher loads and speeds. For these conditions, the nitric oxide concentrations ranged from about 2500 to 4300 ppm.

Figure 17.12 shows the nitric oxide concentrations and the average (during combustion) adiabatic zone gas temperatures as functions of equivalence ratio for the base case. These results are related to the earlier results shown in Figures 17.9 and 17.10. The maximum nitric oxide concentration is for an equivalence ratio of 0.93. For lower or higher equivalence ratios, the nitric oxide concentrations decrease. The highest temperature is for an equivalence ratio near stoichiometric ($\phi = 1.0$), and for equivalence ratios lower or higher than stoichiometric, the temperatures decrease. As the equivalence ratio decreases from stoichiometric, even though the temperature is decreasing, the nitric oxide concentrations increase. This is due to the importance of the increasing oxygen concentrations. For decreases of the equivalence

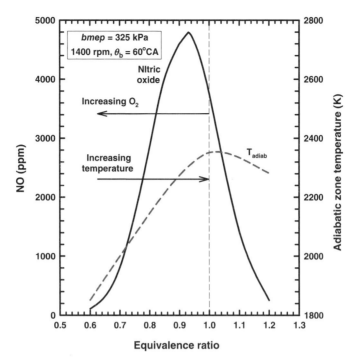

Figure 17.12 The nitric oxide concentrations and the average (one-zone) temperature during combustion as functions of equivalence ratio

ratio less than 0.93, the increasing oxygen concentration is not as important as the decreasing temperature—and therefore, the nitric oxide concentration decreases.

Figure 17.13 shows the nitric oxide concentrations as functions of the relative combustion timing. The nitric oxide concentrations continue to decrease as the timing is retarded from the most advance timing. This is largely a consequence of the gas temperatures decreasing as timing is retarded.

Figure 17.14 shows the nitric oxide concentrations as functions of compression ratio. In general, the effect of compression ratio is modest. For increases of compression ratio from 4 to about 8, the nitric oxide concentration increases slightly. For increases of compression ratio from 8 to 20, the nitric oxide concentration decreases. These slight changes of the nitric oxide concentrations as compression ratio changes are due to complex and subtle changes of items such as the combustion volumes. More details on the effects of compression ratio on nitric oxide concentrations are available elsewhere [13].

Figure 17.15 shows the nitric oxide concentrations as functions of exhaust gas recirculation (EGR).[4] As described elsewhere [14,15] and in Chapter 8, two EGR configurations have been considered. The cooled EGR configuration assumes that the exhaust gas is cooled (external to the thermodynamic system) to the inlet temperature. The adiabatic EGR configuration assumes that the exhaust gas is recirculated to the inlet at the exhaust temperature. These two

---

[4] The effects of EGR on the thermal efficiencies are described in Chapter 8.

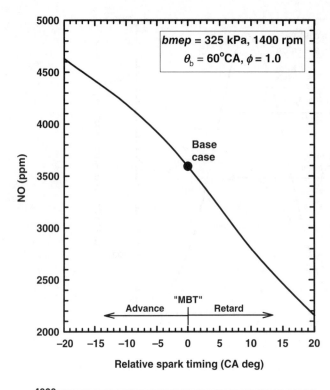

Figure 17.13 The nitric oxide concentrations as functions of the relative combustion timing

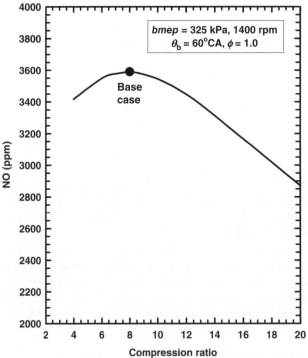

Figure 17.14 The nitric oxide concentrations as functions of the compression ratio

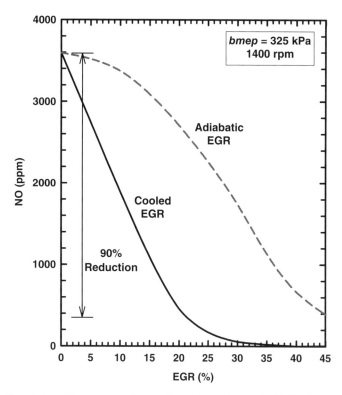

Figure 17.15   The nitric oxide concentrations as functions of the level of EGR for two EGR configurations

configurations bracket the resulting temperatures of actual EGR systems. For both configurations, the use of EGR rapidly decreases the nitric oxide concentrations due to the dilution effect and resulting lower combustion temperatures. Using 20% cooled EGR, the nitric oxide concentration is reduced by about 90%. For 20% adiabatic EGR, the nitric oxide concentration is reduced by about 25%. As mentioned elsewhere, conventional engines may experience longer combustion durations or combustion difficulties for EGR levels above about 20% or 25%.

Figure 17.16 shows the nitric oxide concentrations as functions of the heat transfer multiplier (*htm*). This evaluation is not so much about an engine variable, but rather about the sensitivity of the nitric oxide results to assumptions. As described in the heat transfer case study, the heat transfer correlation is basically an estimate. As the *htm* value increases, the nitric oxide concentration decreases due to the lower gases temperatures from the higher heat transfer. For ±20% of the *htm* value, the nitric oxide values vary by a total of about 13% for these conditions.

17.6   Summary and Conclusions

This case study has employed the thermodynamic engine cycle simulation to investigate the formation and destruction of nitric oxides as functions of engine design and operating

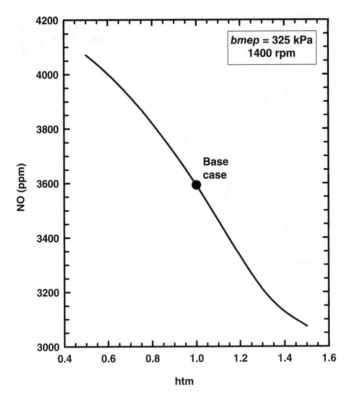

Figure 17.16 The nitric oxide concentrations as functions of the heat transfer multiplier (*htm*) for the base case

variables. A critical feature of the simulation for the nitric oxide computations is the adiabatic zone as part of the combustion submodel. The use of an adiabatic zone allows the capture of the high gas temperatures during combustion which dominate the formation and destruction of nitric oxides.

For comparisons on a similar basis, the start of combustion was adjusted for each case to provide maximum performance (MBT timing); and the inlet pressure (throttle position) was adjusted to maintain a constant load. Specific conclusions and findings include

- For typical engine cylinder temperatures, pressures, equivalence ratios, and residence times, the nitric oxide concentrations must be determined from the detail kinetics.
- The important reaction steps and the sensitivity to the reaction rate expressions were demonstrated.
- For the conditions examined, the exhaust nitric oxide concentrations do not necessarily equal the equilibrium values.
- The importance of the use of the adiabatic zone (and the corresponding higher gas temperature) for computing nitric oxides was demonstrated.
- For the conditions examined, nitric oxide concentrations increased for increasing loads (*bmep*) and speed.

- For a burn duration of 60°CA, the maximum nitric oxide values were noted for an equivalence ratio of about 0.93, and the values decreased for leaner or richer stoichiometry.
- Retarding the spark timing and increasing the EGR levels were two effective ways to reduce nitric oxide values.
- Increasing compression ratio resulted in first increasing and then decreasing nitric oxide concentrations due to a complex combination of several parameters including gas temperatures, cylinder pressure, and combustion gas volumes.

References

1. Caton, J. A. (2003) The use of a three-zone combustion model to determine nitric oxide emissions from a homogeneous-charge, spark-ignition engine, in Proceedings of the 2003 Spring Technical Conference of the ASME Internal Combustion Engine Division, Hellbrunn Palace, Salzburg, Austria, 11–14 May.
2. Caton, J. A. (2002) Detailed results for nitric oxide emissions as determined from a multiple-zone cycle simulation for a spark-ignition engine, in Design, Application, Performance and Emissions of Modern Internal Combustion Engine Systems and Components, ICE-Vol. 39, Proceedings of the 2002 Fall Technical Conference, Wong V. W. (ed.), the ASME Internal Combustion Engine Division, American Society of Mechanical Engineers, paper no. ICEF–2002–491, pp. 131–148, New Orleans, LA, September 08–11.
3. Caton, J. A. (2003) Effects of burn rate parameters on nitric oxide emissions for a spark ignition engine: results from a three-zone, thermodynamic simulation, 2003 SAE International Congress and Exposition, Society of Automotive Engineers, in *Modeling of SI Engines*, SP–1745, paper no. 2003–01–0720, Cobo Hall, Detroit, MI, March 3–6.
4. Heywood, J. B. (1988) *Internal Combustion Engine Fundamentals*, McGraw-Hill Book Company, New York.
5. Lavoie, G. A., Heywood, J. B., and Keck, J. C. (1970) Experimental and theoretical study of nitric oxide formation in internal combustion engines, *Combustion and Science Technology*, **1**, 313–338.
6. Miller, J. A. and Bowman, C. T. (1989) Mechanism and modeling of nitrogen chemistry in combustion, *Progress in Energy and Combustion Science*, **15**, 287–338.
7. Fenimore, C. P. (1971) Formation of nitric oxide in premixed hydrocarbon flames, *Thirteenth Symposium (International) on Combustion*, The Combustion Institute, Pittsburgh, PA, 373–380.
8. Lavoie, G. A. and Blumberg, P. N. (1973) Measurements of NO emissions from a stratified charge engine: comparison of theory and experiment, *Combustion Science and Technology*, **8**, 25–37.
9. Baulch, D. L., Drysdale, D. D., Horne, D. G., and Lloyd, A. C. (1973) *Evaluated Kinetic Data for High Temperature Reactions, Vol. 2 – Homogeneous Gas Phase Reactions of the H2–N2–O2 System*, Butterworths & Co. (Publishers) Ltd., London.
10. Hanson, R. K. and Salimian, S. (1984) Survey of rate constants in the N/H/O system, in *Combustion Chemistry*, Gardiner W. C., Jr. (ed.), pp. 361–422, Springer-Verlag, New York.
11. Dean, A. M. and Bozzelli, J. W. (2000) Combustion chemistry of nitrogen, in *Gas-Phase Combustion Chemistry*, Gardiner W. C., Jr. (ed.), pp. 125–342, Springer-Verlag, New York.
12. Woschni, G. (1968) A universally applicable equation for the instantaneous heat transfer coefficient in the internal combustion engine, Society of Automotive Engineers, *SAE Transactions*, SAE paper no. 670931, **76**, 3065–83.
13. Caton, J. A. (2003) Effects of compression ratio on nitric oxide emissions for a spark-ignition engine: results from a thermodynamic cycle simulation, *Journal of Engine Research*, **4** (4), 249–268.
14. Shyani, R. G. and Caton, J. A. (2009) A thermodynamic analysis of the use of EGR in SI engines including the second law of thermodynamics, *Proceedings of the Institution of Mechanical Engineers, Part D, Journal of Automobile Engineering*, **223** (1), 131–149.
15. Caton, J. A. (2006) Utilizing a cycle simulation to examine the use of EGR for a spark-ignition engine including the second law of thermodynamics, in Proceedings of the 2006 Fall Conference of the ASME Internal Combustion Engine Division, Sacramento, CA, 5–8 November.

18

High Efficiency Engines

18.1 Introduction

This case study is focused on understanding the thermodynamics of high efficiency engines. Engine development work has suggested that certain engine features can dramatically improve engine efficiency: for example, higher compression ratios, lean operation, high levels of EGR, and shorter burn durations. The material in this chapter will quantify the contributions of each of these engine features toward increasing the thermal efficiency, and describe the thermodynamic reasons for the improvements. A key to developing such engines is to manage the combustion process, and this advancement of the combustion process is briefly described next.

A great amount of engine research and development activities have focused on engine combustion processes, and this work has demonstrated the use of novel combustion techniques to achieve low nitric oxide and particulate emissions with concurrent high thermal efficiencies. These activities have used a variety of novel combustion technologies which have included various versions of low temperature, partially premixed combustion. These combustion modes are often called homogeneous charge compression ignition (HCCI), premixed charge compression ignition (PCCI), partially premixed charge (PPC), and many other such names. These developments have included the use of higher compression ratios, lean mixtures, high EGR, multiple fuel injections, two different fuels, variable valve timing, and high inlet pressures (boost). These combustion modes often result in rapid combustion. The design that provides the most effective operation is not yet known, but these types of features are clearly important.

With this in mind, the purpose of this case study is to present a systematic assessment of engine operation with high levels of EGR, lean mixtures, high compression ratios, and rapid combustion. By considering each feature in a step-by-step fashion, the impact of each feature will be quantified. The resulting thermal efficiencies will be reported, and the reasons for the high efficiencies will be determined. Additional information related to this case study may be found in References 1–3.

An Introduction to Thermodynamic Cycle Simulations for Internal Combustion Engines, First Edition.
Jerald A. Caton © 2016 John Wiley & Sons, Ltd. Published 2016 by John Wiley & Sons, Ltd.

18.2   Engine and Operating Conditions

Two versions of the standard automotive, 5.7 liter, V–8 engine were examined in this work: (i) a conventional engine and (ii) a high efficiency engine. For the conventional engine, the heat transfer correlation from Woschni [4] is used with an *htm* of 1.0. For the high efficiency engine, a heat transfer correlation from Chang et al. [5] developed for such engines was used. Another case study in this book (Chapter 13) includes a section on the importance of the selection of the cylinder heat transfer correlation.

The operating condition examined is a moderate load (*bmep* = 900 kPa), moderate speed (2000 rpm) condition. The conventional operating condition includes a compression ratio of 8:1 and stoichiometric mixture. Table 18.1 lists some of the input and output values for this condition. This condition is representative of a conventional, spark-ignition engine, and has been documented in a number of previous studies [1–3, 6]. The high efficiency operating condition is motivated by recent work for low temperature combustion engines. This condition includes a compression ratio of 16:1, an equivalence ratio of 0.7, and 45% EGR (cooled configuration[1]).

The high efficiency engine must operate with a higher inlet pressure than the conventional engine to achieve the 900 kPa *bmep*. Table 18.2 lists some of the input and output values for this condition.

Table 18.1   Conventional engine (case 1)

Item	Value used	How obtained
bmep = 900 kPa, CR = 8, EGR = 0%, MBT timing		
Fuel	C_8H_{18}	Input
Equivalence ratio	1.0	Input
Engine speed (rpm)	2000	Input
Mech frictional mep (kPa)	81.5	From algorithm [11]
Inlet pressure (kPa)	93.2	Input
Exhaust pressure (kPa)	105.0	Input
Start of combustion (°bTDC)	24.0	Determined for MBT
Combustion duration (°CA)	60	Input
Cylinder wall temperature (K)	450	Input
Heat transfer correlation		Woschni [4]
htm	1.0	Input
Gross indicated thermal eff (%)	37.30	Output
Net indicated thermal eff (%)	36.70	Output
Brake thermal eff (%)	33.65	Output
Exergy destruction combustion (%)	20.53	Output
Max press rise rate (kPa/CA)	142	Output

---

[1] Chapter 8 describes two EGR configurations: cooled and adiabatic.

Table 18.2   High efficiency engine (case 6)

Item	Value used	How obtained
bmep = 900 kPa, CR = 16, EGR = 45%, MBT timing		
Fuel	C_8H_{18}	Input
Equivalence ratio	0.7	Input
Engine speed (rpm)	2000	Input
Mech frictional mep (kPa)	115.9	From algorithm [11]
Inlet pressure (kPa)	169.5	Input
Exhaust pressure (kPa)	179.5	Input
Start of combustion (°bTDC)	13.5	Determined for MBT
Combustion duration (°CA)	30	Input
Cylinder wall temperature (K)	450	Input
Heat transfer correlation		Chang et al. [5]
htm	1.0	Input
Gross indicated thermal eff (%)	54.95	Output
Net indicated thermal eff (%)	53.92	Output
Brake thermal eff (%)	47.79	Output
Exergy destruction combustion (%)	24.13	Output
Max press rise rate (kPa/CA)	535	Output

For these computations, the exhaust pressure is not known and will depend on the design of the charging system. For this work, a simple schedule was assumed for the exhaust pressure. For naturally aspirated cases (inlet pressure less than 95 kPa), the exhaust pressure was set at 105 kPa. For cases with inlet pressures above 95 kPa, the exhaust pressure was set at 10 kPa above the inlet pressure.

The following are a few comments concerning the exhaust pressure. Due to the variety of engine configurations and EGR systems, no single assumption for the exhaust pressure is universal. Some engines may use a turbocharger, a supercharger, or both. The exhaust pressure (especially for a turbocharged engine using a high level of EGR) may be expected to be significantly higher than the inlet pressure. The sensitivity of the engine performance on the exhaust pressure is presented elsewhere [7]. The use of a supercharger may not result in a pumping work penalty, but would result in additional mechanical friction. The scope of the current study did not include any detailed evaluation of these types of considerations.

Another item that may be different for different engines is the inlet temperature which will be affected by the degree of EGR cooling. Although the current work has assumed that the EGR is cooled to the nominal inlet temperature, limited results for the adiabatic EGR configuration also are included below for completeness. The sensitivity of the thermal efficiency to the inlet temperature may be found in Reference 7.

18.3 Results and Discussion

The goal of this work is to illustrate the significance of thermodynamics for the development of high efficiency engines. Not only are specific features shown to provide this high efficiency, but the physical reasons for the improvements are provided. The results will include engine performance parameters, thermal efficiencies, nitric oxide (NO) and relative carbon dioxide (CO_2) emissions, and exergy destruction during the combustion process.

As a representative fuel, only isooctane is examined. Although in practice other fuels (and multiple fuels) have been used, the implication of these other fuels to the thermodynamics is modest [6]. As is the case for most of the work reported in this book, for both engines, combustion was assumed successful and proceeds according to the Wiebe function. Again, the focus of this work is not to describe the combustion in any detail. Rather, the focus is to understand the generic thermodynamics based on successful combustion. Of course, successful combustion is a challenge for some conditions, but this aspect is beyond the scope of the current work. The results are presented in four subsections: (i) overall assessment, (ii) individual parameters, (iii) emissions and exergy, and (iv) combustion parameters.

18.3.1 Overall Assessment

This subsection provides detailed, quantitative results for efficiency improvements due to specific features. As described above, a number of operating and design parameters may have a significant effect on the thermal efficiency. Often changing one parameter is not as effective as changing several at the same time. For example, higher compression ratios by themselves may lead to spark knock for a stoichiometric engine. But if the increase of compression ratio is accompanied with lean operation and the use of EGR, spark knock may not be a limitation. So, although the following parameters are discussed individually, the ultimate success is dictated by the final, total design.

The main idea of this study was to complete a systematic assessment of the various features that have been employed in previous investigations that resulted in high efficiencies. By considering each feature in a step-by-step fashion, the impact of each feature can be quantified. Table 18.3 is a description of the cases. For example, going from case 1 to case 2, the compression ratio is increased from 8 to 16, and all other parameters are the same as case 1. Going from case 2 to case 3, the burn duration is decreased from 60°CA to 30°CA, and all

Table 18.3 Description of cases

Case	Description
1 (base)	CR = 8; θ_b = 60°CA; ϕ = 1.0; EGR = 0%; Woschni
2 (CR)	CR = 16; θ_b = 60°CA; ϕ = 1.0; EGR = 0%; Woschni
3 (θ_b)	CR = 16; θ_b = 30°CA; ϕ = 1.0; EGR = 0%; Woschni
4 (ϕ)	CR = 16; θ_b = 30°CA; ϕ = 0.7; EGR = 0%; Woschni
5 (EGR)	CR = 16; θ_b = 30°CA; ϕ = 0.7; EGR = 45%; Woschni
6 (HT)	CR = 16; θ_b = 30°CA; ϕ = 0.7; EGR = 45%; Chang

other parameters are the same as case 2. This continues until all the parameters are set for the high efficiency engine, case 6.

The order that the features were added was arbitrary. A study [2] has examined and quantified the significance of the different orders or sequences of features. In general, the overall trends are the same, but the individual contributions will be slightly different. Obviously, the final result (comparison of case 1 and case 6) will be the same regardless of which sequence is used.

Figure 18.1 shows the gross indicated, net indicated and brake thermal efficiencies for this operating condition for each case as described in Table 18.3. Table 18.4 lists some of the input values for the six cases. The efficiencies increase from case 1 (conventional engine) through case 6 (high efficiency engine). Each of the added features contributes to the final, high efficiency. The brake thermal efficiency increases from 33.7% (case 1) to 47.8% (case 6), and the net indicated thermal efficiency increases from 36.7% (case 1) to 53.9% (case 6). These are major improvements. For the brake thermal efficiency, this is a 1.4 factor (or 140%) increase. The following paragraphs will describe the thermodynamic features that are responsible for these improvements.

The increase of compression ratio, the decrease of equivalence ratio, and the increase of the level of EGR provide the greatest improvements. The shorter burn duration provides a more modest improvement. The implication of each of these features is discussed in more detail in the following.

The use of the heat transfer correlation from Chang et al. [5] instead of the one from Woschni [4] provides a significant improvement, but this is not actually an engine design parameter. The choice of the heat transfer correlation is based on empirical information. Further comments on the choice of the heat transfer correlation are provided in the case study on cylinder heat transfer (chapter 13).

For the addition of lean mixtures (case 4) and high EGR (case 5), the *brake* thermal efficiency increases less than the *indicated* efficiencies. This is due to the need for increases of the inlet pressure (to provide the 900 kPa *bmep*) which increases the mechanical friction for the diluted cases. More discussion of this aspect is provided later in this subsection.

Although no direct verification of the computed results with experimental data is available, a comparison with similar data is useful. Table 18.5 provides a comparison of the results for the high efficiency engine (case 6) from this study with results from experiments reported by Kokjohn et al. [8]. The engine and operating details used by Kokjohn et al. were different than the current work. They used a dual-fueled HCCI and PCCI concept. By the use of in-cylinder

Table 18.4 Detailed description of cases ($bmep$ = 900 kPa, 2000 rpm)

Case	CR	θ_b	ϕ	EGR (%)	HT	θ_o (°bTDC)	p_{in} (kPa)	p_{exh} (kPa)
1 (Base)	8	60	1.0	0	Woschni	24.0	93.2	105.
2 (CR)	16	60	1.0	0	Woschni	20.0	83.7	105.
3 (θ_b)	16	30	1.0	0	Woschni	5.5	82.1	105.
4 (ϕ)	16	30	0.7	0	Woschni	6.0	107.1	117.1
5 (EGR)	16	30	0.7	45	Woschni	7.5	191.1	201.1
6 (HT)	16	30	0.7	45	Chang	13.5	169.5	179.5

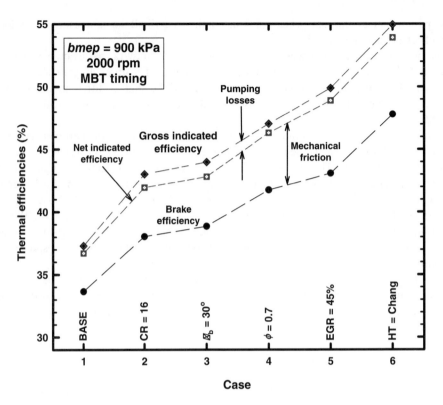

Figure 18.1   The gross indicated, net indicated and brake thermal efficiencies for each case for a *bmep* of 900 kPa and 2000 rpm. *Source:* Caton 2014b. Reproduced with permission from Elsevier

blends of gasoline and diesel fuel they were able to extend the range of operation of the low-temperature combustion. They demonstrated successful operation for several operating conditions. One such case was a 1300 rpm, 11 bar *imep*$_{net}$ condition that resulted in negligible emissions and relatively high thermal efficiencies.

Table 18.5 shows that the work by Kokjohn et al. employed an equivalence ratio of 0.77, an EGR level of 45.5%, and a compression ratio of 16.1. These are similar to the conditions used in this work ($\varphi = 0.7$, EGR = 45%, CR = 16). They required an inlet pressure of 200 kPa, whereas the current work needed an inlet pressure of 170 kPa. They reported an *imep*$_{net}$ of 1100 kPa, whereas the current work resulted in an *imep*$_{net}$ of 1015 kPa. Also, the peak cylinder pressure (12.0 MPa) was the same from the experiment and from the computations.

Of particular relevance, however, is the comparison of the net indicated thermal efficiency: 50% from the experiment, and 53.9% from the computations. The higher computed value is probably due to some of the idealizations included in the simulation such as zero blow-by and zero combustion inefficiency.[2] Also, the actual heat transfer for the experimental engine may not be adequately represented by the Chang et al. [5] correlation. In any case, the two values are reasonably close.

---

[2] Note that features such as blow-by and combustion inefficiency could be included in the simulation if more detailed results were desired.

Table 18.5 Comparisons to results from Reference 8

Item	Reference (Kokjohn et al. [8])	This work (Case 6)
Bore/stroke (mm)	137/165	102/88
Fuels	Gasoline/diesel	Isooctane
Inlet pressure (kPa)	200	169.5
Geometric CR	16.1	16
EGR (%)	45.5	45
Equivalence ratio	0.77	0.7
Speed (rpm)	1300	2000
Results:		
$imep_{net}$ (kPa)	1100	1015
Net indicated efficiency (%)	50	53.92
Peak pressure (MPa)	12	12.0
Nitric oxide (g/kW-h)	0.01	0.005

Although this comparison was not exact, the reasonable agreement of the results from the computations and the experiment provides some confidence that the overall thermodynamics of the high efficiency engine have been captured by the simulation using the features outlined above.

While an exact comparison to the experimental work may be desired, this is often difficult to complete with a high level of fidelity. Often reports of experimental work do not include all of the major parameters needed for the comparison. Some of these parameters are not measured and some are not possible to measure. For example, the cylinder wall temperatures, the cylinder heat transfer and the blow-by rates are rarely reported.

The remainder of this section will report on the details of the above results and discuss the specific importance of various thermodynamic features. The next several figures will show various parameters as functions of the brake thermal efficiency. Typically, the brake thermal efficiency is not used as the independent variable. This is done, however, to better illustrate the effects of the various features as the cases progress from case 1 to case 6 with continually increasing efficiencies.

Figure 18.2 shows the inlet and exhaust pressures as functions of the brake thermal efficiency. As mentioned above, the inlet pressure was selected to obtain the constant *bmep* of 900 kPa. For cases 1–3, the inlet pressure remained below 100 kPa and they were "throttled" cases. For cases 4–6, the inlet pressure was above 100 kPa, and would need to be operated with turbocharging or supercharging. Also as mentioned above and reflected in Figure 18.2, the exhaust pressure was either 105 kPa or 10 kPa above the inlet pressure.

Figure 18.3 shows the mechanical friction mean effective pressure for each case. The mechanical friction increases as a result of higher cylinder pressures (or inlet pressures), and this is largely due to the increases of the reciprocating component of the mechanical friction. The reciprocating component includes the mechanical friction due to the connecting rod bearings, piston skirts, and rings. This is the reason that the brake thermal efficiencies increases are not as great as the indicated thermal efficiencies increases as shown in Figure 18.1.

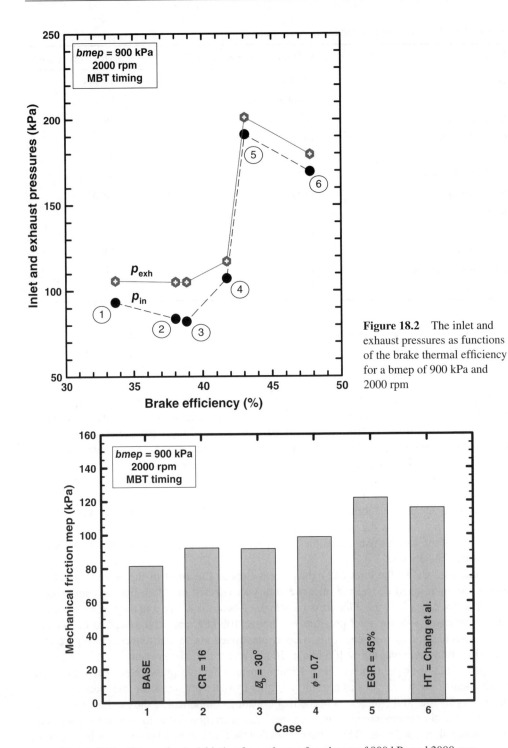

Figure 18.2 The inlet and exhaust pressures as functions of the brake thermal efficiency for a bmep of 900 kPa and 2000 rpm

Figure 18.3 The mechanical friction for each case for a bmep of 900 kPa and 2000 rpm

Figure 18.4 shows the cylinder heat transfer energy and the exhaust sensible gas energy as percentages of the fuel energy as functions of the brake thermal efficiency for each case. To help understand these results, Figure 18.5 shows (i) the one-zone, peak temperature and (ii) the time-average, one-zone temperature during the combustion period (average combustion temperature) as functions of the brake thermal efficiency. As shown in Figure 18.4, for increasing compression ratio (case 1 to case 2), the relative heat transfer slightly increased largely due to a higher surface area to volume ratio for the higher compression ratio. The exhaust energy decreased due partly to the increase of heat transfer, but also because the work output conversion percentage increased. The peak temperature decreased, but the average combustion temperature was fairly unchanged as compression ratio increased.

For shorter burn durations (case 2 to case 3), the relative heat transfer increased slightly, the exhaust energy decreased slightly, and the temperatures increased slightly. The higher temperatures are a result of the higher heat release rate due to the shorter burn duration, and the higher temperatures explain the heat transfer and exhaust energy results.

For the change from stoichiometric to lean operation (case 3 to case 4), the temperatures decreased as expected for the lean combustion. The lower temperatures result in lower relative heat transfer and lower exhaust energy. This is similar for the addition of EGR.

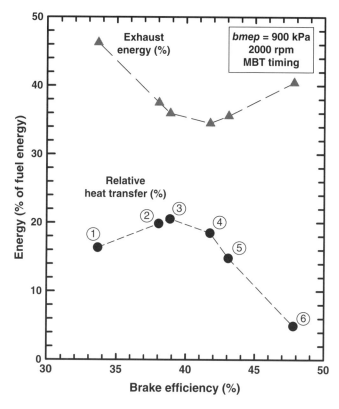

Figure 18.4 The relative heat transfer and exhaust energy as functions of the brake thermal efficiency for a *bmep* of 900 kPa and 2000 rpm

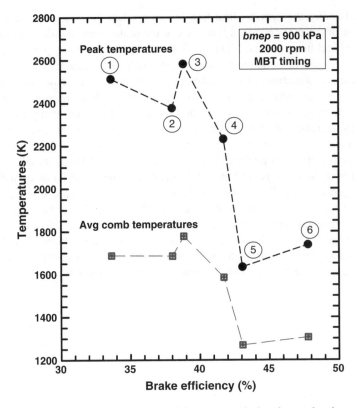

Figure 18.5   The one-zone peak temperatures and the average (during the combustion period) temperatures as functions of the brake thermal efficiency for a *bmep* of 900 kPa and 2000 rpm

For the increase of EGR (case 4 to case 5), the temperatures decrease, and the relative heat transfer decreases. The exhaust gas energy as a percentage of the fuel energy increases slightly.

Finally, for the switch to the Chang et al. [5] heat transfer correlation (case 5 to case 6), the temperatures increase slightly and the exhaust gas energy (as a percentage of the fuel energy) increases as a result of the significant reduction of cylinder heat transfer.

An important feature of these results is the conversion of thermal energy to work which has been shown to be a strong function of the ratio of specific heats [9]. As the ratio of specific heats increases, this conversion increases. Figure 18.6 shows the one-zone average ratio of specific heats (gamma) during the combustion period as a function of the brake thermal efficiency. This value increases dramatically for the change to lean operation and the addition of EGR. Since the work conversion is dependent on the ratio of specific heats as an exponent, small changes in this value have dramatic effects on the work conversion. The increase of this ratio is due to the lower gas temperatures and to a lesser extent due to the change of the mixture composition [9]. Although a more subtle contribution, at least a part of the efficiency gains reflected in Figure 18.1 is due to the increase of the ratio of specific heats [10].

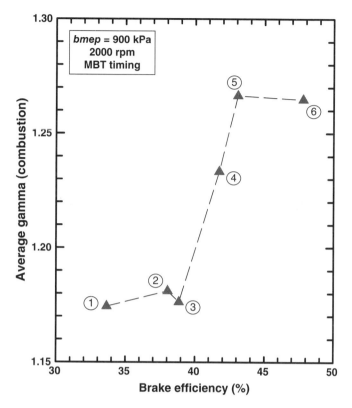

Figure 18.6   The average ratio of specific heats (gamma) during combustion as a function of the brake thermal efficiency for a *bmep* of 900 kPa and 2000 rpm

18.3.2   *Effects of Individual Parameters*

This next section of results will consider the individual contributions of each of the major parameters discussed above.

Lean operation is well known to improve efficiency (as well as lower nitric oxide emissions in certain cases—see related case study on nitric oxides). Figure 18.7 shows the net indicated and brake thermal efficiencies as functions of equivalence ratio (all conditions except for equivalence ratio are the same as for case 6). As the mixture becomes leaner from stoichiometric, the gains are approximately linear for the net indicated efficiency. For the brake efficiency, the leaner cases require higher inlet pressures (to achieve the 900 *bmep*) which results in higher friction. This higher friction is largely due to the connecting rod bearings, piston skirts, and rings [11]. So the gains are somewhat less for the brake efficiency as equivalence ratio decreases. These gains are largely due to the reduced cylinder gas temperatures, the reduced heat losses, and the more favorable thermodynamics (e.g., increased ratio of specific heats). For equivalence ratios greater than stoichiometric (fuel rich mixtures), the efficiencies decrease rapidly due to the excess fuel that is not utilized. Except for special cases (such as cold starts), rich equivalence ratios typically are avoided for normal operation.

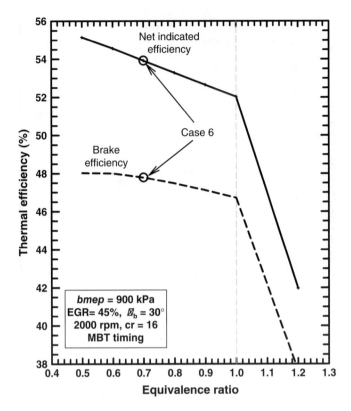

Figure 18.7   The net indicated and brake thermal efficiencies as functions of the equivalence ratio for a *bmep* of 900 kPa and 2000 rpm. *Source:* Caton 2013. Reproduced with permission from ASME

The use of exhaust gas recirculation (EGR) is a very effective option for lowering combustion temperatures, and therefore, nitric oxide formation (see case study on nitric oxides). The use of EGR is also well known to have efficiency benefits. Figure 18.8 shows the net indicated and brake thermal efficiencies as functions of EGR level for cooled and adiabatic EGR configurations. (The results for the adiabatic EGR configuration are included for completeness, but are not part of this particular study.) In general, the use of the cooled EGR configuration results in increasing efficiencies as the EGR level increases. These increases are largely due to reduced heat losses and increases of the ratio of specific heats. The gains are nearly linear for the net indicated efficiency. For the brake efficiency, the increasing friction with higher EGR levels (higher inlet pressures to maintain the constant load comparison) causes the brake thermal efficiency to start to decrease for levels greater than about 45%.

The use of the adiabatic EGR configuration results in decreasing efficiencies as the EGR level increases due to increasing heat losses. As described in Chapter 7, for a throttled (inlet pressures less than atmospheric) operating condition, a special trade-off existed. This trade-off was that for maintaining a constant load, any decreases of engine efficiency due to increasing heat losses were offset by decreasing pumping losses due to increasing inlet pressures. This is not the case for engines operating with inlet pressures above atmospheric such as for the high

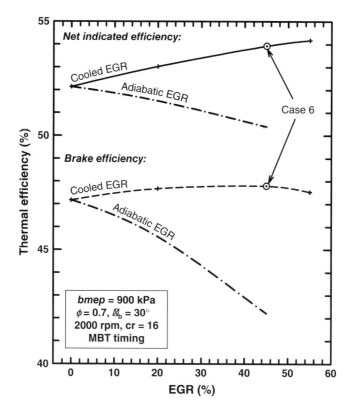

Figure 18.8 The net indicated and brake thermal efficiencies as functions of the EGR for a cooled EGR configuration and an adiabatic EGR configuration for a *bmep* of 900 kPa and 2000 rpm

efficiency engine. In addition, also note that in practice most EGR cooling systems will not achieve the high level of cooling considered here for the cooled configuration.

Since the use of lean mixtures and the use of EGR (cooled configuration) are similar, the relative merits of each approach are of interest. Up to this point, this study has considered the effects of lean operation and EGR separately. In practice, many investigators have suggested that the combined use of both lean mixtures with EGR is the most promising way to maximize thermal efficiency and reduce emissions (e.g., [12]). Previous work [9] has provided a comparison of the two techniques. One conclusion of this work [9] was that the increase of the ratio of specific heats associated with both approaches is the dominant characteristic that leads to higher efficiencies.

The next feature to consider in the pursuit of high efficiency is increasing the compression ratio. Of course high compression ratios have been known since the first engines to provide higher efficiencies. For some spark ignition engines, higher compression ratios result in spark knock and would not be a possible option. For this configuration (using lean mixtures with significant EGR), higher compression ratios are not thought to lead to knock. Figure 18.9 shows the net indicated and brake thermal efficiencies as functions of compression ratio. The gains are greatest for the compression ratio increases up to about 12, and then the gains are

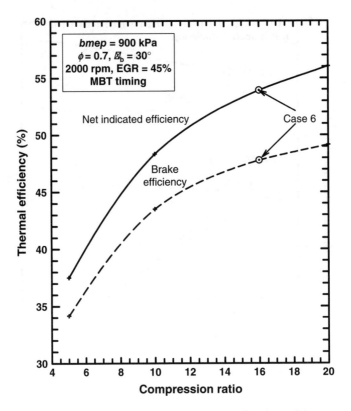

Figure 18.9 The net indicated and brake thermal efficiencies as functions of the compression ratio for a *bmep* of 900 kPa and 2000 rpm. *Source:* Caton 2013. Reproduced with permission from ASME

somewhat less. Again, the gains for the brake efficiencies are slightly less due to the increasing friction.

Short burn durations are also known to lead to high efficiencies since greater advantage of the energy release is obtained during the expansion stroke. For the work reported here, burn duration is the period from 0% to 100% of the fuel mass burned. The disadvantages of short burn durations include possible noise, roughness and harshness due to the rapid energy release. The maximum pressure rise rate can serve as an example of the rapid energy release for the high efficiency engine. For the high efficiency engine and for the conventional engine, the maximum pressure rise rates were 535 and 142 kPa/CA, respectively. In other words, the high efficiency engine has about a 3.8 higher pressure rise rate than the conventional engine. Figure 18.10 shows the net indicated and brake thermal efficiencies as functions of the burn duration. The thermal efficiencies increase as the burn duration decreases. The gains are smaller as the burn duration approaches zero.

All of the preceding results have been obtained for optimal combustion phasing—start of combustion is adjusted for maximum brake torque (MBT). The thermal efficiencies decrease as the timing is retarded or advanced relative to the MBT timing. Further comments on the combustion timing are provided elsewhere [1–3].

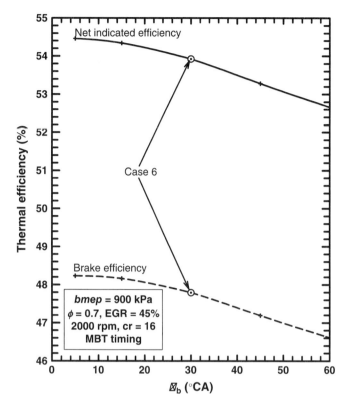

Figure 18.10 The net indicated and brake thermal efficiencies as functions of the burn duration for a *bmep* of 900 kPa and 2000 rpm. *Source:* Caton 2013. Reproduced with permission from ASME

The above features ($\phi = 0.7$, EGR = 45%, CR = 16, $\theta_b = 30°$CA, and MBT timing) provide the parameters of one version of a high efficiency engine. These values and the thermodynamic results appear to be consistent with experimental results. Although the simulation is a powerful tool for understanding the thermodynamics associated with this engine configuration, many other practical aspects will play a role in the final design. For example, combustion has been assumed to be successful and spark knock was not a constraint. Experimental information may indicate these or other assumptions to be invalid. Items such as combustion timing may need to be adjusted to aid in the combustion process and not be adjusted for MBT.

18.3.3 Emissions and Exergy

The thermodynamic cycle simulation may be used to estimate nitric oxide and carbon dioxide emissions. The use of the simulation for estimating nitric oxide emissions is discussed more fully in chapter 17.

Figure 18.11 shows the specific nitric oxide emissions as functions of the brake thermal efficiency for each case. The nitric oxide emissions are affected most by the use of EGR

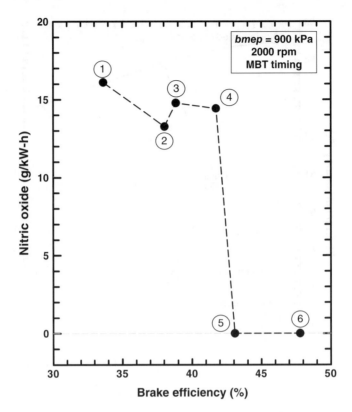

Figure 18.11 The specific nitric oxide emissions as functions of the brake thermal efficiency for a *bmep* of 900 kPa and 2000 rpm

which decreases the nitric oxides to near zero. The other items had modest effects on the nitric oxide levels. Of special interest is the lack of significant reduction by the lean operation (case 3 to case 4). This lack of significant reduction has been reported elsewhere [13]. Basically, lean operation has two major effects—reducing combustion temperatures and increasing the oxygen concentrations. Nitric oxides decrease with decreasing temperatures, but increase with increasing oxygen concentrations. For this change (case 3 to case 4), the oxygen concentration effect roughly equaled the temperature effect. For these conditions, the nitric oxide emission does not decrease significantly until the equivalence ratio decreases below 0.7 [13].

Figure 18.12 shows the relative carbon dioxide (CO_2) emissions as functions of the brake thermal efficiency. Since the carbon dioxide emissions are proportional to the fuel consumption, the fuel consumption has been used as surrogate for carbon dioxide emissions. The relative carbon dioxide emissions are based on the comparisons of the fuel consumption of the conventional engine with the other cases. The conventional engine is assigned a value of 1.0 for this comparison. The relative carbon dioxide emissions decrease monotonically as the cases progress from 1 to 6. The final, high efficiency engine (case 6) has about 70% of the carbon dioxide emissions of the conventional engine (case 1).

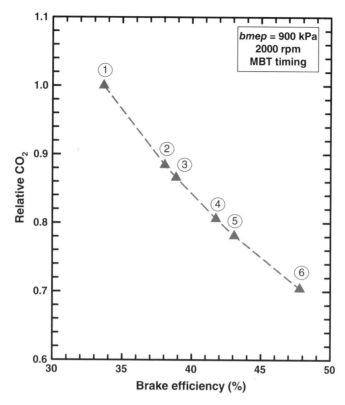

Figure 18.12 The relative carbon dioxide emissions as functions of the brake thermal efficiency for a *bmep* of 900 kPa and 2000 rpm

Figures 18.13 and 18.14 are bar graphs that show the distribution of the energy and exergy for the conventional (case 1) and high efficiency (case 6) engines, respectively. More details concerning exergy analyses may be found in chapter 9 on the second law results. By comparing Figures 18.13 and 18.14, several items are important to note. First, the large increase of the indicated work of the high efficiency engine is accompanied by a significant reduction of the heat loss and a more modest reduction of the exhaust gas energy. From the consideration of exergy, similar statements may be made. In addition, the exergy destruction during combustion is higher for the high efficiency engine compared to the conventional engine. This is due to the lower combustion temperatures associated with the high efficiency engine. In spite of the higher exergy destruction, the high efficiency engine benefits from the reduced temperatures and heat losses, and achieves the higher efficiency. As mentioned below, the higher destruction of the exergy for the high efficiency engine may be considered acceptable in light of the higher final efficiencies.

Figure 18.15 shows the exergy destruction during the combustion period for each case. Exergy destruction during the combustion process is more fully discussed in chapter 9 and in Reference 14. The exergy destruction changes the most with lean operation and the addition

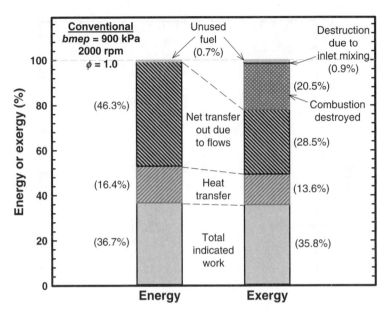

Figure 18.13 The distribution of the energy and exergy for the conventional (case 1) engine.
Source: Caton 2013. Reproduced with permission from ASME

of EGR. This is largely due to the decrease of the combustion temperatures (see Figure 18.5). The exergy destruction increases from 20.5% (case 1, conventional engine) to 24.1% (case 6, high efficiency engine). The increase of the exergy destruction is a trade-off for the higher thermal efficiencies associated with case 6. The idea is to manage the exergy destruction; and

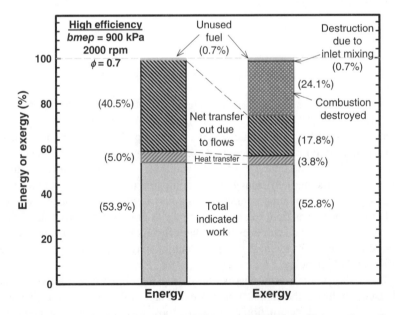

Figure 18.14 The distribution of the energy and exergy for the high efficiency (case 6) engine.
Source: Caton 2013. Reproduced with permission from ASME

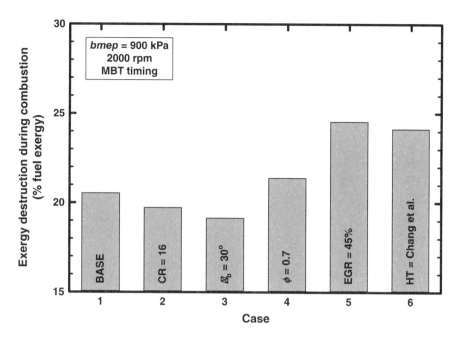

Figure 18.15 The exergy destroyed during the combustion process for each case for a *bmep* of 900 kPa and 2000 rpm. *Source:* Caton 2014b. Reproduced with permission from Elsevier

not necessarily reduce it. As mentioned above, in this situation, the benefits from lower combustion temperatures (lower heat transfer and higher specific heat ratios) are more beneficial than the reduction of the exergy destruction.

18.3.4 Effects of Combustion Parameters

For completeness, the combustion parameters are examined in this subsection. Recall that the cycle simulation uses a Wiebe function to describe the mass fraction burned.

$$x_b = 1 - \exp\left\{-a y^{m+1}\right\}$$

Details of this function and its application to the cycle simulation are provided in section 12.3. This subsection will examine the influence of the parameters "a" and "m" on thermal efficiencies for both the conventional and high efficiency engines. Recall that the values used in the above were $a = 5.0$ and $m = 2.0$.

Figures 18.16 and 18.17 show the brake thermal efficiency as functions of "a" for three values of "m" for the conventional engine (case 1) and for the high efficiency engine (case 6), respectively. For these computations, the combustion timing was adjusted for maximum brake torque (MBT), and the inlet pressure was selected to obtain the constant *bmep* of 900 kPa. For both engines, the thermal efficiency increases with increases in "a" and slightly increases with increases of "m." This result is largely due to the greater influence that "a" has on the maximum rate of burning whereas "m" has a greater influence on the timing of the maximum rate.

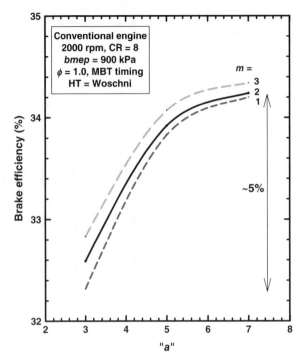

Figure 18.16 The brake thermal efficiency as functions of "*a*" for three "*m*" values for the conventional engine (case 1). *Source:* Caton 2013. Reproduced with permission from ASME

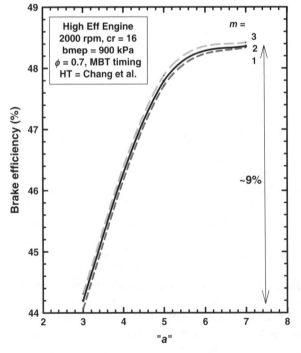

Figure 18.17 The brake thermal efficiency as functions of "*a*" for three "*m*" values for the high efficiency engine (case 6). *Source:* Caton 2013. Reproduced with permission from ASME

For the conventional engine (Figure 18.16), the range of change is about 5% (relative) which represents a change of about 2% actual (absolute) efficiency gain. In contrast, for the high efficiency engine (Figure 18.17), the range of change is about 9% (relative)—or almost 4% actual (absolute) efficiency gain. These numbers for the magnitude of the efficiency change with changes in the Wiebe constants is for a change of the "a" value from 3 to 7. Since the value of "a" for most of the simulation results is normally 5, the change from the base results will be even less.

One other aspect of varying the constants is the impact on the rate of pressure rise. For higher values of "a," the rate of pressure rise increases. For the high efficiency engine, as "a" increases from 3 to 7 (for an "m" of 2), the pressure rise rate increases about 20%. The high efficiency engine conditions appear to be slightly more sensitive to the Wiebe constants.

18.4   Summary and Conclusions

This case study has presented results from both a first law and a second law perspective to better understand the thermodynamics of advanced high efficiency engines. By systematically examining the individual contributions of compression ratio, lean mixtures, fast burn and high EGR, the impact of these features on the thermodynamic efficiency was determined. An automotive engine at 2000 rpm and with a *bmep* of 900 kPa was examined.

- The final case (6) resulted in a net indicated efficiency of 53.9% and a brake efficiency of 47.8% with negligible nitric oxide emissions. To obtain the *bmep* of 900 kPa, the inlet pressure needed to be about 170 kPa for case 6. This will require a turbocharger or supercharger. Since the exhaust gas energy is significantly decreased, the use of a turbocharger may be problematic.
- Increasing the compression ratio from 8.0 to 16.0 provided one of the more effective ways to increase the efficiencies for these conditions.
- Decreasing the equivalence ratio from 1.0 to 0.7, and increasing the exhaust gas recirculation (EGR; cooled configuration) to 45% provided significant efficiency gains largely due to the significant decrease of the relative heat transfer and the increase of the ratio of specific heats during combustion.
- Decreasing the burn duration from 60°CA to 30°CA increased the net indicated efficiency only a modest amount. At least part of the reason for the modest increase was due to the use of MBT timing.
- Some of the gains suggested by the increases of the net indicated thermal efficiency are somewhat mitigated in terms of the brake efficiencies. This is due to the increased cylinder gas pressures (that are necessary to obtain the same load) which increase the mechanical friction component.
- The exergy destruction during the combustion process decreases and increases as the various features are considered due primarily to the increase and decrease of the combustion temperatures. Overall, as an example, exergy destruction was about 20.4% and about 24.0% for cases 1 and 6, respectively. The efficiencies were higher for case 6 compared to case 1 even though the exergy destruction during combustion was higher for case 6.
- The lower combustion temperatures for the final configuration (case 6) result in negligible nitric oxides, and the carbon dioxide emissions for the high efficiency engine are about 70% of the emissions for the conventional engine (due to the reduction of fuel consumption).

- For a range of values for the Wiebe constants, only modest changes of the thermal efficiencies were obtained. The high efficiency engine appears to be slightly more sensitive to the values of the Wiebe constants.

References

1. Caton, J. A. (2010) An assessment of the thermodynamics associated with high-efficiency engines, in Proceedings of the ASME 2010 Internal Combustion Engine Division Fall Technical Conference, paper no. ICEF2010–35037, San Antonio, TX, September 12–15.
2. Caton, J. A. (2012) The thermodynamic characteristics of high efficiency, internal-combustion engines, *Energy Conversion and Management*, **58**, 84–93.
3. Caton, J. A. (2014) Thermodynamic considerations for advanced, high efficiency IC engines, *Journal of Engineering for Gas Turbines and Power*, **136** (10) 101512–101512–6; also in Proceedings of the 2013 Fall Technical Conference of the ASME Internal Combustion Engine Division, paper no. ICEF2013–19040, Dearborn, MI, October 13–16, 2013.
4. Woschni, G. (1968) A universally applicable equation for the instantaneous heat transfer coefficient in the internal combustion engine, *SAE Transactions*, SAE paper no. 670931, **76**, 3065–3083.
5. Chang, J., Guralp, O., Filipi, Z., and Assanis, D. (2004) New heat transfer correlation for an HCCI engine derived from measurements of instantaneous surface heat flux, Society of Automotive Engineers, SAE Paper No. 2004–01–2996.
6. Caton, J. A. (2010) Implications of fuel selection for an SI engine: results from the first and second laws of thermodynamics, *Fuel*, **89**, 3157–3166, November.
7. Caton, J. A. (2014) A thermodynamic comparison of external and internal exhaust gas dilution for high efficiency engines, accepted for publication, *International Journal of Engines*, October 01, 2014.
8. Kokjohn, S L., Hanson, R. M., Splitter, D. A., and Reitz, R. D. (2009) Experiments and modeling of dual-fuel HCCI and PCCI combustion using in-cylinder fuel blending, Society of Automotive Engineers, SAE paper no. 2009–01–2647.
9. Caton, J. A. (2013) A comparison of lean operation and exhaust gas recirculation: thermodynamic reasons for the increases of efficiency, 2013 SAE International Congress and Exposition, Society of Automotive Engineers, SAE paper no. 2013–01–0266, Cobo Hall, Detroit, MI, April 16–18.
10. Caton, J. A. (2014) On the importance of specific heats as regards efficiency increases for highly dilute IC engines, *Journal of Energy Conversion and Management*, **79**, 146–160.
11. Sandoval, D., and Heywood, J. B. (2003). An improved friction model for spark-ignition engines, Society of Automotive Engineers, SAE paper no. 2003–01–0725.
12. Tabata, M., Yamamoto, T., and Fukube, T. (1995) Improving NO_x and fuel economy for mixture injected SI engine with EGR, Society of Automotive Engineers, SAE paper no. 950684.
13. Caton, J. A. (2012) Effect of equivalence ratio on nitric oxides for conventional and high efficiency engines, in proceedings of the 2012 Central States Section, Combustion Institute, Technical Meeting, Dayton, OH, April 22–24, 2012.
14. Caton, J. A. (2010) The destruction of exergy during the combustion process for a spark-ignition engine, proceedings of the ASME 2010 Internal Combustion Engine Division Fall Technical Conference, paper n6. ICEF2010–35036, San Antonio, TX, September 12–15.

19

Summary: Thermodynamics of Engines

As closure to this book, this chapter provides a brief summary of the book, and then highlights some of the major results presented throughout the book. These results are a good summary of some of the main, fundamental thermodynamic foundations of IC engines. As described in the introduction, each chapter in this book has contributed to the understanding of engine cycle simulations and to the results that such simulations can provide.

19.1 Summaries of Chapters

Chapter 1: Introduction. This chapter described the role of engine simulations relative to designing and understanding IC engines.

Chapter 2: Overview of engines and their operation. This chapter provided a brief overview of engine fundamentals, terminology, components, operation, and performance parameters.

Chapter 3: Overview of engine cycle simulations. This chapter was a description of the evolution of thermodynamic engine cycle simulations. The description began with the simple ideal (air-standard) cycle analyses, and continued with the historical development of the thermodynamic simulations. Brief comments were included on quasi-dimensional thermodynamic simulations, and multi-dimensional (CFD) simulations.

Chapter 4: Properties of the working fluids. This chapter was a comprehensive presentation on the development of the algorithms needed to determine the compositions and properties of the unburned and burned mixtures. Compositions for combustion products are described for both frozen species and for chemical equilibrium. The chapter ends with results for various properties for typical engine conditions as functions of temperature, pressure and equivalence ratio.

An Introduction to Thermodynamic Cycle Simulations for Internal Combustion Engines, First Edition.
Jerald A. Caton © 2016 John Wiley & Sons, Ltd. Published 2016 by John Wiley & Sons, Ltd.

Chapter 5: Thermodynamic formulations. This chapter included the details for developing the governing thermodynamic relations for the cylinder contents. The chapter ends with a concise summary of a set of differential equations for the cylinder pressure, temperatures, volumes, and masses.

Chapter 6: Items and procedures for solutions. This chapter specified all the required items to solve the governing differential equations developed in Chapter 5. These items included cylinder volumes, fuel burning rates, heat transfer, flow rates, friction, and other sub-models as needed. The chapter ends with comments on solution procedures and convergence to final answers.

Chapter 7: Basic results. This chapter was focused on presenting detailed, instantaneous results for the base case condition largely as functions of crank angle for parameters such as cylinder pressures, temperatures, species, and energy terms.

Chapter 8: Performance results. This chapter provided overall global engine performance metrics as functions of engine operating and design parameters. Overall performance metrics included power, torque, thermal efficiency, and mean effective pressure. Engine operating and design parameters included speed, load, combustion variables, compression ratio, equivalence ratio, EGR, inlet pressure, and combustion timing.

Chapter 9: Second law results. This chapter presented the development of the aspects needed to complete second law assessments of IC engines. This included the property *exergy* which allows the irreversibilities of the engine processes to be quantified. Results for entropy and exergy were presented as functions of engine operating and design parameters.

Chapter 10: Other engine combustion processes. This chapter was devoted to providing a brief overview of alternative combustion processes. These combustion processes included those for diesel engines, stratified charge engines, and low temperature combustion engines.

Chapters 11–18: Case studies. These chapters provided seven case studies using the engine cycle simulations. These cases include studies of combustion, cylinder heat transfer, fuels, oxygen-enriched reactants, overexpanded engines, nitric oxide emissions, and advanced, high efficiency engines.

19.2 Fundamental Thermodynamic Foundations of IC Engines

Much of this book has focused on the thermodynamics of engines. This final section is aimed at highlighting and summarizing many of the previous results, and in particular, the fundamental thermodynamic concepts related to IC engines. Since the results and conclusions of this work are often quite detailed, the general and "universal" findings are sometimes not obvious. The following then are specific aspects of the thermodynamics of engines that may be quantified by use of engine cycle simulations. For completeness, some of these comments are based on well-known understandings.

Item 1: Heat Engines versus Chemical Conversion Devices

One fundamental thermodynamic concept that is often misunderstood is whether the "Carnot" efficiency is an appropriate limit for the IC engine. The Carnot efficiency applies to a specific, ideal heat engine that receives and rejects energy to thermal reservoirs. This is an appropriate ideal model for steam power plants or other power plants based on the use of energy from sources that may be approximated as thermal reservoirs. The IC engine, however, is more correctly described as a chemical conversion device [1,2]. The IC engine fails the definition of a heat engine on several levels. First, the IC engine does not go through a thermodynamic cycle since the working material is exhausted after each cycle and new material is inducted for the next cycle. Further the energy input is from a chemical conversion and not from a thermal reservoir. Finally, the energy input is a complex process at variable temperatures and pressures. For these reasons, then, the IC engine is not limited by the "Carnot" efficiency. The IC engine is, however, constrained by the second law of thermodynamics as described throughout this book.

Item 2: Air-Standard Cycles

Air-standard cycles were introduced in the late 1800s to help understand the thermodynamics of IC engines. These ideal models are based on many simplified assumptions and approximations which are not close to the reality of actual IC engines. One of the most troublesome approximations is the use of heat addition to simulate combustion. As mentioned above, this simplification is not thermodynamically consistent with the actual combustion process in IC engines. From an introductory point of view, the use of ideal air-standard cycles provides some advantages, but the lack of rigor needs to be kept in mind.

Other assumptions and approximations of the ideal, air-standard cycles include the use of pure air, adiabatic processes, no gas exchange processes, and no friction. These assumptions and approximations lead to relationships that result in thermal efficiencies that are roughly twice as high as actual results. Even though the absolute numbers are not realistic, the simple models result in relationships that preserve the correct trends with respect to compression ratio and specific heats. More extensive evaluations of the weaknesses and strengths of the ideal cycles are presented elsewhere [1, 3].

The following describes results from an air-standard cycle as functions of compression ratio and specific heats. If the heat release is assumed to occur at constant volume at TDC ("Otto" cycle), the resulting expression for the thermal efficiency (for constant properties) is

$$\eta = 1 - r^{(1-\gamma)}$$

where r is the compression ratio and γ is the ratio of specific heats.

Using this relationship for the "Otto" cycle, Figure 19.1 shows the thermal efficiency as a function of compression ratio for two values of the ratio of specific heats. The thermal efficiency increases with increases of compression ratio and with increases of the ratio of specific heats. Actual engine data also demonstrate similar trends [1]. These results lead to the next two items.

Item 3: Importance of Compression Ratio

As mentioned in *item 2*, increases of compression ratio provide increases of thermal efficiency and overall performance. This feature has been recognized since the earliest engine

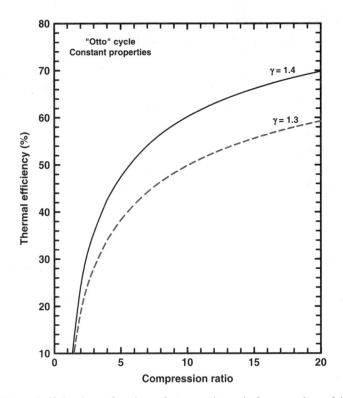

Figure 19.1   Thermal efficiencies as functions of compression ratio for two values of the ratio of specific heats using the "Otto" cycle with constant properties

development in the 1800s. In the early 1900s, IC engines were limited to compression ratios less than about 5 due to spark knock. As the fuel quality improved and anti-knock fuel additives were discovered, compression ratios rapidly increased to values of 8 or higher. Engines with these higher compression ratios provided higher efficiencies and performance. Much detailed work has been completed to demonstrate the importance of higher compression ratios [1].

Figure 19.2[1] shows the net indicated and brake thermal efficiencies as function of compression ratio for one set of operating conditions for the standard engine configuration used in this book. The indicated thermal efficiency increases rapidly at first and then at a slower rate as the compression ratio increases. The brake thermal efficiency is lower and increases at a slower rate due to the increasing friction for the higher compression ratios. Depending on the specific friction and heat transfer characteristics, a compression ratio that provides the maximum brake thermal efficiency is often in the range of 12–16 for automotive engines.

The fundamental thermodynamic benefit of higher compression ratios is the potential of the higher expansion ratios. In general, as compression ratio increases, the compression work

---

[1] The data for Figure 19.2 are the same as those used for Figure 8.16.

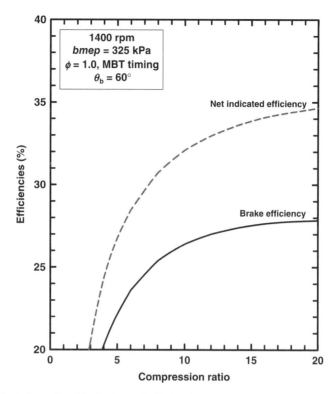

Figure 19.2 Net indicated and brake thermal efficiencies as functions of compression ratio for a case with an engine speed of 1400 rpm and a *bmep* of 325 kPa using the cycle simulation. *Source:* Caton 2007. Reproduced with permission from ASME

increases, but the expansion work increases more such that the final result is a net increase of the work output (and therefore, for the same fuel input, an increase of the thermal efficiency).

Item 4: Importance of the Ratio of Specific Heats

As demonstrated by the simple "Otto" cycle efficiency (*item 2*), increases of the ratio of specific heats increase the thermal efficiency. The qualitative effects of the ratio of specific heats on actual engine performance are well understood [1], but until recently [4], these effects have not been quantified for actual engines.

As part of appendix A in Chapter 3, the effects of the specific heats on efficiency were illustrated for a simple "Otto" cycle. These results showed that the efficiency increases significantly for modest increases of the ratio of specific heats for two compression ratios. Similar results from the engine cycle simulation are difficult to obtain due to the connections between the properties, the work output, and cylinder heat transfer. Caton [4] has attempted to separate these various effects and isolate the contribution of the increase of the specific heats toward efficiency increases. These results have demonstrated the dominating contribution that increases of the ratio of specific heats have on efficiency [4].

Item 5: Cylinder Heat Transfer

Cylinder heat transfer is intimately connected to the thermodynamics of the IC engine. The case study on cylinder heat transfer (chapter 13) contains descriptions of many of these aspects. As described, simple global heat transfer correlations from several sources provide markedly different results for the detailed heat transfer. Fortunately, for many engine performance parameters, the results are somewhat insensitive to these variations. Nevertheless, better understandings of the cylinder heat transfer would aid in more precise engine computations and in the design of more efficient engines.

As might be expected, reductions of cylinder heat transfer may result in some increase of the engine's thermal efficiency. For conventional engines, this increase is often small. An important thermodynamic reason for this is that the conversion of thermal energy to work is difficult. This conversion is governed by the value of the specific heats (see item 4). Increases of the specific heats (e.g., due to increases of temperatures) result in decreases of this conversion. For decreases of heat transfer, the gas temperatures generally increase, and therefore, this conversion decreases.

Another important aspect of the cylinder heat transfer is the relation to engine speed. Although increases of engine speed result in increases of the heat transfer rate, the higher speed means that less "real" time is available for the process. This often results in decreasing importance of cylinder heat transfer with increases of engine speed.

Other such insights on the role of cylinder heat transfer are provided throughout the book. One such insight is described next regarding low heat rejection engine concepts.

Item 6: The Potential of a Low Heat Rejection Engine

The concept of an engine with reduced heat losses has been explored at least since the early 1900s. Some of these activities were described as the development of an "adiabatic" engine. Of course, no practical engine could ever reduce all heat losses (adiabatic)—so a better term is "low heat rejection." In general, such an engine should have a higher thermal efficiency, may require a smaller (or no) cooling system, and could have other advantages.

This concept has been considered most applicable for diesel engines since the reduced heat losses would result in higher gas temperatures which could result in more potential for spark knock for spark-ignition engines. Many concepts have been developed and tested to reduce the cylinder heat losses. These concepts included the use of ceramic and other low-conducting materials to "insulate" the cylinder contents and minimize cylinder heat losses. Most of these materials are less ductile than the cast iron and steel components that they replace. Due to this characteristic, these materials suffered low durability and often were not able to sustain the harsh environment of the engine combustion chamber.

Although major research and development activities have been completed on this topic, low heat rejection engines have not been commercialized. The general consensus has been that efficiency gains, if any, are small. As demonstrated in the heat transfer case study (chapter 13), at least one major reason for this is the adverse change of the specific heats due to the higher gas temperatures. One possibility that may minimize this negative feature is the combination of the low heat rejection concept with low temperature combustion engines.

Item 7: Lean Operation and the Use of EGR

Fundamental studies of engine performance have demonstrated the importance of engine operation with lean mixtures and using exhaust gas recirculation (EGR). Both of these approaches have been studied, and have proven useful for increasing thermal efficiencies. These "dilute" engines must operate with increased inlet pressures to achieve the same output as engines using stoichiometric mixtures.

These dilute engines possess several features that result in the higher efficiencies. First, the high dilution results in lower combustion temperatures which in turn results in lower cylinder heat transfer. Second, the lower temperatures result in lower specific heats (higher ratio of specific heats) and this increases the conversion of thermal energy into work.

A related feature that is unique for lean mixtures (whereas EGR does not provide this benefit) is due to the change of composition for lean mixtures. This composition change results in lower specific heats (even for no temperature change) due to the change of composition.

For spark-ignition engines, these dilute engines also may benefit due to a reduction of throttle losses since the engine must be operated at higher inlet pressures for the same engine load. This benefit is only possible for original operating conditions at or below an inlet pressure near atmospheric.

Item 8: Insights from the Second Law of Thermodynamics

As described in Chapter 9, the second law of thermodynamics provides important and unique insights concerning the thermodynamics of engines. Although this short section cannot capture the breadth of these insights, the following statements are intended to highlight these ideas. Perhaps one of the most important of these insights relates to the conversion of thermal energy to work. The first law of thermodynamics provides only an upper limit to this conversion. From the first law, basically the amount of work may not exceed the thermal energy (energy conservation). The second law, on the other hand, states that not all of thermal energy may be converted to work. Further, the second law provides the quantitative expressions based on engine parameters and thermodynamic properties, to determine this conversion.

Examples of the above have been provided throughout this book. For instance, for reductions of cylinder heat transfer, the increase of work output is only a fraction of the energy associated with this reduction. For some cases, the increase of the work output could be near zero. The second law provides this insight which is not available otherwise.

Another example is regarding the potential to produce work with exhaust gases. For cases examined in this book, the exhaust gases may possess on the order of 40% of the fuel energy. But the work potential (exergy) of these gases only represents about 25% of the fuel energy. Further, only a portion of this 25% may be converted to work due to friction, heat losses, and other real effects. Again, without the second law insights, the work potential of the exhaust gases could be greatly overestimated.

A final comment relates to the destruction of exergy during the combustion process. This item is not part of any first law analyses, and can only be appreciated by the use of the second law. Destruction is always present in combustion processes, and quantifying this destruction is useful for a full understanding of the engine operation.

Item 9: Timing of the Combustion Process

The cylinder pressure which results from the combustion process provides the maximum work output for a specific timing of the combustion process. This is a fundamental consequence of the engine geometry and kinematics, and is easily demonstrated by the engine cycle simulation. Other aspects of the combustion process which relate to the thermal efficiency of the engine are the mass fraction burn rate and the duration of the combustion. The thermal efficiencies increase for shorter combustion durations, but the impact of the mass fraction burn rate is modest for reasonable descriptions of the combustion process.

Item 10: Technical Assessments of Engine Concepts

As illustrated by the case studies, thermodynamic engine simulations are an excellent approach for examining the technical merits of various engine concepts and technologies. The case studies on oxygen enriched inlet air and on overexpanded engines are two good examples of this approach. The engineer can quantify any advantages of the engine concept in terms of efficiency, power density, or other performance parameters. These results can be balanced by any disadvantages either with respect to the thermodynamics or with respect to the overall design (complexity, cost, ease of manufacturability, material constraints, ...)

19.3 Concluding Remarks

As described in the introduction, the main purpose of writing this book was to document the development and use of thermodynamic engine cycle simulations. To illustrate the value of these simulations, representative results were provided for a large range of conditions. In addition to this main purpose, a secondary purpose was to stimulate the interest and excitement of using fundamental thermodynamic principles to understand a complex device. As demonstrated in the book, many phenomena related to engine operation and design may be understood in a more complete fashion by focusing on the fundamental thermodynamics.

The internal combustion engine is a fantastic engineering achievement. For further improvements of these engines, fundamentally precise understandings are required. The author hopes that some of the information in this book will be useful for this fundamental foundation. He is confident that he will continue to witness extraordinary advances of the IC engine.

References

1. Heywood, J. B. (1988). *Internal Combustion Engine Fundamentals*, McGraw-Hill Book Company, New York.
2. Lauck, F., Uyehara, O. A., and Myers, P. S. (1963). An engineering evaluation of energy conversion devices, Society of Automotive Engineers, *SAE Transactions*, **71**, 41–50.
3. Shyani, R. G., Jacobs, T. J., and Caton, J. A. (2011). Quantitative reasons that ideal air-standard engine cycles are deficient, *International Journal of Mechanical Engineering Education*, **39** (3), 232–248.
4. Caton, J. A. (2014). On the importance of specific heats as regards efficiency increases for highly dilute IC engines, *Journal of Energy Conversion and Management*, **79**, 146–160.

Index

adiabatic engines (see low heat rejection engines)
air, 38
air standard cycles (see ideal cycles)
air-fuel ratio, stoichiometric, 41, 254
alcohol fuel (see methanol and ethanol)
Annand heat transfer correlation, 230
Atkinson
   cycle, 21, 295-297
   James, 295
availability (see exergy)
available energy (see exergy)

blow-by, 64
blow-down, 20, 108, 159
boundary layer, 21, 22, 72-73,103, 106, 107
brake parameters, 14
burned gas fraction, 38
burned gas mixture
   composition, 42-45
   equilibrium, 43-45
   low temperature, 42-43

CA_{50}, 206-221
CA_{pp}, 101, 129, 206-221
carbon monoxide
   properties, 41, 254
   results, 255-268, 273
Carnot efficiency, 3, 20, 356, 357
case studies, descriptions of, 187-190
chemical conversion device, 3, 356, 357

chemical equilibrium, 43-45
clearance volume, 80
combustion
   duration
      definitions, 82-84
      effects on performance, 135, 346-347
   efficiency, 17
   phasing, 205-222
compression ratio
   definition, 80
   effects related to exergy analyses, 170-171
   effects on performance, 131-133, 345-346
   effects on nitric oxide emissions, 327-328
   importance of, 357-358
computational fluid dynamics (CFD), 23-24, 25
cosine law, 83-84
crank-angle numbering scheme, 13
cycles
   Atkinson, 21
   Diesel, 20
   dual, 20
   Miller, 295
   Otto, 20, 29-34
cylinder
   rate of change of volume, 81, 102, 104
   surface area, 81-82
   volume, 80,102-103

An Introduction to Thermodynamic Cycle Simulations for Internal Combustion Engines, First Edition.
Jerald A. Caton © 2016 John Wiley & Sons, Ltd. Published 2016 by John Wiley & Sons, Ltd.

dead state, 156
Diesel
    cycle, 20
    engines, 2, 20, 179-181
    Rudolf, 20
discharge coefficient for valves, 86
displaced volume, 80
dual cycle, 20

efficiency
    combustion, 17
    fuel, 17
    mechanical, 17
    thermal, 16
    volumetric, 18, 129, 136, 241, 244
Eichelberg heat transfer correlation, 230
energy
    as function of crank angle, 114-115
    distribution, 101, 114-115, 121-134,
        146-149, 245, 269-273, 302
engine
    characteristics, 11
    classifications, 10
    combustion processes, 179-184
    components, 12
    design goals, 9
    high efficiency, 34, 218-221,
        333-354
    kinematics, 80-81
    simulations (see thermodynamic
        simulations)
    specific power, 286
enthalpy
    property relations, 46
    results, 59, 108-109
entropy
    balances, 155
    generated, 155
    property relations, 46-47
    results, 60, 159-160
equilibrium composition (see chemical
    equilibrium)
equivalence ratio
    definition, 44
    effects on performance,
        133-135

effects on nitric oxide emissions,
        324-327
    effects related to exergy analyses, 167-
        170, 343-344
    importance of, 361
essergy, 153
ethanol
    properties, 41, 254
    results, 255-268, 270
exergy,
    balances, 156-157
    definitions, 153-154, 156-158
    destroyed
        by heat transfer, 174
        by inlet mixing, 163, 167-168
        during combustion, 162-173, 256,
            266-268, 304, 308, 349-351
    distribution, 114, 162-165, 169-173,
        269-273, 349-350
    function of crank angle, 162
exhaust gas temperatures
    energy averaged, 140-142
    enthalpy averaged, 261-262, 265
    mass averaged, 140
    time averaged, 140-142, 261
exhaust gas recirculation (EGR)
    definition, 38
    effects on performance, 142-145, 215,
        218-221, 247-250, 334-339,
        344-345, 361
    effects on nitric oxides, 327, 329
    effects related to exergy analyses,
        173-174, 349-350
exhaust pressure, effects on performance,
        136-139, 339-340
expansion ratio, 21, 295-309

flow rates
    computation, 86-89
    exhaust, 108, 165, 173, 283, 300-303,
        306-307
    intake, 108, 283, 300-303, 306-307
    sonic, 87
    subsonic, 87
four-stroke cycle, 3, 11, 12
friction, 89-93

fuels
    heating values, 41, 254
    efficiency, 17
    properties. 41, 254
    results, 255-273
fuel cell, 2, 174

gamma (γ)
    average during combustion,
        245-246, 249
    burned mixtures, 55
    definition, 29
    unburned mixtures, 51
gas constant, 56-58, 102, 110, 112
Gibbs free energy, 158

heat engine, 3, 356
heat release
    apparent, 192, 197
    ideal cycle, 20
    rate, 197-201
    results, 197-206
    schedule, 191-192
heat transfer, cylinder
    correlations, 227-230
    effects on performance, 119-126, 131,
        133-134, 230-250, 260, 263,
        286-287, 302, 341
    effects related to exergy analyses,
        160-176
    general, 85-86, 225-228, 254, 360
hexane,
    properties, 41, 254
    results, 255-268, 271
high efficiency engine (see engine, high
        efficiency)
history of engine simulations, 5, 21-24
Hohenberg heat transfer correlation, 230
homogeneous charge compression ignition
        (HCCI), 181, 333
hydrogen
    properties, 41, 254
    results, 255-268, 272

ideal cycles, 5, 19-21, 29-35, 357
ideal gas, 46

ignition delay, 129-130
indicated parameters, 13-14
internal energy
    property relations, 46
    results, 57-58, 108-109
isooctane
    property relations, 47
    properties, 41, 254
    results, 255-268, 272

KIVA code, 24-25
knock, 120, 132, 180, 241

load, engine (as bmep)
    effects on performance, 121-123,
        213-215
    effects on nitric oxide emissions, 326
    effects related to exergy analyses,
        163-168
    effects related to fuels, 263-266
low heat rejection (LHR) engines,
        241-246, 360
low temperature combustion (LTC),
        181-184, 246, 334, 337-338

mass fraction burned, 64, 82-85, 183,
        192-205, 208, 351
MBT timing, 99
mean effective pressure (mep), 15-16
methane
    properties, 41, 254
    results, 255-268, 270
methanol
    properties, 41, 254
    results, 255-268, 269
Miller cycle, 295
mole fractions
    burned mixtures, 51-53
    unburned mixtures, 48-49
molecular mass
    air, 45
    burned mixtures, 39, 53-54, 110-111
    exhaust gases, 316
    fuel, 38, 41, 254, 259
    unburned mixtures, 39, 49-50,
        110-111

nitric oxides (NO)
    chemical formation mechanisms,
        312-313
    rate constants, 314-315
    results, 289-291, 317-330,
        347-348
nitrogen oxides (NO_X), 311
numerical methods, 94-96
Nusselt number, 227

Otto
    cycle, 20
    Nikolaus, 20
overexpanded cycle, 295-309
overview of book, 6
oxygen
    oxygen containing fuels, 40
    oxygen enriched air, 275-277
    results for oxygen enriched air,
        277-291

part load, 99, 136, 148, 297
partially premixed combustion (PPC),
        182, 333
polytropic compression/expansion
        relation, 103
power, 14
premixed charge compression ignition
        (PCCI), 333, 337
pressure, peak cylinder, 101, 129
pressure-crank angle diagrams, 102-104,
        106-107, 193, 208-211, 280
pressure-volume diagrams, 13, 16, 102, 105,
        299-300, 302-303, 305-306
propane
    properties, 41, 254
    results, 255-268, 271
pumping
    friction, 90-91
    loop, 102
    mean effective pressure, 90-91
    work, 13, 102, 114, 121, 145-149,
        257-258, 335

reactivity controlled compression ignition
        (RCCI) 182, 183

residual mass fraction
    definition, 37
    results, 101, 102, 136, 137, 256, 281
Reynolds number, 227, 228

second law analyses, 154-158
second law of thermodynamics, 153, 154,
        184, 236, 361
slider-crank mechanism, 296
spark assisted compression ignition (SACI),
        182
spark timing
    effects on nitric oxide emissions, 323-324
    effects on performance, 129-131, 362
    effects related to exergy analyses,
        171-173
    MBT definition, 99
specific fuel consumption (sfc)
    brake, 17
    indicated, 17
    results, 127, 129
specific heats
    importance of, 35, 246, 249, 342-345,
        359
    property, 29, 50-51, 55-56
speed, engine
    effects on heat transfer, 238-241
    effects on performance, 119-120,
        123-129, 213
    effects on nitric oxide emissions, 324-326
    effects related to exergy analyses,
        164-168
    effects related to fuels, 259-262
stratified charge engines, 181

temperatures, gas,
    as functions of CA, 103-107, 232-234, 244,
        280-282, 300-301, 322, 326-327
    combustion, 128-129
    exhaust, 140-142, 286-287,
    inlet, 135-136, 142-143
    peak, 130, 171-172, 244, 341-342
temperature-entropy diagrams, 159
thermodynamic engine simulations
    approximations and assumptions, 64
    commercial products, 24-25

formulations,
    one-zone, 65-67
    three-zone, 72-76
    two-zone, 67-72
  history, 5, 21-24
  quasi-dimensional, 22-23
  zero-dimensional, 21-22
thermodynamic properties
  descriptions, 46-47
  results, 47-60
timing (see spark timing)
torque, 15
tuning, intake/exhaust, 120
two-stroke cycle engines, 3, 11, 12

unburned gas mixture composition, 37-41

valves
  choked flow, 87
  curtain area, 87
  discharge coefficient, 86
  lift, 87
  overlap, 87-88
volumetric efficiency, 18, 129, 136,
    241, 244

water gas-shift reaction, 42-43
wave effects, 120, 129
Wiebe function
  definition, 82
  effects on performance, 201-205
  effects on mass fraction burned,
    194-196
work, 12-14, 116, 147-148
working fluids, 37-60
Woschni heat transfer correlation,
    228-230

Zeldovich mechanism, 289,
    311-315